Skin 3-D Models and Cosmetics Toxicity

Aditya Bhushan Pant • Ashish Dwivedi •
Ratan Singh Ray • Anurag Tripathi •
Atul Kumar Upadhyay • Shiv Poojan
Editors

Skin 3-D Models and Cosmetics Toxicity

Springer

Editors
Aditya Bhushan Pant
CSIR – Indian Institute Of Toxicology Research
Lucknow, Uttar Pradesh, India

Ashish Dwivedi
Photobiology Division
CSIR – Indian Institute Of Toxicology Research
Lucknow, Uttar Pradesh, India

Ratan Singh Ray
Photobiology Division
CSIR – Indian Institute Of Toxicology Research
Lucknow, Uttar Pradesh, India

Anurag Tripathi
Food, Drug & Chemical Toxicology
CSIR – Indian Institute Of Toxicology Research
Lucknow, Uttar Pradesh, India

Atul Kumar Upadhyay
Department of Biotechnology
Thapar University
Patiala, Punjab, India

Shiv Poojan
bioTox Solution Pvt. Ltd.
Institute of Engineering and Technology
Lucknow, Uttar Pradesh, India

ISBN 978-981-99-2803-3 ISBN 978-981-99-2804-0 (eBook)
https://doi.org/10.1007/978-981-99-2804-0

This Springer imprint is published by the registered company Springer Nature Singapore Pte Ltd.
The registered company address is: 152 Beach Road, #21-01/04 Gateway East, Singapore 189721, Singapore

Preface

Skin is the body's largest organ, with an average area of 1.8 sqm. It is an organ that people notice first in our body, even from afar, so everyone wants it to be healthy, beautiful, and appealing. Aside from acting as a barrier against the invasion of microorganisms and environmental chemicals into the inner body parts—bones, muscles, blood, etc., the skin performs various other critical functions such as thermoregulation, the first line of immune defense, and so on. Protecting the skin against any xenobiotics exposure to keep it healthy is as important as any other organ in the body. Therefore, a proven safety, efficacy, and quality assurance assessment of all cosmetics and their ingredients must be ensured before commercialization.

Despite the fact that concerted efforts to develop 3D skin-mimicking model systems, animal-based testing and 2D monolayer cultures of independent skin cell types were the only options for screening the safety of cosmetics and their ingredients until 2013. After many countries banned animal testing in 2014, there has been a surge in the development of human skin-specific 3D models for toxicity and efficacy testing of cosmetics and their ingredients. The ultimate goal was to simulate the morphology and functionality of human skin in culture conditions outside the body.

In fact, the preface to developing such human-specific sophisticated 3D-skin models for cosmetic testing was written way back in the late 1930s when a "new and improved eyebrow and lash dye" called "Lash Lure" was introduced in the US market. This product was discovered to have severe to fatal effects on users' faces, eyelids, and eyes. The untested chemical compound in the product, paraphenylenediamine, was to blame for the adverse reactions. As a result, in 1938, the US Congress passed the Federal Food, Drug, and Cosmetic Act, which established strict regulations for cosmetic products. It has since become an essential component of product development in the cosmetic industry.

Throughout the twentieth century, many animal models for testing cosmetic and personal care products were developed and widely used. However, growing ethical concerns prompted regulators and lawmakers to limit the use of laboratory animals and develop new alternative systems for cosmetic efficacy and toxicity testing. In the 1990s, Europe took the lead and restricted animal testing for cosmetic products and their ingredients. The European Union banned animal testing for cosmetics and their ingredients in 2013. It ensured that no such product derived from animal testing would be available on the European market. Following that, a ban was enacted in dozens of other countries, including India, in 2014. Because of consumer concerns and global trade barriers, the use of animals in cosmetics testing is declining in other countries where it is still permitted.

Under this new scenario, all stakeholders are working hard to create unique and more predictive alternatives to animal model systems as rapid and high-throughput tools for testing cosmetics and their ingredients. The advancements in technology in the cultivation, bulk production, differentiation, and cytosolic and genetic transformation of immortalized and primary cells derived from human and animal origin in 2D and 3D cultures have given us an advantage in developing high throughput screening models for testing chemicals and products, including cosmetics and their ingredients. This mission has been accelerated by the unlimited proliferation and differentiation potential of stem cells and induced pluripotent stem cells (iPSCs).

One of the success stories began with the development of a total thickness human skin equivalent 3D model for studying wound healing in a hydrogel using dermal fibroblasts and keratinocytes. The model, however, lacked vascularization. Recognizing the limitations, decellularized porcine jejunum was used to provide vascularization in the scaffold. Following that, this vascularized 3D-skin model has the added benefit of allowing researchers to investigate the concentration and amount of cosmetic products or ingredients that enter blood vessels after topical application. Attempts are also being made to broaden the applicability of a 3D skin-based model like the melanoma assessment model by integrating it with cancer cells and adding immune cells to assess skin irritation/sensitization and other immunological responses.

The OECD has already approved and adopted human reconstruct models of epidermis/cornea-like epithelium as an alternative tool for testing cosmetics and their ingredients. Many industries have developed human-skin equivalent 3D model systems in collaboration with academia and research organizations over the last two decades. These models are primarily created by combining various skin representative cells such as keratinocytes, fibroblasts, Langerhans cells, and melanocytes. Attempts are also being made to develop these terminally differentiated functional cells in 3D cultures from human stem cells or human fibroblast-derived iPSCs to improve predictability in human trials. Because 3D-skin equivalents resemble human organ architecture, they are more predictive than 2D experimental model systems. Recent studies show that the 3D human skin reconstructs model can be used to assess the antiaging activity of test compounds (cosmetic products) by measuring the increased levels of collagen production by fibroblasts in the extracellular matrix. Reduced collagen content is known to weaken the bonding between the dermis and epidermis, contributing to wrinkle formation in the skin. In addition, by measuring pro- and anti-inflammatory cytokines, immunohistochemical changes, and so on, these 3D skin models are being used to evaluate cosmetics' anti-inflammatory responses, such as reduction in epidermal swelling, one of the hallmark endpoints of inflammation.

Industry and academia joint efforts have also culminated in the less complex, unique 3D well-characterized spheroid epidermis model using two major cell types present in the epidermis, i.e., keratinocytes and melanocytes. This human epidermis mimicking model was suitable for studying the complete melanogenesis pathway and the transfer of melanin from melanocytes to physiologically differentiated keratinocytes within the 3D spheroids microenvironment.

With the pace of working, it is anticipated that full-thickness skin models consisting of the basement membrane, dermis, and epidermis and displaying the mimicking morphology and functionality of human skin will be available shortly. Once such a perfect model is available, it will parallel the human body conditions and minimize/eliminate the reliance on animal data. It will further reduce rodents' sacrifices, which are extensively required in xenograft assays for testing cosmetics and their ingredients, and oncology experiments. There are enormous applications of human 3D-skin models that still need to be validated, such as testing of cytotoxic responses of cosmetics ingredients and products, pigmentation, depigmentation, anti-skin-aging, hair-growth pattern studies, mechanistic insight studies of skin functionality, physiology under normal and diseased conditions, production of engineered skin grafts for significant burns and wound healings, 3D bioprinting of miniaturized form of organ for more efficacious drug and cosmetics screening, etc. Of course, 3D-skin models developed so far and currently being developed are primarily directed towards screening cosmetics and their ingredients only, as regulatory agencies demand it. Still, indeed very soon, these will apply to biomedical research, pharmaceuticals, and hundreds of other applications.

The book *Skin 3-D models and cosmetics toxicity* provides an overview of skin cell culture approaches, as well as recent achievements and future challenges, while also providing a thumbnail sketch of how modern technologies have evolved in recent times to innovative human three-dimensional skin equivalent model systems for efficacy and safety testing of cosmetics and their ingredients. The book is divided into a total of 14 chapters that discuss the progress made thus far, the challenges and roadblocks encountered, and the future of developing and validating 3D models for testing cosmetic and personal healthcare products. The book began with an editorial followed by 13 chapters that spoke volumes about the significant progress made thus far, challenges, and prospects for creating 3D reconstruct models. The overall content flow extends from an overview of the path of significant achievements that underpin the birth of this new discipline of 3D model systems to a discussion of the future of 3D model systems' future.

This book may be the first to reveal the previously unknown journey of this discipline's current status and the various developments, achievements, and lessons learned. Issues addressed that have shaped the current face of research in using 3D model systems in the safety/toxicity of cosmetics and personal care products. This book will be an invaluable resource for researchers, academia, cosmetic industries, regulators, policymakers, graduate and postgraduate students, PhD, and postdoctoral fellows working in the fields of toxicology/biosafety of cosmetics and personal healthcare products, with case studies, technical and applied approaches, pictorial representations, informative tables, and simple language. The book will also instill in readers a practical understanding of the discipline of 3D culture systems, leaving them in awe of the incredible efforts made by the scientific community to contribute to understanding the complexities the topic presents. Furthermore, the content has been organized to reach out to the general public, the non-scientific fraternity, to develop a basic understanding of the subject with greater relevance in everyday life.

Lucknow, Uttar Pradesh, India Aditya Bhushan Pant
Lucknow, Uttar Pradesh, India Ashish Dwivedi
Lucknow, Uttar Pradesh, India Ratan Singh Ray
Lucknow, Uttar Pradesh, India Anurag Tripathi
Patiala, Punjab, India Atul Kumar Upadhyay
Lucknow, Uttar Pradesh, India Shiv Poojan

About This Book

Toxicology research is currently at a crossroads, transitioning from traditional animal testing studies to mechanistic understanding using 'omics' methodologies that allow for greater confidence in subsequent risk evaluations. Following the complete prohibition on animal testing in cosmetics and personal healthcare goods ingredients, there has been a significant effort to develop and validate non-animal methodologies. The OECD has previously acknowledged the work and approved reconstructed human epidermis/corneal-like epithelium as a human skin-mimicking model for skin corrosion, skin irritation, and eye irritation. Nonetheless, no data on genotoxicity, phototoxicity, absorption, cytotoxicity, or other effects utilizing such in vitro models have been received by any of the validating authorities so far. As a result, research organizations and cosmetic enterprises worldwide are attempting to address the issue by developing more human-mimicking 2D and 3D model systems with more exact and targeted endpoints. The book contains a series of subtopics devoted to progress made thus far, obstacles and roadblocks encountered, and the future ahead of developing and validating 3D models for testing cosmetic and personal healthcare products. The book begins with an editorial followed by 13 chapters that are themed to speak volumes about the significant progress made thus far, challenges, and prospects for creating 3D reconstruct models. The overall flow of content extends from an overview of the path of important achievements that underpin the birth of this new discipline of 3D model systems to a discussion of the future of 3D model systems.

This book may be the first of its kind to unravel the untold journey of this discipline's current status and the various developments, achievements, and lessons learned. Issues addressed that have ultimately shaped the current face of research in the application of 3D model systems in the safety/toxicity of cosmetics and personal care products. With case studies, technical and applied approaches, pictorial representations, informative tables, and simple language, this book will be an invaluable resource for researchers, academia, cosmetic industries, regulators, policymakers, graduate and postgraduate students, PhD, and post-doctoral fellows working in the fields of toxicology/biosafety of cosmetics and personal health care products. The book will also instill in the readers a practical awareness of the discipline of 3D culture systems, leaving them in awe of the amazing efforts made by the scientific community to contribute to comprehending the intricacies the topic presents. Furthermore, the content has been organized to reach out to the general public, the non-scientific fraternity, to develop their rudimentary grasp of the subject with better relevance in everyday life.

Contents

Editors and Contributors

About the Editors

A. B. Pant is a seasoned toxicologist with over 30 years of active research career. Presently serving as "Senior Principal Scientist" at CSIR-Indian Institute of Toxicology Research, Lucknow, and a "Professor" of the Academy of Scientific and Innovative Research, an Institution of National Importance set up by the Act of Parliament, Government of India. In the scientific fraternity, Professor Pant is renowned for his elegant research on applying human stem cells and iPSCs in developmental neurotoxicity and establishing the number of in vitro model systems as an alternative to laboratory animals for biomedical research. He also founded a new discipline of "In-Vitro Toxicology" in the country. Dr Pant also dedicated himself to fostering contemporary science in postgraduate, students, researchers, and university teachers. He is a lead GLP inspector for the Government of India and is an expert for regulatory bodies, such as BIS, CDSCO, NABL, FSSAI, NGCMA, ICMR, DST, and Pharmacovigilance. He has won several awards and honors, including Vigyan Ratna Award 2010 (CST UP), Shakuntala Amir Chand Prize 2007 (ICMR), National Bioscience Award 2012 (DBT), Prof. KT Shetty Memorial Oration Award 2017 (Indian Academy of Neurosciences), Toxicology Promotion Award 2018 (National Academy of Sciences, Allahabad), STOX/ASAW Gold Medal (STOX, India), etc. He has been elected Fellow of the Academy of Toxicological Sciences, USA; Society of Toxicology, India; Indian Academy of Neurosciences, Academy of Sciences for Animal Welfare, India; Academy of Environmental Biology, India; etc. In view of outstanding credibility in toxicology, Professor Pant has been included as a global toxicologist in the "European and United Kingdom Register of Toxicologists and has also been conferred the DSc (h. c.) degree by HNB Medical Education University, Dehradun, Uttarakhand, India." He currently serves as Chairman and Member of several National and International Committees/Boards of Government Institutes and Bodies. He also serves as a member of the editorial board of several reputable journals.

Ashish Dwivedi graduated from CSIR-Indian Institute of Toxicology Research, Lucknow, India, in the area of Photobiology and Phototoxicology. Thereafter, he enriched himself in the same research domain as a Post-Doctoral Fellow at the Israel Institute of Technology, Israel; Colorado University, USA; and Banaras Hindu University, Varanasi, India. Based on his outstanding research experience, he got an opportunity to return to his parenteral research organization-CSIR-IITR, where he is currently serving as a Scientist in the System Toxicology and Health Risk Assessment Group. In a decade-long research career, Dr Dwivedi has established himself as one of the pioneers in phototoxicity/photosafety with special reference to the photosafety of therapeutic agents, nanotized phytochemicals, environmental pollutants, etc. Besides the laboratory work, he has also fostered contemporary science to postgraduate, PhD students, researchers, and university teachers. Dr Dwivedi's research efforts have been rewarding in terms of the quantifiable number of research publications in journals of high repute. He has also edited three books for Springer Nature in the area of Photoprotection, Skin Aging, and Neuroprotection. He is a member of various scientific societies, to name a few are—Society of Toxicology, USA; American Society for Photobiology, USA; Indian Photobiology Society, India; Society of Toxicology, India, etc.

Ratan Singh Ray is a Chief Scientist and Head Photobiology Division at CSIR-Indian Institute of Toxicology Research, Lucknow, India. He is one of the founders who established a new discipline of Photobiology/Phototoxicity in the country. His active research career of over three decades in Photobiology made him instrumental as a resource person in the Bureau of Indian Standards for the formulation/revision of several standards for cosmetics and personal healthcare products. His research group is credited for creating the complete spectrum of UV-A, UV-B, and UV-C radiation falling on the earth's surface at latitude-longitude 26.8467° N, 80.9462° E for over 20 years. Such dedicated and sustained efforts in creating this database have created immense wealth for the scientific fraternity working in this area of research, industries involved in the production and R & D of cosmetic and personal healthcare products and regulatory agencies. He has supervised more than a dozen students for PhD, several post-doctoral fellows and postgraduate students. He also renders his intellectual input and expertise to various scientific societies and government agencies, such as the Bureau of Indian Standards, India; American Society for Photobiology, USA; Indian Photobiology Society, India; and Society of Toxicology, India. His research is highly cited in the scientific literature, as evidenced by a Google Scholar Score of over 2000 with an H-Index of 28. He is currently coordinating several investigator-driven research grants from different national funding agencies.

Anurag Tripathi works as Principal Scientist at CSIR-Indian Institute of Toxicology Research. He did his PhD from the School of Biotechnology, Banaras Hindu University, on the Immunomodulatory effects of Prolactin and Growth hormone on macrophages. His work contributes to an in-depth understanding of the immunomodulatory and immunosuppressive effects of common food additives, drugs, and cosmetic ingredients. He has also worked on the tumorigenic potential of mycotoxins and the molecular mechanisms leading to skin cancer. Dr Tripathi has over 16 years of research experience and published 50 articles in peer-reviewed international journals such as Immunology,

Molecular Immunology, Molecular Aspects of Medicine, Toxicology Letters, ACS Nano, Toxicology and Applied Pharmacology. His publications are well-cited, with an H-index of 25. He is a life member of Indian Nanoscience Society. He serves as an editorial board member in several prestigious international journals, such as the Journal of Immunology and Serum Biology, SM Journal of Nanotechnology and Nanomedicine, Journal of Toxicology and Forensic cases, and Austin Immunology.

Atul Kumar Upadhyay received his PhD in Life Sciences from National Centre for Biological Sciences (NCBS), Bangalore; Tata Institute of Fundamental Research (TIFR), Mumbai, in 2015. His primary research area encompasses in-depth plant genomic analysis and computer simulations for developing database management systems and bioinformatics tools. He started his teaching career as Assistant Professor at Lovely Professional University, Jalandhar, Punjab, India, in 2016. He is an Assistant Professor at the Department of Biotechnology, Thapar Institute of Engineering and Technology (TIET), Patiala, Punjab, India. His areas of expertise are comparative genomics and bioinformatics, computational biology, molecular modeling of biomolecules, and drug designing. Based on his research, he has come up with a good number of research publications in national and international journals of high repute. Besides the research publications, he has also edited one book, *Bioinformatics for Agriculture: High-throughput Approaches* for Springer Nature. He is an author/co-author in several book chapters in his area of expertise. Presently, he is also an editorial member in a few good journals in his area of research.

Shiv Poojan currently holds the Scientist position at Sidney Kimmel Cancer Center, Thomas Jefferson University, Philadelphia, PA, USA. He is a research graduate from CSIR-Indian Institute of Toxicology Research, Lucknow, India. His early research focused on deciphering the mechanistic incites of arsenic-induced developmental toxicity, genetic alteration, and carcinogenicity in a mouse stem cell-derived 3D reconstructed skin model. Then he moved to the National Cancer

Center (NCC), South Korea, as a post-doctoral fellow, where he culminated a task of developing the target-specific stable transformations in healthy and cancerous iPSCs to read across the customized changes in epigenetic profiling and genomic sequences, hence could be utilized to identify the drug resistance genes. For several years, he has worked on a rare genetic disease, epidermolysis bullosa, to identify suitable inhibitors to reduce cancer development in RDEBSCC patients and to understand RDEB squamous cell carcinoma development. In acknowledgment of his outstanding research acumens, he has been awarded "Carl Storm International Diversity Fellowship" at the Gordon Research Conference (GRC), Italy (2013), and "Rambhau Kulkarni Best Scientist Award" at Indian Association of Cancer Research (2010). He has come up with a good number of research papers in diversified journals of high impact, to name a few are—Nature's Scientific Reports, Experimental and Molecular Medicine, PLoS-ONE, Mutagenesis, Rheumatology, Cell Biology International, Histology, Histopathology, etc.

Contributors

Michael H. Alexander Department of Dermatology and Cutaneous Biology, Thomas Jefferson University, Philadelphia, PA, USA

Saroj Kumar Amar Photobiology Laboratory, Systems Toxicology & Health Risk Assessment Group, CSIR-Indian Institute of Toxicology Research (CSIR-IITR), Lucknow, Uttar Pradesh, India
Environmental Lab, US Army Engineer Research and Development Center, Vicksburg, MS, USA
School of Bioscience and Bioengineering, Lovely Professional University, Phagwara, Punjab, India

Anthony Atala Wake Forest Institute for Regenerative Medicine, Winston-Salem, NC, USA
Wake Forest School of Medicine, Medical Center Boulevard, Winston-Salem, NC, USA
Atrium Health Wake Forest Baptist, Medical Center Boulevard, Winston-Salem, NC, USA

Madhu Bala Photobiology Laboratory, Systems Toxicology & Health Risk Assessment Group, CSIR-Indian Institute of Toxicology Research (CSIR-IITR), Lucknow, Uttar Pradesh, India

Academy of Scientific and Innovative Research, AcSIR Headquarters, CSIR-HRDC Campus, Ghaziabad, Uttar Pradesh, India

Tanmayee Basu Department of Biotechnology, Thapar Institute of Engineering & Technology, Patiala, Punjab, India

Ashis Kumar Bera Department of Biomedical Engineering, Indian Institute of Technology Hyderabad, Kandi, Telangana, India

Deepti Bhat bioTox Solution Pvt. Ltd., Institute of Engineering and Technology, Lucknow, Uttar Pradesh, India

Anugya Bhatt Division of Thrombosis Research, Department of Applied Biology, Biomedical Technology Wing, Sree Chitra Tirunal Institute for Medical Sciences and Technology, Thiruvananthapuram, Kerala, India

Deepti Chopra Photobiology Laboratory, Systems Toxicology and Health Risk Assessment Group, CSIR-Indian Institute of Toxicology Research (CSIR-IITR), Lucknow, Uttar Pradesh, India
Photobiology Laboratory, Systems Toxicology and Health Risk Assessment Group, CSIR-Indian Institute of Toxicology Research, Lucknow, Uttar Pradesh, India
Department of Biochemistry, School of Dental Sciences, Babu Banarasi Das University, Lucknow, Uttar Pradesh, India

Rashi Chugh Department of Biotechnology, Thapar Institute of Engineering & Technology, Patiala, Punjab, India

Snehal Desai bioTox Solution Pvt. Ltd., Institute of Engineering and Technology, Lucknow, Uttar Pradesh, India

Divya Dubey Department of Molecular Biosciences, The Wenner-Gren Institute, Stockholm University, Stockholm, Sweden

Prakriti Gaur Photobiology Division, CSIR-Indian Institute of Toxicology Research, Lucknow, Uttar Pradesh, India
Photobiology Laboratory, Systems Toxicology and Health Risk Assessment Group, CSIR-Indian Institute of Toxicology Research (CSIR-IITR), Lucknow, Uttar Pradesh, India

Anurag Kumar Gautam Department of Pharmaceutical Sciences, Babasaheb Bhimrao Ambedkar University (A Central University), Lucknow, Uttar Pradesh, India

Anastasiya Gorkun Wake Forest Institute for Regenerative Medicine, Winston-Salem, NC, USA
FSBSI Institute of General Pathology and Pathophysiology, Moscow, Russia
Sechenov First Moscow State Medical University, Institute for Regenerative Medicine, Moscow, Russia

Shruti Goyal bioTox Solution Pvt. Ltd., Institute of Engineering and Technology, Lucknow, Uttar Pradesh, India

Ranjit Singh Gujjar Division of Crop Improvement, Indian Institute of Sugarcane Research, Lucknow, Uttar Pradesh, India

Alan Jacobson Wake Forest Institute for Regenerative Medicine, Winston-Salem, NC, USA

Adam M. Jorgensen Wake Forest Institute for Regenerative Medicine, Winston-Salem, NC, USA
Wake Forest School of Medicine, Medical Center Boulevard, Winston-Salem, NC, USA

Mohd Danish Kamar Photobiology Laboratory, Systems Toxicology & Health Risk Assessment Group, CSIR-Indian Institute of Toxicology Research (CSIR-IITR), Lucknow, Uttar Pradesh, India
Academy of Scientific and Innovative Research, AcSIR Headquarters, CSIR-HRDC Campus, Ghaziabad, Uttar Pradesh, India

Varun R. Kasula Wake Forest Institute for Regenerative Medicine, Winston-Salem, NC, USA

Malini Kotak ITC Life Sciences and Technology Center, ITC Ltd., Bangalore, Karnataka, India

Vipin Kumar Department of Pharmaceutical Sciences, Babasaheb Bhimrao Ambedkar University (A Central University), Lucknow, Uttar Pradesh, India

Naresh Mahajan Wake Forest Institute for Regenerative Medicine, Winston-Salem, NC, USA

Bhartendu Nath Mishra Department of Biotechnology, Institute of Engineering & Technology, Lucknow, Uttar Pradesh, India

Krishan Mistry Translational and Clinical Research Institute, Newcastle University, Newcastle upon Tyne, UK

Renjith P. Nair Division of Thrombosis Research, Department of Applied Biology, Biomedical Technology Wing, Sree Chitra Tirunal Institute for Medical Sciences and Technology, Thiruvananthapuram, Kerala, India

Aditi Pandey Graduate Institute of Nanomedicine and Medical Engineering, Taipei Medical University, Taipei, Taiwan

Anuj Pandey CSIR-Indian Institute of Toxicology Research (CSIR-IITR), Lucknow, Uttar Pradesh, India

Ruchi Pandey School District of Philadelphia, Philadelphia, PA, USA

A. B. Pant CSIR-Indian Institute of Toxicology Research (CSIR-IITR), Lucknow, Uttar Pradesh, India

Academy of Scientific and Innovative Research (AcSIR), Ghaziabad, Uttar Pradesh, India

Sunil Kumar Patel Photobiology Laboratory, Systems Toxicology & Health Risk Assessment Group, CSIR-Indian Institute of Toxicology Research (CSIR-IITR), Lucknow, Uttar Pradesh, India
Academy of Scientific and Innovative Research, AcSIR Headquarters, CSIR-HRDC Campus, Ghaziabad, Uttar Pradesh, India
Photobiology Laboratory, Systems Toxicology and Health Risk Assessment Group, CSIR-Indian Institute of Toxicology Research, Lucknow, Uttar Pradesh, India

Diksha Pathania Photobiology Laboratory, Systems Toxicology and Health Risk Assessment Group, CSIR-Indian Institute of Toxicology Research (CSIR-IITR), Lucknow, Uttar Pradesh, India
Academy of Scientific and Innovative Research (AcSIR), CSIR-IITR Campus, Lucknow, Uttar Pradesh, India

Falguni Pati Department of Biomedical Engineering, Indian Institute of Technology Hyderabad, Kandi, Telangana, India

Shiv Poojan bioTox Solution Pvt. Ltd., Lucknow, Uttar Pradesh, India
Bangalore, India
Bangalore, India

Gaurav Prajapati Photobiology Laboratory, Systems Toxicology & Health Risk Assessment Group, CSIR-Indian Institute of Toxicology Research (CSIR-IITR), Lucknow, Uttar Pradesh, India
Academy of Scientific and Innovative Research, AcSIR Headquarters, CSIR-HRDC Campus, Ghaziabad, Uttar Pradesh, India

Priyanka Rana Photobiology Laboratory, Systems Toxicology and Health Risk Assessment Group, CSIR-Indian Institute of Toxicology Research (CSIR-IITR), Lucknow, Uttar Pradesh, India
Academy of Scientific and Innovative Research (AcSIR), CSIR-IITR Campus, Lucknow, Uttar Pradesh, India

Havisha M. H. Rao bioTox Solution Pvt. Ltd., Institute of Engineering and Technology, Lucknow, Uttar Pradesh, India

N. S. Remya Division of Toxicology, Department of Applied Biology, Biomedical Technology Wing, Sree Chitra Tirunal Institute for Medical Sciences and Technology, Thiruvananthapuram, Kerala, India

Sana Sarkar CSIR-Indian Institute of Toxicology Research (CSIR-IITR), Lucknow, Uttar Pradesh, India

Rama Shankar Sahu Package Reliability at Central Quality, Micron, Taichung, Taiwan

Saumya Shukla Photobiology Laboratory, Systems Toxicology and Health Risk Assessment Group, CSIR-Indian Institute of Toxicology Research (CSIR-IITR), Lucknow, Uttar Pradesh, India
Department of Biochemistry, School of Dental Sciences, Babu Banarasi Das University, Lucknow, Uttar Pradesh, India

Shakti Vinay Shukla Fragrance & Flavour Development Centre, Kannauj, Uttar Pradesh, India
Fragrance & Flavour Development Centre, Ministry of MSME, Govt. of India, Kannauj, Uttar Pradesh, India

Vinay Shakti Shukla Fragrance & Flavour Development Centre, Kanauji, Uttar Pradesh, India

Amita Singh Department of Pharmaceutical Sciences, Babasaheb Bhimrao Ambedkar University (A Central University), Lucknow, Uttar Pradesh, India

Shay Soker Wake Forest Institute for Regenerative Medicine, Winston-Salem, NC, USA
Wake Forest School of Medicine, Medical Center Boulevard, Winston-Salem, NC, USA
Atrium Health Wake Forest Baptist, Medical Center Boulevard, Winston-Salem, NC, USA

Swastika Suvirya Department of Dermatology, Venereology and Leprosy, King George's Medical University, Lucknow, Uttar Pradesh, India

Atul Kumar Upadhyay Department of Biotechnology, Thapar Institute of Engineering & Technology, Patiala, Punjab, India
Division of Crop Improvement, Indian Institute of Sugarcane Research, Lucknow, Uttar Pradesh, India

S. Usha bioTox Solution Pvt. Ltd., Institute of Engineering and Technology, Lucknow, Uttar Pradesh, India

M. R. Vijayakumar Department of Pharmaceutical Sciences, Babasaheb Bhimrao Ambedkar University (A Central University), Lucknow, Uttar Pradesh, India

Apeksha Vikram Photobiology Laboratory, Systems Toxicology & Health Risk Assessment Group, CSIR-Indian Institute of Toxicology Research (CSIR-IITR), Lucknow, Uttar Pradesh, India
Academy of Scientific and Innovative Research, AcSIR Headquarters, CSIR-HRDC Campus, Ghaziabad, Uttar Pradesh, India

Grayson A. Wagner Wake Forest Institute for Regenerative Medicine, Winston-Salem, NC, USA

Kelsey Willson Wake Forest Institute for Regenerative Medicine, Winston-Salem, NC, USA

Wake Forest School of Medicine, Medical Center Boulevard, Winston-Salem, NC, USA

Akhilesh Kumar Yadav Regulatory Toxicology, CSIR-Indian Institute of Toxicology Research, Lucknow, Uttar Pradesh, India

Ashok Kumar Yadav Department of Pharmaceutical Chemistry, Hygia Institute of Pharmaceutical Education and Research Lucknow, Lucknow, Uttar Pradesh, India

Artificial Skin Models for Animal-Free Testing: 3D Skin Reconstruct Approach, a Journey in the Past Two Decades

1

Ruchi Pandey and Shiv Poojan

Abstract

Growing ethical concerns regarding the use of animals in research have managed to the creation of several alternative procedures based on the refinement, reduction, and replacement, which Russell and Burch initially introduced in 1959. After that, since 2013, as animal experimentation ethics have been supported in the European Union, artificial skin models have fascinated interest as an alternative to use of animal model for testing for establishing the efficacy and toxicity of products. In addition to concerns for animal welfare, using animals in experiments should be reduced and avoided. Research on cosmetics and the skin is particularly important to objections to animal testing. Due to numerous constraints, including the fact that human and animal skins have distinct immune systems and anatomical makeup, investigations on animals may not correctly anticipate results in people. This chapter's major goal is to provide an overview of the strategies for creating 3D skin models, along with their advantages and disadvantages, as well as how new methods can be used to create constructions that are truly physiologically accurate and useful for preclinical innovation. In vivo animal testing for evaluating efficacy and safety in the beauty in the field of pharmaceutical sectors can be replaced with artificial skin models that closely resemble human skin. The primary investigations on cell-to-cell interactions, cell-matrix interactions, tissue creation, and development can also benefit from using 3D skin construct models. An integrated application of these approaches would give insight into the minimum use of animals in scientific experiments.

R. Pandey
School District of Philadelphia, Philadelphia, PA, USA

S. Poojan (✉)
bioTox Solution Pvt. Ltd., Institute of Engineering and Technology, Lucknow, Uttar Pradesh, India

A. B. Pant et al. (eds.), *Skin 3-D Models and Cosmetics Toxicity*,
https://doi.org/10.1007/978-981-99-2804-0_1

Keywords

3Rs · Artificial skin · 3D skin construct · European Union · Cell–matrix interaction

1.1 Introduction

The preamble of sophisticated 3D skin model development that is now being used by Global cosmetic research began more than 80 years ago. In the USA, in 1933, "an eyebrow and lash dye, Lash Lure" was introduced to the market. At first, no one thought an innocuous cosmetic product could have acute and even lethal repercussions. A chemical paraphenylenediamine, which was present in the product, was a substance that has not been thoroughly studied and can have significant effects on the face, eyelids, and eyes (McCally et al. 1933), and the impact of that compound was very intense. After using the dye, more than a dozen women lost their eyesight, and one of them contracted a fatal bacterial infection. After that, the U.S. Food and Drug Administration (FDA) authority has taken over cosmetics testing. Also, in 1936 publication introduced by "American Chamber of Horrors: The Truth about Food and Drug," Lamb (1936) emphasized the many examples where consumer goods show any change in terms of injury or even death. Even in 1938, the Federal Food, Drug and Cosmetic Act was passed by the U.S. Congress, which requires stricter regulations for cosmetic products (U.S. Congress 1934). Since then, cosmetic testing has been a crucial component of product development due to the possibility of adverse health consequences that could be severe due to high and frequent exposure. Following these instances that prompted consumer protection, animal testing quickly became required under American law (Zurlo et al. 1994). Due to this, the Scientific Committee on Consumer Safety (SCCS) decided to publish ad continuously update the SCCS Notes of Guidelines for Testing of Cosmetic Ingredients and their Safety Assessment (Bernauer et al. 2019). Since then, pharmacological and cosmetic product screening for skin research has been frequently done on animals (such as mice and pigs). Hence, the initial animal rights movement was established when animal testing became necessary.

1.2 Approach Toward an Animal Alternative

To overcome the problems with animal research and steer clear of unethical practices, alternative models to animal testing have been offered. In 1959, Russell and Burch published the first description of the fundamentals of human experimental research (Tannenbaum and Bennett 2015), which are elegantly referred by Russell and Burch as follows:

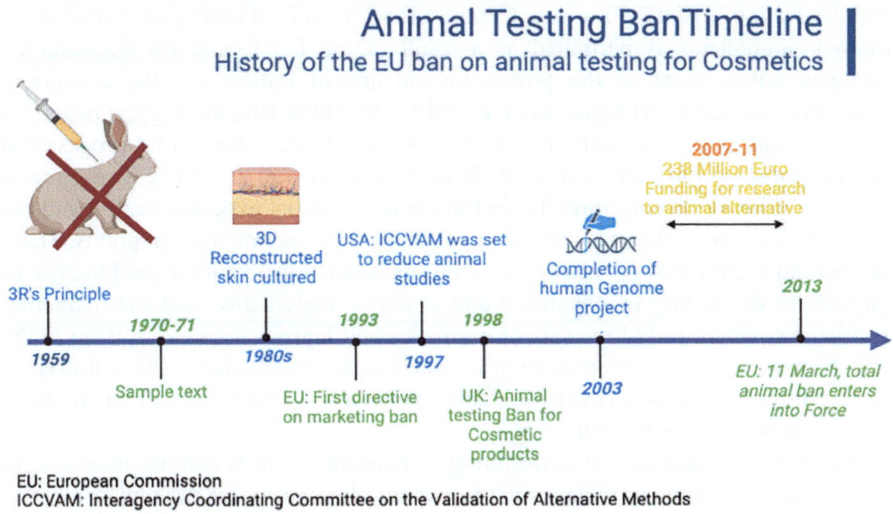

Fig. 1.1 EU timeline history for animal testing ban on cosmetics

First method is refinement, which involves the adoption of sophisticated techniques to prevent animal suffering or distress.

The second method is reduction, which involves employing methods to get similar amounts of data from fewer animals.

Third is substitute trials that do not include animals when possible. Three key factors for the development of non-animal testing emerged when the 3Rs were viewed from the current angle: ethical reasons, the absence of practical extrapolation, and economic considerations. When animal experimentation was eliminated, the search for human-relevant data at an affordable cost has been driven by the ethical consideration that they wanted to avoid (Silva and Tamburic 2022). The weight of ending animal testing has been placed on the cosmetics industry by animal welfare activism and public opinion, which has ultimately led to a fourth and crucial driver—animal testing restrictions. The result of these partnerships between researchers, NGOs, decision-makers, and the cosmetics business is the field of "New Approach Methodologies," which is constantly developing (NAMs).

Nonetheless, during the twentieth century, using animal testing models was common for determining the safety and effectiveness of novel medications or cosmetic elements (Semlin et al. 2011), yet because of growing worries about what happens to lab animals and because of ethical and scientific considerations (Ferdowsian and Gluck 2015), European legislators were compelled to strictly control and ultimately outlaw the use of animals in cosmetics testing. The laws governing animal testing have tightened up since the 1990s. A partial ban was enacted in the European Union (EU) in 1993 after the European Commission (EU) published the first legislation to restrict the use of animals in the cosmetic sector, whereas, in 1998 (Fig. 1.1), the UK became the first country in the world to completely outlaw cosmetics tested on animals, and 20 years later, in 2013, the EU followed suit (CREDIT: European Commission). No animal testing on formulations or their ingredients is permitted as a result of the EU Cosmetics Regulation's stringent enforcement of the prohibition on animal testing and the concurrent marketing ban as of 2013 (Suhail et al. 2019; EC 2009). But, the biggest negatives were the high cost, time commitment, and, most critically, the painful process that animals must go through (Ahn et al. 2010; Cheluvappa et al. 2017). These issues were the primary driving force behind the ban on animal experimentation and the hunt for alternatives that may be used to evaluate the experimental endpoints. There are two different kinds of restrictions: I testing restrictions, which forbid the use of animals in the testing of cosmetics and cosmetic ingredients, and (ii) marketing restrictions, which forbid the sale of cosmetics and ingredients used in them in the EU. Numerous other nations have also issued bans. Nonetheless, it is still used in some nations' cosmetic markets, including those in the United States, where using animals in research is not outlawed.

Nevertheless, also its use is declining as consumers are becoming increasingly critical. These regulatory changes led to various alternatives for the replacement of in vivo animal tests, and together pharmaceutical and cosmetics companies started to find suitable skin models that could be used to access new formulations and other topical products (Jung et al. 2014; Catarino et al. 2018). The fundamental justification for seeking an animal-free option for skin research, aside from ethical considerations, is that animals do not exhibit the same physiological, structural, and biochemical behavior as human skin, which leads to high drug attrition rates in later phases of study. Organization for Economic Co-operation and Development (OECD) approved the toxicological testing strategies for skin sensitization, hazard assessment without animal testing. The first OECD Guide Line (GL) to include defined approaches (DAs) for skin sensitization study is (GL) No. 497, a new type of (non-test) guideline that employs combined data from non-animal methods to provide toxicological information for hazard and potency evaluations (EC 2002; OECD 2021a).

1.3 Merits/Demerits of Elective Methods

Replacement could be the best way to dodge animal torment and endure. Although it is hard to completely replace animal research, combining two or three in vivo procedures should be required as an elective method. Investigational projects are consistently applied with limited funds and time. Selecting elective non-animal experiments can uniquely spare both fetched and time. Besides, in a few cases, elective methods are way well suited than the test in animal since they permit higher quality tests to be required. Animal and human skins are endlessly distinctive regarding architecture and immune response; conjointly, animal skin often has a significantly lower life expectancy than human skin (Van Gele et al. 2011). Therefore, it is outlandish to know precisely what happens within the experimental animal body, and elective in vitro methodology can more obviously uncover the overall mechanisms involved. The comprehensive immune response of animal experiments plays the most crucial role. Ironically, elective methods held back on this front, which is why animals are quiet often used for different dermatological assessment and research. Here is the development of other optional plans by focusing on the use of elective an animal alternative model for cosmetic research.

Biochemical methods, 2D and 3D cell cultures, genomics research, and the creation of in bioinformatics-based in silico simulations of skin models are some specific/other potential alternatives to animal models (Nakamura et al. 2018; Yun et al. 2018). Different methods, such as cell culture or artificial skin models, have been developed to produce and construct biological skin models that replicate the human skin's very complex and stratified structure. These models use synthetic or natural biomaterial-based scaffolds (Przekora 2020); even though they have numerous drawbacks, their careful and skillful handling makes these methods indispensable for futuristic dermatology research. EpiDerm® (MatTek Corporation, USA), EpiSkin® (L'Oreal, France), epiCS® (CellSystems, Germany), Holoderm®, and SkinEthic® (SkinEthics, France) were the most popular epidermis models based on the utilization of human skin cells (Mao et al. 2003; Whang et al. 2005; You et al. 2012). More recently, a few advanced skin models were commercialized, including NeoDerm® (Tego Science, Korea), Phenion® (Henkel, Germany), Genoskin Ex Vivo (USA) (summarized in Table 1.1). A collagen matrix comprising human fibroblasts and an epidermal overlay made of human keratinocytes serve as the foundation for sophisticated skin models (Ackermann et al. 2010; Kano et al. 2010). The representation of the full thickness of human skin's structural makeup is crucial in engineered skin models and the skin's cellular components. Complex human skin models with the proper cell compositions and matrix structure can be made using a number of techniques, such as electrospinning, three-dimensional (3D) bioprinting, and microfluidic systems (Kempf et al. 2011; Koch et al. 2012; Atac et al. 2013). A significant tool for understanding cell–cell, cell–matrix, and dermal–epithelial interactions in dermatology, as well as for assessing the safety of novel drug formulations or cosmetic elements, is the 3D engineered skin model (Suhail et al. 2019; Lee et al. 2014a, b).

1.4 Skin Equivalent: 3D Skin Model, a Valuable Alternative to Animal Tests

There has been an enormous increase in dermatology research using skin equivalents for fundamental and industrial research in the past several years and for clinical applications (Choudhury and Das 2021). Tissue-engineered human skin equivalents have been produced to regenerate the skin's main structural and functional behavior in vitro. These synthetically built skin substitutes, made up of epidermal and dermal layers, are known as skin equivalents (Dellambra et al. 2019). They are made from primary cells (keratinocytes, fibroblasts, and/or stem cells) and extracellular matrix (ECM) components in a manner that closely resembles natural skin, allowing for the study of their effect. They have been broadly utilized for skin homeostasis studies and its alterations and also for generating therapeutic tools, which can be used for

Table 1.1 Commercially available reconstructed skin models

Name of reconstructed skin model	Cells used for culture	Structures	OECD approved	Provider	Place
EpiSkin™	Keratinocytes	Cultured on a matrix	TG439, TG431	EpiSkin SNC	Lyon, France
EpiDerm™	Keratinocytes	Based on tissue culture inserts	TG431, TG439	MatTek	Ashland, MA, USA
SkinEthic™ RHE	Keratinocytes	Cultured on polycarbonate filter	TG431, TG439		EPISKIN Labs, France
epiCS™	Keratinocytes	Cultured on tissue culture inserts	TG431	Cell Systems	Troisdorf, Germany
SkinEthic™ RHPE	Keratinocytes Melanocytes	Pigmented epidermis			
T-Skin™	Keratinocytes Fibroblasts	Full-thickness model			
EpiDerm™ FT	Keratinocytes Fibroblasts	Full-thickness model			
MelanoDerm™	Keratinocytes Melanocytes	Pigmented epidermis			
epiCS™-M	Keratinocytes Melanocytes	Pigmented epidermis			
Ex Vivo Skin	From surgical discard	Full-thickness model			
Naïve Ex Vivo Skin	Donated surgical skin	Full-thickness model		Genoskin, USA	

chronic skin lesions (Martínez-Santamaría et al. 2012). Right now, the finest substitute tool for animal research is three-dimensional (3D) reconstructed living human skin analogs. They have been extensively used to study a variety of dermatological research because these can reproduce close resemblance of structural and functional properties with natural human skin. With the aid of a computer-controlled 3D printer, tissues and organs can be created by precisely positioning living cells, biological components, and biochemicals (Fig. 1.2). Therefore, 3D printing method has developed as a propitious tool to assemble structural matrix and attracted the attention of skincare companies. There are several methods to mimic skin models, including electrospinning, 3D printing, and even microfluidic devices, which show the human skin's structure in much more detail. Table 1.2 enlightens the various procedures used for the fabrication of artificial skin model using different biomaterials and their applications in cosmetic research (Yun et al. 2018). When used to create tissue-engineered constructs, 3D bioprinting imparts high accuracy, reproducibility, and good control over a scaffold's internal structure and external shape (Koch et al. 2010, 2012).

The skin is known to be a multilayered structure containing various cell types, and thus, 3D bioprinting could provide the opportunity to deposit cells in this

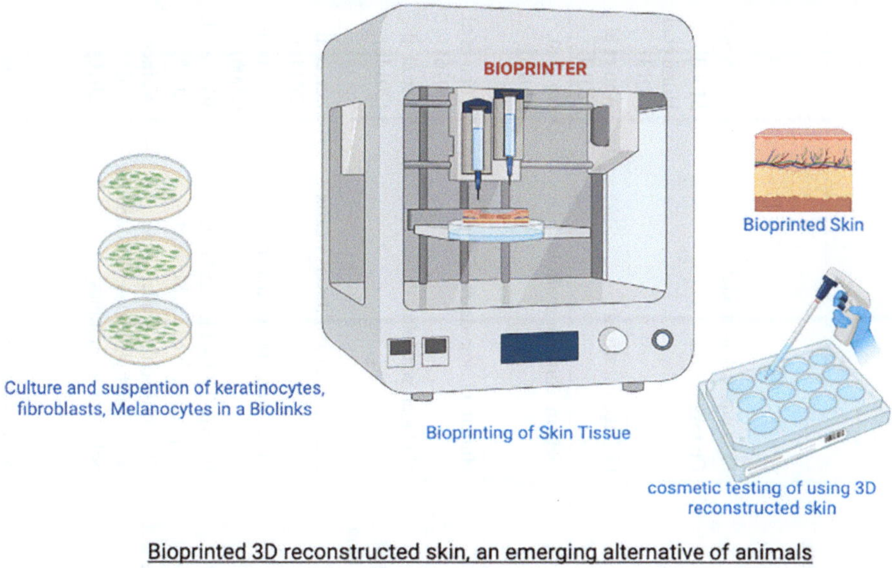

Fig. 1.2 Bioprinting used for bioprinted 3D reconstructed skin

arrangement. Multilayer artificial skins were created by depositing a collagen type I from rat tail hydrogel precursor, fibroblasts, and keratinocytes using a layer-by-layer printing process. Lee et al. (2014b) described the creation of 3D manufactured skin models using collagen and human skin cells assembled, layer by layer. Primarily, collagen layer was printed, followed by both the cells' (keratinocytes and fibroblasts) deposition on the top of every particular collagen matrix. In the reconstructed skin model construct, cell viability was great (>94%), and at 14 days following air–liquid interface culture, completely mature skin tissue displayed 3–7 different cell layers in the epidermis. In a different study, Ng et al. (2016) created skin constructs using a gelatin–chitosan bioink with good printability and antibacterial properties. The dermal region and portions of the outer epidermal layer were printed in three dimensions at around 400 μm. The fibroblasts from human foreskin showed a spindle-like morphology on the 5% gelatin–chitosan, and more viable cells were seen on hydrogels.

Creating complex epidermis and dermis structures via 3D printing offers hope for skin tissue engineering, but many technical obstacles still exist. The difficulties in fabrication of 3D skin constructions with high resolution and printability include biomaterials restrictions due to biocompatibility, biodegradability, and physico-chemical qualities (Zhu et al. 2016). Additionally, the conditions of 3D printing should be tuned to reduce stress-related cellular and biological component damage during the deposition phase (Patra and Young 2016).

Table 1.2 Various skin reconstruct fabrication methods, model, and their application

Methods	Cell type used for culture	Structure	Use in cosmetic industry/ dermatology research	Skin equivalent model	References
Freeze dying	Primary human dermal fibroblasts	Dermis	Skin tissue engineering	Chitosan sheet	Mohd Hilmi et al. (2013)
	Human foreskin fibroblasts Human keratinocytes	Dermis epidermis	Skin tissue engineering	Collagen scaffold	Ahn et al. (2010)
3D printing	Human neonatal fibroblasts	Dermis epidermis	Skin tissue engineering Bioink	Gelatin–chitosan scaffold	Ng et al. (2016)
	Human fibroblasts human keratinocytes	Dermis epidermis	Bioink	Human plasma-derived fibrin matrix	Cubo et al. (2016)
	Human fribroblasts HaCaT cells	Dermis epidermis	Bioink drug transdermal test skin regeneration	Collagen scaffold	Lee et al. (2014b)
Electrospinning	Normal human fibroblasts	Dermis	Skin tissue engineering	Silk fibroin nanofibrous matrix	Lee et al. (2014a)
	Normal human fibroblasts Normal human keratinocytes	Dermis epidermis	Skin tissue engineering	Chitin nanofibrous matrix	Noh et al. (2006)
Electrospinning/ freeze-drying	–	Dermis epidermis	Antibacterial wound dressing	Bilayer TiO_2-chitosan/human adipose-derived ECM sheet	Woo et al. (2015)
Microfluidic device	Human fibroblasts and dermal keratinocytes Blood vessel endothelial cellsHuman umbilical vein endothelial cells (HUVECs), HaCaT cells	Dermis epidermis vascular layer	Drug toxicity test	Skin-on-a-chip model using polydimethylsiloxane (PDMS) and polyester membranes	Wufuer et al. (2016)
	Human keratinocytes human dermal fibroblasts			Skin chip model using PDMS and collagen hydrogel	Lee et al. (2017)

1.5 Skin Equivalents As Far

1.5.1 In Vitro Reconstituted Epidermis

In the 1980s, collagen gel and multilayered human epidermal keratinocytes (isolated and serially reproduced in vitro from a tiny skin biopsy) were used to culture human fibroblasts for 3D culture for the first time (Bell et al. 1981). After that, as an alternative, essential reconstructed skin culture comprising cultured keratinocytes on a mesh is started to be utilized for assessing the safety of home and personal care products (Triglia et al. 1991). Though the more enhanced performance of in vitro reconstitutes, the epidermis can be obtained using biological matrices for keratinocyte seeding, such as a fibrin substrate that permits keratinocyte stem cell conservation and growth factor delivery (Hynds et al. 2018; Wang et al. 2017). However, as demonstrated by several clinical trials, a genuine dermis is necessary for improvement in the histological quality of the newly regenerated skin and cell engraftment (Tavakoli and Klar 2021).

1.5.2 Development of Full-Thickness Skin Equivalents (FTSE)

The reconstituted epidermis has its own limitations; therefore, the 3D organotypic models have been produced to overcome these. FTSE also promotes cellular communication between the dermal and epidermal layers, making it possible to use it as a more complicated model to study processes like skin formation, infection, or wound healing. Epidermal models still offer highly standardized conditions for risk assessment (Reuter et al. 2017). By seeding primary keratinocytes on a de-epidermized dermis, full-thickness skin equivalents (FTSEs) are created (Zhang and Michniak-Kohn 2012; Singh et al. 2020; Reijnders et al. 2015) or on biodegradable polymer substrates (natural origin hydrogels or synthetic hydrogels) incorporating human dermal fibroblasts. The basal layer keratinocytes also rigorously control the proliferation. Although hydrogel production can be optimized, no perfect bioink exists to produce a hydrogel that mimics the structural, mechanical, and biochemical characteristics of native skin in a medically meaningful manner. Thus, an additional focus on hydrogel composition is vital. Another approach that may be considered is utilizing the biochemical pathways for physiological polymer formation to create bioink materials with increased physiological pertinence (Randall et al. 2018). It accurately reflects the physiological aspect, the structural and mechanical elements of 3D skin creations, including appendages and macrostructures like glands and vasculature, which are crucial. Additionally, integrating technologies like MESW and 3D bioprinting open up new possibilities for merging synthetic and natural matrix that produce the tissue environment required for cell survival while offering structural support. To prevent dermal framework shrinkage in long-lasting cultures, fibrin or a composite silk–collagen matrix should be preferred (Janani et al. 2019). At the confluence point of air and liquid, the skin equivalents are present in an uncovered position to cultivate entire dermis differentiation and to facilitate full epidermal differentiation and stratification (Roger et al. 2019). However, these skin equivalents are comparatively more porous than native human skin (Bouwstra and Ponec 2006). At the epidermal basal layer, co-seeded keratinocytes and melanocytes merge as a single entity to regenerate their physiological distribution (Cichorek et al. 2013). Full-thickness skin model (human skin equivalent HSE, FTSE) offers several benefits: (1) Since most models are composed of primary human cells, inter-species extrapolation is avoided; (2) repeated application of formulations can be performed in contrast to ex vivo human skin for at least several weeks; (3) as it is ready to use, it does not require advanced knowledge of cell culture technique; (4) It has become the great alternative to animal models for research and development applications in regulatory toxicology and in the cosmetic industry (Zhang and Michniak-Kohn 2012; Brohem et al. 2011; Groeber et al. 2011). Besides these applications, models of skin have been employed frequently to investigate the cellular and molecular mechanisms governing cutaneous disease.

1.5.3 Most Recent Skin Equivalents

Unfortunately, with regards to the production of 3D skin, analysts confront numerous issues that vexed researchers in the mid-twentieth century, i.e., complicated structures like glands and tactile corpuscles, physiological oxygen, and nutrient delivery via a perfused vasculature, as well as easily available and repeatable 3D model for use in research laboratories. A significant barrier to furthering our understanding of the skin is the absence of skin appendages in skin grafts. Therefore, skin appendages are being incorporated into full-thickness skin equivalents to produce the latest generation skin equivalents. Adipocytes at the time of maturity are well co-cultured with fibroblasts and keratinocytes. It improves the stability between epidermal growth and differentiation and develops a more competent epidermal barrier (Vig et al. 2017; Yang et al. 2019). As of late, an "endogenous" HSE was produced by employing a active whirligig culture of fibroblasts fixed in a temporary matrix made of gelatin microspheres, which is steadily damaged. In contrast, fibroblasts congregate with the extracellular matrix (ECM) and lead to improved epidermal barrier function (Tracy et al. 2016). The main drawback of skin equivalents is their lack of a functional vascular system. Vascularizing skin structures for therapeutic applications is vital since it is required for good and long-lasting structure and function. However, vascular skin equivalents are not helpful as models to study the main features of diseased skin, i.e., leukocyte trafficking across vascular endothelium or testing the skin's ability to absorb an intravenously supplied chemical. Black et al. created the first skin analog with a capillary-like architecture in 1998 (Black et al. 1999). The main use has been to improve the graft uptake in a clinical setting or to examine angiogenic and angiostatic drugs (Veith et al. 2019; Gradin et al. 2021; Shahin et al. 2020). The vascularized tissue generation was performed using decellularized porcine small bowel segments due to their tendency to have a collagen matrix scaffold showing the structures of native vascular network and is repopulated with endothelial progenitor cells. With its characteristic endothelial differentiation structures, it is able to form a vascular network (Schanz et al. 2010). The attainment of typical skin architecture can accelerate because of such system, which supports the 3D skin under immersed conditions and at the air-lift liquid interface (Groeber et al. 2011). As of late, dermal fibroblasts have been utilized for neural stem cell generation through direct reprogramming and to obtain a neuroimmune-cutaneous system. These have been added to skin substitutes made of silk collagen that contain immune and adipose cells. A major setback is cultivated primary skin cells' low proliferation potential and scarcity of these skin analogs. On the other hand, using induced pluripotent stem cells (iPSCs) is very crucial as these can differentiate into different skin cells, with dermal papilla and sensory neuron cells (Abaci et al. 2018). These have also been used to generate FTSE. Therefore, it could be an incredible source for massive-scale generation of distinctive skin cell sorts, with improved reproducibility.

1.6 Future Perspective: Next-Generation Skin Equivalents, a More Advanced Way Ahead!

With so many advancements in the skin research field, it is tough to optimize a method where an organotypic model can be developed, summarizing human skin's whole complexity and similarity. Once immune-competent cells are effectively integrated into hydrogels and inside a circulating vasculature tissue culture, modeling of skin disease will become a reality (Randall et al. 2018). Integrating various biofabrication techniques, such as electrospinning and bioprinting, will target both appendages to increase the possibility of producing a functional skin and the consolidation of immune cells within the skin model for specific formulations and other preclinical applications (Fig. 1.2). The next step toward developing the skin equivalent within microfluidic may provide much better models, which can mimic skin function even more efficiently (van den Broek et al. 2017). In previous days, silicone microfabrication and micromachining techniques were used to produce microfluidic devices (Preetam et al. 2022). On the other hand, a comparatively cheaper and easier way to develop microfluid devices is to use biocompatible silicone rubber poly(dimethylsiloxane) (PDMS) (Torino et al. 2018). Due to perfused vascular structure of microfluidic platforms, it became easier to mimic in vivo physical force applied by blood flow (shear stress), which is necessary to regulate endothelial cell gene expression, morphology, proliferation, and apoptosis (D'Arcangelo et al. 2016; Osaki et al. 2018). Skin analogs, skin biopsies, or explants of individual hair follicles have all been cultured in a dynamically perfused bioreactor based on chip chamber by subjecting them to varying mechanical shear stress. Long-term composite skin equivalents can be maintained, and multiple tests can be performed without device disassembly/tissue disruption.

Furthermore, histological procedures and other analyses can be performed after removing the tissue from the device. Epidermal stratification, differentiation, and barrier functions can be improved by allowing dynamic perfusion and a finely controlled region that is exposed to air movement and gas composition in microfluidic systems (Sriram et al. 2018). Microfluidic devices play an important role to create skin immuno-competent models. To represent human dendritic cells, a keratinocyte cell line HaCaT has been cultured as an epidermis barrier model on one side using a bi-channel device and on the other side using a human leukemic monocyte lymphoma cell line (U937) (Ramadan and Ting 2016). The effects of UV irradiation are evaluated by measuring an integrated magnetic-bead immunological test and trans-epithelial electrical resistance.

Similar to this, a model based on three microfluidic channels was developed to simulate skin inflammation and edema for drug testing (Wufuer et al. 2016), and to further improve skin equivalent complexity, 3D bioprinting technology has been applied. Indeed, to build a similar structure as native human skin, deposition of various cell types and biomaterials has been permitted by this fully automated system (Ng et al. 2016; Lee et al. 2017). Using 3D bioprinting, Abaci and colleagues created a 3D skin with perusable, made of both primary and iPSC-derived endothelial cells (Abaci et al. 2016). They also worked on a pumpless "skin-on-a-chip" model (Abaci et al. 2015). A HUVEC-coated nylon wires were used to produce perfusable skin equivalent when inserting within the dermal compartment and using 3D bioprinting. The benefit of this model is that, after medication delivery, it demonstrates effective percutaneous penetration in the endothelialized tubes (Mori et al. 2017). The organization of skin appendage age niches in mini-organoids may be recreated using 3D bioprinting technology. The difference in density, anatomy, and function of different fibroblast subpopulations can easily be observed in the native dermis. However, the difference in the composition of extracellular matrices is also predominant (Sriram et al. 2015). By fusing various fibroblast subpopulations with various extracellular matrix elements, 3D bioprinting technology can be an effective tool for recreating the dermal natural composition. Altogether, 3D bioprinting technology can be very supportive to summarize the local dermal composition by combining diverse fibroblasts with various extracellular matrix components and can achieve a way to make a connection between in vitro models of different tissues and the skin "human-on-a-chip" system for drug screening. Recently, skin has been connected with organs such as kidney, liver, and intestine (Risueño et al. 2021).

Some pigmentation experiments utilize human melanocytes in the base layer of multifaceted epidermal keratinocytes. Skin aging-related studies about wrinkles and elasticity of skin employ some full-thickness skin models including keratinocytes and fibroblasts. OECD records a few skin constructs as options to animal experiments for chemical testing in their technical guidelines (TG): TG431, a skin corrosion test, and TG439 (OECD 2019), a skin irritation test (OECD 2021b).

However contrary, 3D skin models cannot be used for drug penetrability tests because of lipid proportions in these 3D remodeled skin models that are not accurate compared to in ex vivo human skin ex plant and thus exhibit an increase in drug penetrance up to 5–50 fold in these models. The pivotal restriction of 3D skin cultures is their confinement of having a beneficial barrier and competent immune response. In some studies of 3D reconstructed skin models, the incorporation of immune cells has been thorough. Duval et al. (2003) have studied skin aging to access UV-induced skin damage and skin modifications and, for this, have used reconstructed skin containing Langerhans cells. In another study by Pageon et al. (2017), reconstructed skin containing monocytes was utilized to evaluate the glycation reaction, which is a partial reason for skin aging.

There are several advantages of 3D skin equivalents for both cosmetic and pharmaceutical industries. Each new substance/drug had to undergo various in vitro safety checks before each clinical study, and therefore, the cosmetic industry researchers may evaluate the medicines/chemicals by using 3D bioprinter-fabricated skin models, whereas, prior to any kind of marketing of cosmetic formulations, it is of utmost importance to evaluate the potential toxic and allergic effect of the same (Sarkiri et al. 2019). All of these needs and their ethical approach make 3D bioprinted skin a great tool to execute evaluation and screening of pharmaceutical and cosmetic products. Additionally, 3D skin bioprinting may be used to examine how well drugs and other active substances penetrate and absorb through skin. Global giants in the cosmetics industry, like L'Oreal and Proctor and Gamble, were interested in this technology and invested in the study and creation of 3D bioprinted skin models.

1.7 Conclusion

A PubMed search revealed references to "alternative to animal model" in every subject, indicating the significant and growing interest in studies utilizing alternatives to animals. The number of non-animal research has increased from 628 in 2007 to 212 in 1997 and 1219 in 2017 respectively. The skin equivalent strategy appears to be the most effective method now available, having advanced from systems that just consisted of keratinocytes seeded on a medium to more intricate cell and matrix combinations. An improvement in recreating the skin's structural, functional, and molecular network features is made possible by 3D bioprinting and microfluidic tools. The corresponding models replicate skin architecture and blood flow effects more precisely. The difficult task of collecting a more precise understanding of biological systems and appropriately resolving issues of cost, time, and ethics calls for the further development of in vitro skin systems. Forthcoming substitute technologies should ideally be able to simulate skin inside the framework of an artificial body, simulating certain connections with other organs. There is so much work to be done, but that will be very worthwhile. The straightforward hand-poured hydrogel matrix will also become obsolete in the twenty-first century with the adoption of 3D printing for usage in biological procedures providing a new benchmark for creating 3D tissue constructs. The necessary physiologically relevant skin components, such as the ECM and microbiome, can currently be produced using bioengineering techniques. Still, future advancements in these techniques and the creation of completely new ones will allow the cost-effective and repeatable in vitro production of physiological skin. A distinct physiological matrix and microenvironment, the addition of extra specific type of cells, and the simplicity of manufacture using novel fabrication processes are crucial components to creating a more accurate 3D skin model. Future researchers would not have to worry about choosing sources and techniques, and all skin aging research can be carried out utilizing comprehensive skin simulation models. It is critically necessary to collect reliable analyses of human data and conduct extensive and thorough sampling to realize this ideal model.

References

Uncategorized References

Abaci HE et al (2015) Pumpless microfluidic platform for drug testing on human skin equivalents. Lab Chip 15(3):882–888

Abaci HE et al (2016) Human skin constructs with spatially controlled vasculature using primary and iPSC-derived endothelial cells. Adv Healthc Mater 5(14):1800–1807

Abaci HE et al (2018) Tissue engineering of human hair follicles using a biomimetic developmental approach. Nat Commun 9(1):5301

Ackermann K et al (2010) The Phenion® full-thickness skin model for percutaneous absorption testing. Skin Pharmacol Physiol 23(2):105–112

Ahn S et al (2010) Designed three-dimensional collagen scaffolds for skin tissue regeneration. Tissue Eng Part C Methods 16(5):813–820

Atac B et al (2013) Skin and hair on-a-chip: in vitro skin models versus ex vivo tissue maintenance with dynamic perfusion. Lab Chip 13(18):3555–3561

Bell E et al (1981) Living tissue formed in vitro and accepted as skin-equivalent tissue of full thickness. Science 211(4486):1052–1054

Bernauer U, Bodin L, Chaudry Q, Coenraads PJ, Dusinka M, Ezendam J et al (2019) The SCCS notes of guidance for the testing of cosmetic ingredients and their safety evaluation: 10th revision. Publications Office

Black AF et al (1999) A novel approach for studying angiogenesis: a human skin equivalent with a capillary-like network. Cell Biol Toxicol 15(2):81–90

Bouwstra JA, Ponec M (2006) The skin barrier in healthy and diseased state. Biochim Biophys Acta 1758(12):2080–2095

Brohem CA et al (2011) Artificial skin in perspective: concepts and applications. Pigment Cell Melanoma Res 24(1):35–50

Catarino CM et al (2018) Skin corrosion test: a comparison between reconstructed human epidermis and full thickness skin models. Eur J Pharm Biopharm 125:51–57

Cheluvappa R, Scowen P, Eri R (2017) Ethics of animal research in human disease remediation, its institutional teaching; and alternatives to animal experimentation. Pharmacol Res Perspect 5(4): e00332

Choudhury S, Das A (2021) Advances in generation of three-dimensional skin equivalents: pre-clinical studies to clinical therapies. Cytotherapy 23(1):1–9

Cichorek M et al (2013) Skin melanocytes: biology and development. Postepy Dermatol Alergol 30(1):30–41

Cubo N et al (2016) 3D bioprinting of functional human skin: production and in vivo analysis. Biofabrication 9(1):015006

D'Arcangelo C et al (2016) Wear properties of dental ceramics and porcelains compared with human enamel. J Prosthet Dent 115(3):350–355

Dellambra E et al (2019) Non-animal models in dermatological research. ALTEX 36(2):177–202

Duval C et al (2003) The use of reconstructed human skin to evaluate UV-induced modifications and sunscreen efficacy. Exp Dermatol 12(s2):64–70

EC (2002) Twenty-sixth Commission Directive 2002/34/EC of 15 April 2002 adapting to technical progress Annexes II, III and VII to Council Directive 76/768/EEC on the approximation of the laws of the Member States relating to cosmetic products

EC (2009) Regulation (EC) no 1223/2009 of the European parliament and of the council on cosmetic products

Ferdowsian HR, Gluck JP (2015) The ethical challenges of animal research. Camb Q Healthc Ethics 24(4):391–406

Gradin R et al (2021) Quantitative assessment of sensitizing potency using a dose-response adaptation of GARDskin. Sci Rep 11(1):18904

Groeber F et al (2011) Skin tissue engineering–in vivo and in vitro applications. Adv Drug Deliv Rev 63(4–5):352–366

Hynds RE, Bonfanti P, Janes SM (2018) Regenerating human epithelia with cultured stem cells: feeder cells, organoids and beyond. EMBO Mol Med 10(2):139–150

Janani G et al (2019) Insight into silk-based biomaterials: from physicochemical attributes to recent biomedical applications. ACS Appl Bio Mater 2(12):5460–5491

Jung KM et al (2014) KeraSkin™-VM: a novel reconstructed human epidermis model for skin irritation tests. Toxicol in Vitro 28(5):742–750

Kano S et al (2010) Utilization of reconstructed cultured human skin models as an alternative skin for permeation studies of chemical compounds. Altern Anim Test Exp 15(2):61–70

Kempf T et al (2011) GDF-15 is an inhibitor of leukocyte integrin activation required for survival after myocardial infarction in mice. Nat Med 17(5):581–588

Koch L et al (2010) Laser printing of skin cells and human stem cells. Tissue Eng Part C Methods 16(5):847–854

Koch L et al (2012) Skin tissue generation by laser cell printing. Biotechnol Bioeng 109(7): 1855–1863

Lamb RD (1936) American chamber of horrors

Lee OJ et al (2014a) Development of artificial dermis using 3D electrospun silk fibroin nanofiber matrix. J Biomed Nanotechnol 10(7):1294–1303

Lee V et al (2014b) Design and fabrication of human skin by three-dimensional bioprinting. Tissue Eng Part C Methods 20(6):473–484

Lee S et al (2017) Construction of 3D multicellular microfluidic chip for an in vitro skin model. Biomed Microdevices 19(2):22

Mao J et al (2003) Study of novel chitosan-gelatin artificial skin in vitro. J Biomed Mater Res A 64(2):301–308

Martínez-Santamaría L, Guerrero-Aspizua S, Del Río M (2012) Skin bioengineering: preclinical and clinical applications. Actas Dermosifiliogr (Engl Ed) 103(1):5–11

McCally AW, Farmer AG, Loomis EC (1933) Corneal ulceration following use of lash-lure. JAMA 101:1560–1561

Mohd Hilmi AB et al (2013) In vitro characterization of a chitosan skin regenerating template as a scaffold for cells cultivation. Springerplus 2(1):79

Mori H et al (2017) Endothelial barrier protein expression in biodegradable polymer sirolimus-eluting versus durable polymer everolimus-eluting metallic stents. JACC Cardiovasc Interv 10(23):2375–2387

Nakamura M et al (2018) Alternative test models for skin ageing research. Exp Dermatol 27(5): 495–500

Ng WL, Yeong WY, Naing MW (2016) Polyelectrolyte gelatin-chitosan hydrogel optimized for 3D bioprinting in skin tissue engineering. Int J Bioprint 2(1). https://doi.org/10.18063/IJB.2016.01.009

Noh HK et al (2006) Electrospinning of chitin nanofibers: degradation behavior and cellular response to normal human keratinocytes and fibroblasts. Biomaterials 27(21):3934–3944

OECD (2019) Test no. 431: In vitro skin corrosion: reconstructed human epidermis (RHE) test method

OECD (2021a) Guideline no. 497: defined approaches on skin sensitisation

OECD (2021b) Test no. 439: in vitro skin irritation: reconstructed human epidermis test method
Osaki T et al (2018) In vitro microfluidic models for neurodegenerative disorders. Adv Healthc Mater 7(2)
Pageon H et al (2017) Glycation stimulates cutaneous monocyte differentiation in reconstructed skin in vitro. Mech Ageing Dev 162:18–26
Patra S, Young V (2016) A review of 3D printing techniques and the future in biofabrication of bioprinted tissue. Cell Biochem Biophys 74(2):93–98
Preetam S et al (2022) Emergence of microfluidics for next generation biomedical devices. Biosens Bioelectron X 10:100106
Przekora A (2020) A concise review on tissue engineered artificial skin grafts for chronic wound treatment: can we reconstruct functional skin tissue in vitro? Cell 9(7):1622
Ramadan Q, Ting FCW (2016) In vitro micro-physiological immune-competent model of the human skin. Lab Chip 16(10):1899–1908
Randall MJ et al (2018) Advances in the biofabrication of 3D skin in vitro: healthy and pathological models. Front Bioeng Biotechnol 6:154
Reijnders CMA et al (2015) Development of a full-thickness human skin equivalent in vitro model derived from TERT-immortalized keratinocytes and fibroblasts. Tissue Eng Part A 21(17–18): 2448–2459
Reuter C, Walles H, Groeber F (2017) Preparation of a three-dimensional full thickness skin equivalent. In: Koledova Z (ed) 3D cell culture: methods and protocols. Springer, New York, NY, pp 191–198
Risueño I et al (2021) Skin-on-a-chip models: General overview and future perspectives. APL Bioeng 5(3):030901
Roger M et al (2019) Bioengineering the microanatomy of human skin. J Anat 234(4):438–455
Sarkiri M et al (2019) Bioengineered skin intended for skin disease modeling. Int J Mol Sci 20(6): 1407
Schanz J et al (2010) Retracted: vascularised human tissue models: a new approach for the refinement of biomedical research. J Biotechnol 148(1):56–63
Semlin L et al (2011) In vitro models for human skin disease. Drug Discov Today 16(3):132–139
Shahin H et al (2020) Vascularization is the next challenge for skin tissue engineering as a solution for burn management. Burns Trauma 8:tkaa022
Silva RJ, Tamburic S (2022) A state-of-the-art review on the alternatives to animal testing for the safety assessment of cosmetics. Cosmetics 9(5):90
Singh S et al (2020) Long-term and clinically relevant full-thickness human skin equivalent for psoriasis. ACS Appl Bio Mater 3(10):6639–6647
Sriram G, Bigliardi PL, Bigliardi-Qi M (2015) Fibroblast heterogeneity and its implications for engineering organotypic skin models in vitro. Eur J Cell Biol 94(11):483–512
Sriram G et al (2018) Full-thickness human skin-on-chip with enhanced epidermal morphogenesis and barrier function. Mater Today 21(4):326–340
Suhail S et al (2019) Engineered skin tissue equivalents for product evaluation and therapeutic applications. Biotechnol J 14(7):e1900022
Tannenbaum J, Bennett BT (2015) Russell and Burch's 3Rs then and now: the need for clarity in definition and purpose. J Am Assoc Lab Anim Sci 54(2):120–132
Tavakoli S, Klar AS (2021) Bioengineered skin substitutes: advances and future trends. Appl Sci 11(4):1493
Torino S et al (2018) PDMS-based microfluidic devices for cell culture. Inventions 3(3):65
Tracy LE, Minasian RA, Caterson EJ (2016) Extracellular matrix and dermal fibroblast function in the healing wound. Adv Wound Care (New Rochelle) 5(3):119–136
Triglia D et al (1991) In vitro toxicity of various classes of test agents using the neutral red assay on a human three-dimensional physiologic skin model. In Vitro Cell Dev Biol Anim 27(3): 239–244

U.S Congress (1934) United States code: federal food, drug, and cosmetic act, 21 U.S.C. §§ 301–392 Suppl. 5. [Periodical] Retrieved from the Library of Congress. https://www.loc.gov/item/uscode1934-006021009/

van den Broek LJ et al (2017) Progress and future prospectives in skin-on-chip development with emphasis on the use of different cell types and technical challenges. Stem Cell Rev Rep 13(3): 418–429

Van Gele M et al (2011) Development of a 3D pigmented skin model to evaluate RNAi-induced depigmentation. Exp Dermatol 20(9):773–775

Veith AP et al (2019) Therapeutic strategies for enhancing angiogenesis in wound healing. Adv Drug Deliv Rev 146:97–125

Vig K et al (2017) Advances in skin regeneration using tissue engineering. Int J Mol Sci 18(4):789

Wang Z et al (2017) Novel biomaterial strategies for controlled growth factor delivery for biomedical applications. NPG Asia Mater 9(10):e435–e435

Whang KK et al (2005) Comparative treatment of giant congenital melanocytic nevi with curettage or Er:YAG laser ablation alone versus with cultured epithelial autografts. Dermatol Surg 31(12): 1660–1667

Woo CH et al (2015) A bilayer composite composed of TiO2-incorporated electrospun chitosan membrane and human extracellular matrix sheet as a wound dressing. J Biomater Sci Polym Ed 26(13):841–854

Wufuer M et al (2016) Skin-on-a-chip model simulating inflammation, edema and drug-based treatment. Sci Rep 6(1):37471

Yang R et al (2019) Epidermal stem cells in wound healing and their clinical applications. Stem Cell Res Ther 10(1):229

You HJ et al (2012) Treatment of diabetic foot ulcers using cultured allogeneic keratinocytes–a pilot study. Wound Repair Regen 20(4):491–499

Yun YE et al (2018) Artificial skin models for animal-free testing. J Pharm Investig 48(2):215–223

Zhang Z, Michniak-Kohn BB (2012) Tissue engineered human skin equivalents. Pharmaceutics 4(1):26–41

Zhu W et al (2016) 3D printing of functional biomaterials for tissue engineering. Curr Opin Biotechnol 40:103–112

Zurlo J, Rudacille D, Goldberg AM (1994) Animals and alternatives in testing: history, science, and ethics. Mary Ann Liebert, New York

Fabrication of Ready-to-Use Ex Vivo Human Skin Models for Chemical Testing: Current Status and Challenges

2

Anastasiya Gorkun, Naresh Mahajan, Kelsey Willson, Adam M. Jorgensen, Grayson A. Wagner, Varun R. Kasula, Alan Jacobson, Anthony Atala, and Shay Soker

Abstract

The skin is the largest organ of the body, composed of the epidermis, dermis, and subcutaneous tissue, each with unique functions. The development of *ex-vivo* human skin models for chemical testing is a current challenge in skin research. While 3D printing technology has been used to develop bioprinted skin, few studies have included 3D printed sebaceous glands, making it challenging to create a fully functional skin model. The ideal biomaterial for skin bioprinting should have mechanical properties similar to those of native skin, support high cell viability, have adequate biodegradation rate, provide a suitable

Anastasiya Gorkun and Naresh Mahajan contributed equally to this work.

A. Gorkun
Wake Forest Institute for Regenerative Medicine, Winston-Salem, NC, USA

FSBSI Institute of General Pathology and Pathophysiology, Moscow, Russia

Sechenov First Moscow State Medical University, Institute for Regenerative Medicine, Moscow, Russia

N. Mahajan · G. A. Wagner · V. R. Kasula · A. Jacobson
Wake Forest Institute for Regenerative Medicine, Winston-Salem, NC, USA

K. Willson · A. M. Jorgensen
Wake Forest Institute for Regenerative Medicine, Winston-Salem, NC, USA

Wake Forest School of Medicine, Medical Center Boulevard, Winston-Salem, NC, USA

A. Atala · S. Soker (✉)
Wake Forest Institute for Regenerative Medicine, Winston-Salem, NC, USA

Wake Forest School of Medicine, Medical Center Boulevard, Winston-Salem, NC, USA

Atrium Health Wake Forest Baptist, Medical Center Boulevard, Winston-Salem, NC, USA
e-mail: ssoker@wakehealth.edu

A. B. Pant et al. (eds.), *Skin 3-D Models and Cosmetics Toxicity*,
https://doi.org/10.1007/978-981-99-2804-0_2

microenvironment for skin cell functionality, and be highly biocompatible. Natural biomaterials are commonly used in skin bioprinting, but they lack stable mechanical properties and have low gelation levels. Synthetic materials have controllable mechanical and chemical properties, but low biocompatibility and biodegradability. Composite natural and synthetic biomaterials can help balance the biological and mechanical features and provide more stable bioink. The development of bioprinted skin models will help to advance skin research and provide a customizable approach to the development of skin tissue. In summary, both skin bioprinting and organoid technology have revolutionized the field of tissue engineering and modeling. Skin bioprinting has shown promising results in the fabrication of skin substitutes for wound healing and has the potential to transform the cosmetic industry. Organoids have broad applications in disease modeling, drug testing, and the development of treatment strategies for various genetic and infectious diseases. While animal models remain the gold standard, organoids provide a closer recapitulating system of human organs and have the advantage of being easily cultured, genetically modified, and cryopreserved while maintaining their phenotype. Overall, these technologies offer new possibilities for research, dermatopathology, wound healing, and drug and vaccine development.

Keywords

Skin bioprinting · Organoid technology · Tissue engineering · Wound healing and drug testing

2.1 Part 1: Introduction

As the largest organ of the body, skin weighs in at 16% of an adult's total weight. It is comprised of three layers: epidermis, dermis, and subcutaneous (hypodermal) tissue. The epidermis is the outermost layer, consisting primarily of keratinocytes (KCs), without blood vessels. The epidermis can be further segregated into five layers. Beginning with the innermost layer, these are the stratum basale followed by the stratum spinosum, the stratum granulosum, the stratum lucidum, and the stratum corneum on the uppermost outer part (Kanitakis 2002). These epidermal sub-layers harbor a number of cells such as KCs, melanocytes (MCs), Langerhans cells, and Merkel cells.

Following the epidermis is the dermis, a significantly thicker section residing in the middle of the skin. Much like the epidermis, the dermis can also be subcategorized, divided into papillary dermis and reticular dermis. The papillary layer lies below the epidermis and contains dermal papillae that project into the epidermis. These papillae help anchor the two layers together. In comparison, the reticular layer is deeper within the dermis and contains fibroblasts (FB), collagen, and elastin fibers, which give skin its strength and elasticity. Both layers are inundated with blood vessels and nerves, which serve as conduits for nutrition and sensation. In addition, the dermis contains important sub-structures, which provide functionality to the skin including sweat glands, hair follicles (HF), and sebaceous glands (Marques et al. 2017).

The final layer of the skin is the subcutaneous tissue or hypodermis. This layer is composed of adipocytes (ACs) surrounded by collagen. Despite its simple makeup, the hypodermal layer has many important protective roles, including thermal equilibrium, providing padding to protect deeper tissues from blunt trauma, and buoyancy. In addition, the subcutaneous tissue can be reabsorbed as an energy reserve and in times of need can function as an endocrine system. Together, these three layers protect internal tissues from external forces, providing a physical barrier to micro-organisms/external materials. It prevents the loss of fluid, allows temperature regulation, acts as a moderator for the immune system, and allows each individual to sense the external world (Marques et al. 2017).

2.1.1 Skin Appendages

Internal structures within the skin include HFs, sebaceous glands, and sweat glands. While it is readily acknowledged that these are important for fully functional skin, bioengineered models of these appendages are scarce. This review is focused on the use of 3D printing as a technology for furthering the development of bioprinted skin. To date, very few studies have included 3D-printed sebaceous glands. Instead, the field has focused on the development of appropriate organization (including the three layers discussed above) and the addition of hair follicles/sweat glands. HFs reside in the dermal layer of the skin. These structures consist of hair papillae, hair matrix, root sheath, and hair bulges. The base of the HF is the papilla, a large structure of primarily connective tissue with a capillary loop. The papilla acts as a control center for the HF, determining many characteristics of the hair, including length, hardness, and the overall growth cycle of the follicle. Surrounding this is a root sheath, a dual-layered covering containing bother internal and external sheaths. The outmost sheath contains the hair bulge. This is also the point of insertion of the arrector pili muscle (Buffoli et al. 2014). The hair bulge houses several types of stem cells with superior clonogenicity and proliferative capacity, which supplies the entire HF with new cells and assists in healing any epidermal injuries. HFs are complex but vital sub-structures within the skin, which have been shown to aid antibacterial abilities and inhibit scar formation (Weng et al. 2021), making them of extreme interest to those developing bioengineered skin.

The second structure of popular interest is sweat glands, particularly eccrine sweat glands. These are found across all skin surfaces but are especially prevalent on the palms and soles of the feet. Playing an important role in the regulation of body temperature, eccrine sweat glands contain a coiled secretory tubule, which is connected to the exterior of the epidermis via a long duct. These glands are activated via changes in temperature or emotion, resulting in the excretion of sweat, complete with proteolytic enzymes and interleukin-1. These exuded factors are believed to play a part in the overall barrier function of skin (Lee et al. 2009). In addition, sweat contains urea, lactic acid, and creatine, which contribute to the inhibition of bacterial growth.

Alongside the structural subcomponents of skin, there is a nonstructural feature that is of extreme interest to those trying to engineer replicates: color. Skin color, if primarily developed through MCs, located in the basal layer of the epidermis. These cells produce melanin/melanosomes, which, when deposited into the extracellular matrix (ECM), result in pigmentation. Possibly more important than the visual pigmenting, these melanosomes are utilized by KCs to aid in protection from ultraviolet (UV) damage. Studies have shown that the MCs and KCs have two-way communication, while the MCs provide protections, growth factors from KCs aid in the proliferation, differentiation, and migration of MCs (Weng et al. 2021). Besides HFs and sweat glands, skin color is one of the important skin parameters. Skin color is mainly related to MCs, which are located in. This makes the development of appropriate pigmentation in bioengineered skin vital not only from a cosmetic point, but also through ensuring adequate UV protection.

2.2 Part 2: Reconstructed Human Epidermis (RHE) and Full-Thickness Skin (FTS) Models

There are several in vitro skin models commercially available, the two most common types being full-thickness skin (FTS) models and reconstructed human epidermis (RHE). FTS models are typically defined as an epidermal and dermal layer, where FBs are seeded onto a scaffold and KCs are seeded on top. RHE models differ in that there is no base dermal component, rather only KCs embedded on a scaffold (Camarena et al. 2020; Catarino et al. 2018).

2.2.1 Reconstructed Human Epidermis (RHE)

The first attempt of in vitro model of RHE was carried out by a group of researchers in France, culturing human KCs on a dermal equivalent with an air–liquid interface to recapitulate a functional epidermis (Asselineau et al. 1986). A primary question posed in this work was whether any epidermis obtained in vitro could be considered as "normal," able to recapitulate native in vivo epidermal functions with a focus on the epidermis' role as a barrier. While still an ongoing question in the skin community, 4 years later, in 1990 Rosdy and Clauss successfully obtained a terminal epidermal differentiation of human KCs grown onto inert filters via air–liquid interface in a chemically defined medium (Rosdy and Clauss 1990). RHEs, based on these early studies, mimic solely the epidermis and typically consist of normal human KCs. Fabrication of a RHE begins with KCs that are first expanded in culture, then seeded onto a scaffold, and finally cultured using an air–liquid interface to promote differentiation and maturation.

Since their development in 1986, RHEs have become a useful tool for researchers, especially those interested in toxicology. As an incomplete model of FTS, RHEs have been developed and validated as in vitro skin models. Importantly, these have been validated as alternatives for conventional animal models, according to criteria/guidelines outlined by the Organization for Economic Co-Operation and Development (OECD). These regulations remain the standard for studies aiming to improve upon in vitro skin models. In 2010, Liao et al. developed a new RHE model utilizing the OECD guidelines. The epithelium, developed at Industrial Technology Research Institute (ITRI), and name EPiTRI (epithelium-ITRI), was validated using an OECD-approved skin irritation test (SIT). Briefly, EPiTRI was tested with 20 reference chemicals with known Irritant Index and the results showed an accuracy of irritation response of 96%, that comparable to animal and in vitro reference models, meeting the OECD criteria for screening irritating chemicals in vitro (Liao et al. 2021). There are currently seven RHE models that are considered validated reference methods for in vitro skin irritation testing viz. EPiSkin™ (VRM), EpiDerm™ SIT (EP-200) (VRM), SkinEthic RHE™, LabCyte EPI-Model24 SIT, epiCS®, Skin+®, and KeraSkin™ SIT (OECD 2021). All seven models adhere to the performance standards in OECD TG439. OECD TG431 includes five of these as validated for corrosion testing—EPiSkin™ (SM), EpiDerm™ SCT (EPI-200), epiCS®, and LabCyte EPI-Model24 SCT (OECD n.d.).

Despite these validated RHE models, ongoing research still shows that RHEs are incomplete models of FTS. Catarino et al. (2018) compared novel RHE models to FTS, monitoring their responses when subjected to OECD skin corrosion assays. The results of the study showed higher cell viability of the FTS model compared to the RHE model. This indicates that the FTS maintained an improved barrier function, following the exposure to the substances test on the corrosion assays (including 2-phenylethyl bromide, benzylacetone, lactic acid, and octanoic acid), compared to the single-layer RHE. In addition, the RHE models were found to be significantly more permeable than ex vivo human skin, while their FTS counterparts were been found to have enhanced barrier function (Catarino et al. 2018). This study emphasizes the need for continued development of truly physiologically relevant skin models for in vitro use, which better mimic the in vivo situation for the toxicological detection of substances (Catarino et al. 2018), while RHE models are the only commercial models verified to be used in irritation and corrosion tests (Catarino et al. 2018). It is important to note that without the representation of a dermal layer, the use of RHE models may not fully represent the human skin response in irritation and corrosion tests as the interaction between the epidermal and dermal layers has been shown to affect skin homeostasis.

2.2.2 Full-Thickness Skin (FTS)

FTS models differ from RHEs by including a second layer that mimics the dermal layer in human skin. Typically, the reconstructed dermis is formed using proteins that are found in the human ECM. Collagen type 1 and human FBs are commonly used and provide the basal layer that is then embedded with normal human KCs to form the top epidermal layer (Catarino et al. 2018). Mok et al. demonstrated the formation of a reconstructed human skin equivalent (RSE) with a self-assembled dermal layer. This model consisted of dermal and epidermal layers, making use of the FB ability to secrete their own ECM. The model was developed according to OECD TG439 and was evaluated for toxicity. During 4 weeks of culture, primary dermal FBs formed a dermal FB sheet by secreting ECM. Human KCs were subsequently embedded into this dermal FB sheet. This model was able to closely mimic native human skin structure with a stratified epidermis (Mok et al. 2022).

2.2.3 Future Developments of FTS and RHE Models

While models of FTS and RHEs exist, they are by no means perfect substitutions for in vivo skin. As such, there is continued research improving and redesigning these systems. This is not solely focused on the development of individual cell layers, but also the refinement of the scaffolding material used to provide structure to the models. A recent study by Camarena et al. (2020) demonstrated that FTS and RHE models can be created using novel electrospun scaffolds. They used synthetic polymers instead of animal protein-based materials to create electrospun polymer mats that served as a base for seeding FBs and KCs. PET, PBT, and N6/6 are among the tested synthetic polymers that could be used in place of the typical scaffold materials (i.e., polycarbonate filters or collagen) (Camarena et al. 2020). The ability to alter the growth matrix for these systems may lead to exciting new developments in altering cellular interactions and development to create truly physiological in vitro skin models.

2.3 Part 3: Bioprinting of Skin Constructs

The models discussed above were created through conventional cell seeding methods, and while they have created a solid foundation for skin models in the laboratory, they can be time-consuming and inimitable to develop in large numbers. The shift from hand-crafted to high-speed fabrication of tissues and organs became closer to reality with development of the first bioprinter by Dr. Thomas Boland. Developed through the modification a standard HP inkjet printer to place layer of cells on top of one another in the early 2000s, this marked an important step toward rapid manufacturing of cellularized constructs (Thayer et al. 2018). The rapid advancement of printing technologies and computer-aided design (CAD) has transformed bioprinting into a premium manufacturing platform. Capable of generating custom tissues with defined deposition of living cells, biomaterials, and growth factors at micro- and macro-scales, bioprinting is characterized by both its high customizability and repeatability in the generation of new tissues (Murphy and Atala 2014). These traits have remained true as 3D bioprinting has expanded to include a variety of different printing methods such as laser-assisted bioprinting, inkjet-based bioprinting, pressure-assisted bioprinting, and electrohydrodynamic jetting (Ng et al. 2016; Yan et al. 2018).

3D bioprinting holds several advantages over traditional methods of tissue engineering in skin construct preparation: (1) Computer scanning/imaging technology can be utilized to allow for rapid development of custom skin models matching the shape and depth of the wound surface; (2) the availability of multiple bioinks combined with the ability to deposit them independently within structures can provide skin similar morphology and physiology; (3) the development of in situ printing at the wound surface can be used in clinical treatment; and (4) 3D bioprinting allows for the generation of large, porous constructs providing cell support, gas, and nutrition exchange (Weng et al. 2021). Constant advancement of bioprinting techniques and biomaterials continues to expand this list, with recent studies aiming to add fabrication of complex vasculature and the skin appendages as HFs and sebaceous glands to the 3D printing repertoire.

2.3.1 Biomaterials for Skin Bioinks

The composition of specific bioinks plays a key role in skin bioprinting. The ideal biomaterial should retain mechanical properties similar to those of native skin, support high cell viability and adequate biodegradation rate, a suitable microenvironment for skin cell functionality, good adaptability to printing, and high biocompatibility. The commonly used biomaterials can be divided into natural and synthetic materials (Boland et al. 2003).

Among natural biomaterials, the leading position in skin bioprinting belongs to collagen, an essential ECM component of skin. Other substances present in skin ECM are also used including gelatin, hyaluronic acid, and acellular dermal matrix. Biomaterials sourced from outside the skin are used as well, among them fibrin, agarose, alginate, chitosan, and silk fibroin. Natural biomaterials show high biocompatibility but lack stable mechanical properties and have low gelation levels, making them difficult to handle during and after printing. In contrast, synthetic materials such as polylactic acid (PLA), polycaprolactone (PCL), polyethylene glycol (PEG), and gelatin methacrylate (GelMA) have controllable, and highly reproducible, mechanical, and chemical properties, but low biocompatibility and biodegradability (Yan et al. 2018). Choosing the "right" biomaterial for skin bioprinting is an amorphous challenge, complicated by the many criteria defining "right," and the need to balance both biological and mechanical features. The development of composite natural and synthetic biomaterial can help address the different biological requirements while improving the stability of the bioink. Structural materials such as collagen, alginate, and chitosan can aid in cell adhesion, proliferation, and differentiation post-printing. In contrast, fugitive and support materials, not containing cells but acting as sacrificial materials, can be rapidly dissolved to create voids and channels within 3D structures (fugitive materials) or to improve physical strength and integrity of bioink (support materials), providing material transport and appropriate internal architecture for a print. Examples include polyurethanes (PUs), PCL, and PLGA. Functional materials are also included in composite bioinks, with molecules such as heparins and GAGs used to stimulate cell behavior and development through signaling and binding with growth factors (Manita et al. 2021). With a field composed of so many uniquely diverse options, the use of bioinks can assist in driving appropriate architecture within skin constructs while also being tuned to aid in appropriate cell maturation in bioengineered skin models.

2.3.2 Advances in Skin Bioprinting

In 2009, Lee et al. successfully bioprinted a multi-layered skin substitute using human skin FBs and KCs using freeform fabrication on collagen matrix (Lee et al. 2009). The authors observed cell proliferation in both planar and nonplanar surfaces in their in vitro model and suggested the feasibility of using 3D printing as an on-demand skin graft fabrication method. Later, in 2010, Binder et al. validated the potential of in situ 3D bioprinting for wound healing. In the study, they used human FBs and KCs in a fibrin and collagen matrix directly in full-thickness wounds on immunodeficient mice using inkjet printing approach. The authors observed decreased contraction and better wound healing compared to controls, untreated allogeneic implant, and hydrogel matrix (Binder et al. 2010).

In 2013, Michael et al. placed FBs and KCs on top of a stabilizing matrix (MatriDerm®) using laser-assisted bioprinting (LAB) in the fabrication of skin for the first time. Maturation of the fabricated constructs was monitored in vitro, with samples maintained at an air–liquid interface, and in vivo, where samples were implanted in the dorsal skin fold chamber of nude mice. The results of this study showed that LAB fabricated skin was able to integrate post-implantation, forming a multi-layered differentiating epidermis in vivo. This epidermis demonstrated basal keratinocyte proliferation, primarily in supra-basal layers, typical of native skin. Interestingly, the in vitro constructs also exhibited the formation of a multi-layered epidermis; however, a less matured version with the basal proliferating keratinocytes was present in all (Michael et al. 2013). This study not only highlighted the enormous impact culture conditions can have on the formation of skin constructs, which replicate native tissue, but also showed the value in utilizing bioprinting to accurately layer cells, positioning them for the appropriate development of multi-layered, functional epidermis.

One of the strengths of bioprinting is the ability to combine technologies. A hybrid 3D cell printing system was developed by Kim et al. (2017), allowing the use of extrusion and inkjet modules at the same time. The extrusion module was used to develop a collagen-based construct embedded with a PCL mesh, designed to prevent the contraction of collagen during tissue maturation. The inkjet system was used simultaneously to distribute KCs uniformly across the surface, developing an epidermal layer on top of the engineered dermis (Kim et al. 2017).

3D printing can not only be combined through multi-printing modalities, but also with other clinical technologies. This was the case in the development of BioMask, combining 3D printing with computed tomography (CT) data to develop custom, patient-specific, models. In short, CT images were used to develop placement patterns for both cellularized hydrogels and a wound dressing material. These were then fabricated using an extrusion printer. The final model (BioMask) contained a porous PU layer, a KC-laden hydrogel layer, and a FB-laden hydrogel layer. The printed construct was then implanted on a mouse and monitored for skin regeneration. Histological assays showed that BioMasks aided in the regeneration of multi-layered skin tissue, consisting of both epidermis and dermis, in complex wounds (Seol et al. 2018).

While BioMask focused on smaller, complex wounds, others have pursued larger wounds with the goal of not only developing patient-specific prints but also being able to print these constructs directly into the patient. In 2019, Albanna et al. described a novel, mobile skin bioprinting system, meant for the treatment of extensive wounds through in situ printing. Using integrated wound imaging technology, the group scanned a wound and then delivered either dermal FBs or epidermal KCs directly to the injury (Albanna et al. 2019). This replicated the layered skin structure without the use of secondary support materials or the need to transfer the print from a build plate to the wound site, acting as proof-of-concept in the validation of a mobile, patient-specific in situ bioprinter. The following year, another group was able to use bioprinted skin to recreate an epidermal barrier in a full-thickness wound model, complete with normal, non-scarring, collage remodeling (Jorgensen et al. 2020). The results of these studies, taken in combination, highlight a just a few of the very exciting opportunities that 3D bioprinting could exploit to develop new and complete models of skin for both the bench and the clinic.

Bioprinting has shown many strengths in differentiation itself as a rapid manufacturing technique. However, in biology, it is not the process that reigns supreme—it is the final product. In moving forward with a new manufacturing technique, it is important to compart the fabricated constructs not only to the goal (FTS) but also to prior models, to understand alterations and where researchers might expect differences from previously obtained results. In 2018, a group of researchers from the Singapore Centre for 3D bioprinting did just this, comparing 3D-bioprinted pigmented skin constructs with pigmented skin constructs fabricated using a conventional manual casting approach. The group completed an in-depth characterization of these models, concluding that the 3D-printed pigmented models more closely resembled the native skin control. This was true not only for the development of macro-architecture (dermal and epidermal layers) but also for micro-architectures, including the development of a continuous basement membrane, which was not present in the manually cast samples. The group concluded that the 3D-printed constructs were an improvement over conventionally manufactured pigmented skin models with potential for toxicology testing and furthering fundamental cell biology research on the bench (Ng et al. 2018).

2.3.3 Bioprinting Limitations and New Prospective

The introduction of 3D-bioprinted skin has enabled the development of customizable skin constructs for patients, printed either onto a conventional build plate or directly onto wound sites. However, while the strengths of bioprinting have been highlighted above, it is important to consider the limitations of the techniques as well. These include significant lead time required to develop enough cells to print (3–4 weeks/m^2 according to Cubo et al. 2016), highly trained and specialized personnel to run the printers and develop the 3D models, and high production costs (driven by the price of cell expansion, bioink development, and bioprinter costs). Despite these drawbacks, there is a continually growing demand for artificial skin. These facts together point the skin regeneration field in the direction of automation, standardization, and overall system reduction for both cost and production time. Addressing these points will allow researchers to truly bring RHE and FTS models into the clinic in a meaningful way. New trends in skin bioprinting are focused on these areas while still pressing forward to replicate in vivo skin through the incorporation of stem cells in the skin substitute to develop microvasculature (Abaci et al. 2016) and sweat glands (Yao et al. 2020) as well as combining bioprinting techniques to fabricate skin appendages during the printing process, streamlining the fabrication of physiological skin (Abaci et al. 2018).

2.4 Part 4: Micro-Tissue Equivalents (Organoids) Models

This review has primarily focused on the development of flat models of skin, developed for both research and clinical purposes. When removing the clinical aspect, researchers aim for the closest recapitulating system, which, as with many organs, is an animal model. Animal models remain the gold standard for replicating the functional and cellular interactions of human tissues. They can be used to predict the development of diseases and the efficacy of treatments. They do have deficiencies driven differences in species biology or sensitivity. In addition, animal models can be very expensive in both monetary value and personnel time, which can lead to lower throughput than may be ideal (Hartung 2008; Shanks et al. 2009). In response to this, 3D organoid cultures have emerged. While these systems do not replicate the macro-structure of organs, they exquisitely mimic the micro-structure and functionality of human organs (Li and Izpisua Belmonte 2019).

Organoids are 3D cell structures made up of = cell mixtures appropriate to the organ being modeled, which better mimic cell–cell and cell–matrix interactions compared to 2D cultures (Bates et al. 2000, p. 200). The 3D microenvironment allows to mimic cellular heterogeneity observed in vivo in different contexts, developing both structural and functional similarities for their in vivo counterparts (Weiswald et al. 2015). Studies have shown that organoids can provide excellent platforms for scientific and clinical applications, recapitulating human physiology and positioning themselves as a contender to replace current models in biological/biomedical research (Bell et al. 1981). When compared to 2D culture, organoids can be cultured for longer time periods, easily cryopreserved, and genetically modified while maintaining their phenotype (Clevers 2016; Drost and Clevers 2018). These features have allowed the use of organoids in various research applications. This has included utilizing organoids as a platform to gain new understanding of organ-specific physiology and to investigate disease-specific modeling in comparison with cell lines. When compared to animal models, organoid culture is advantageous with its high throughput and reduces cost, of particular interest to groups interested in screening large numbers of novel drugs (Weiswald et al. 2015). Organoids can be formed from various stem cells including adipose-derived mesenchymal stromal cells AD-MSCs), embryonic stem cells (ESCs), induced pluripotent stem cells (iPSCs), and patient-derived tumor tissue cells, making them applicable to a wide range of systems (Clevers 2016).

While many cell types can be used in fabricating organoids, cancer cells are the most widely used today, due to their easy fabrication in vitro, which is done by embedding cancer cells in a specific ECM, alongside medium niche factors, and additional cells (Ruiz-Garcia et al. 2020; Dominijanni et al. 2020). In 2015, Skardal et al. successfully created liver-based cell organoids in a rotating wall vessel bioreactor. In addition, they were able to combine the manufactured organoids with colon carcinoma cells, developing liver tumor organoids, which acted as in vitro models of liver metastasis. The authors documented that the in vitro 3D liver tumor organoid model replicated tumor responses to current and newly discovered drugs (Skardal et al. 2015). Mazzocchi et al. (2019) created hydrogel-based models to create lung cancer organoids using a single-cell source, pleural effusion aspirate, from multiple lung cancer patients. The authors observed that the cells isolated from the patient, assembled into anatomically relevant structures when seeded into organoids, exhibited behavior specific to lung cancers (Mazzocchi et al. 2019). This application of patient-specific organoids was expanded upon by Forsythe et al. (2022). The group used patient tumor organoids (PTOs) to model rare malignancy "Merkel cell carcinomas" in patient-specific trials. The models were exposed to chemotherapy or immunotherapy agents and monitored for viability after exposure. The authors observed 66% response to chemotherapy in 4/6 specimens with cisplatin and doxorubicin, while immunotherapy was not effective in the immune PTO (iPTO) sets, indicating that these systems could be used to screen for the ideal patient treatment plan (Forsythe et al. 2022). Recently, a group of researchers from Wake Forest developed a novel immune-enhanced tumor organoid (iTO) system to study factors affecting immune checkpoint blockade response (Shelkey et al. 2022).

Unfortunately, these models do not fully mimic human biology. One of the major barriers preventing this is the lack of skin appendages, such as HFs and sweat glands. In addition, the minimalistic approach to skin models has neglected the addition of skin-related cells including dermal fat, sensory cells, and neurons. These deficiencies highlight areas of improvement, which could be targeted to further understand skin through in vitro models (Lee and Koehler 2021). Pushing forward with these 3D models is key, as, 2D cell culture models are less likely to reflect physiological responses than their 3D counterparts (Sun et al. 2006). Many researchers are working on this front, as showcased by the recent development of a skin organoid model that not only uses an air–liquid interface but also included stromal cells, which acted as a source of vital growth factors. Others have focused on the inclusion of new cell types, developing immunocompetent and tumor skin models through the addition of macrophages, T-lymphocytes, melanoma cells, and epithelial carcinoma cells (Gaviria Agudelo and Restrepo 2022).

2.4.1 Organoid Models of Disease

In addition to their application in understanding organ development and drug testing, 3D organoids have applications in disease modeling. Organoids have been used to establish disease models in several tissues. In lung, Wang et al. (2019) used small-cell cancer organoids to investigate the antitumor effect of an irreversible pan-HER receptor tyrosine kinase inhibitor pyrotinib. The study reported various driver gene mutations in lung cancer (Wang et al. 2019). In 2020, Dieterich et al. developed organoid modeling celiac disease using patient-specific cells and reported dissimilarities in phenotypes between the study groups (Dieterich et al. 2020). Dijkstra et al. in 2021 developed a gastroenteropancreatic neuroendocrine carcinoma model, fabricating organoids from stomach/colon cells used to test drugs for this difficult-to-treat tumor (Dijkstra et al. 2021).

Viral infections can also be studied with organoids. This was showcased through the development of human gastric organoids, which modeled the viral infection of Helicobacter pylori in the stomach (Pompaiah and Bartfeld 2017). Zika virus was modeled in brain organoids. (Sutarjono 2019) and then used to test a variety of chemicals mitigating the hypomorphic effect of zika virus (Xu et al. 2016). Multiple intestinal infections have been modeled including norovirus and rotavirus. These have been successfully cultured in human intestinal organoid models (Ettayebi et al. 2016; Finkbeiner et al. 2012). Major intestinal bacterial pathogens, Salmonella typhi and Clostridium difficile, have likewise been cultured in intestinal organoids (Engevik et al. 2015; Heo et al. 2018).

During the recent global pandemic, organoids proved a valuable research tool for those combating SARS-CoV-2. SARS-CoV-2 was shown to be able to infect and propagate in multiple organ systems including primary human liver–gut organoids PSC-derived blood vessel and kidney through experiments done with organoids (Lamers et al. 2020; Monteil et al. 2020). Skin organoids, fabricated from human-induced pluripotent stem cells (hiPSC), were also utilized with the virus, acting as a pathophysiological model of the infection (Ma et al. 2022). In addition, organoids were able to verify COVID-19 pathogenesis, leading researchers to the discovery of mechanism through which SARS-CoV-2 enters host cells. The angiotensin-converting enzyme 2 was proven to not only assist in initial COVID-19 infection but also in transference to tissues beyond the lungs (Hoffmann et al. 2020).

Human organoids are able to reproduce host–pathogen interactions in vitro. In vitro skin model engineering with optimized interaction with the microbiome may help to understand skin microbial ecology and host-related disease mechanisms. In comparison with their 2D counterparts, organoids have been shown to mimic organ pathologies, acting as effective models for human translational studies. This allows them to be used as development platforms for treatment strategies, applying scientific discoveries to a wider range of human diseases. The miniaturized models allow researchers to recreate complex systems for high-throughput studies. This holds true for skin, where it is important to model not just the cellular makeup, but also the ecology of the skin, complete with bacteria, fungi, and viruses. This can be considered analogous to the microbiome in our gut, which plays an essential role in protecting against pathogens (Belkaid and Segre 2014; Scharschmidt and Fischbach 2013). Disruption of this balance leads to inflammation (Costello et al. 2009; Dekio et al. 2005). In 2018, a model of HF induction was developed using cells derived from interfollicular epidermis (IFE) and HFs in canines (Wiener et al. 2018). Wang and coworkers in 2021 developed a method for the establishment and expansion of human primary epidermal organoids for testing antifungal drugs under chemically defined conditions (Wang et al. 2021). Jung et al. optimized the skin organoid platform using air–liquid interface (ALI) to model atopic dermatitis by Staphylococcus aureus (SA) colonization and infection and observed a disrupted skin barrier and increased production of inflammatory cytokines (Jung et al. 2022). These studies help showcase how organoid technology can be used to understand mutations and potential therapeutic strategies for clinical management of genetic diseases.

Alternatively, approaches for studying genetic diseases have also been used with organoid models. Schwank et al. (2013) fabricated the first gene-corrected intestinal organoids from patients with cystic fibrosis. Biopsies were taken and then processed with CRISPR-Cas9 technology to alter the homozygous CFTR F508 deletion (Schwank et al. 2013). Later studies used patient-derived cells, their homozygous mutations corrected by CRISPR, to develop iPSCs. The resulting gene-corrected iPSC-derived organoids were able to into airway epithelium with normal CFTR expression and function (Firth et al. 2015). Similar technology has been used to assess contractile function in engineered heart tissue. Yang et al. (2018) used iPSC-derived organoids to model abnormal contractile functions in patient-specific organoids from those suffering from familial cardiomyopathy. The myosin heavy-chain 7 mutation (E848G) was modeled effectively and the researchers showed that gene correction was possible in dystrophin mutations, showcasing proof of concept for gene application in the treatment of tissue replacement therapy (Yang et al. 2018).

2.5 Part 5: Chemical Irritation, Corrosion, and Sensitization Testing Using Skin Organoids

Skin is the first barrier of the body and the main target for disruptive and hazardous agents of different origin. Exposure to different substances can lead to either reversible (irritation) or non-reversible (corrosion) skin damages. As a result of exposure, humans may experience contact dermatitis—acute inflammation in the skin as both allergic and non-allergic reactions. Recent analysis showed that allergic contact dermatitis (ACD) accounts for 20% among all cases of dermatoses, and rising level of spreading ACD demands new reliable tests to identify new hazardous agents. The OECD formulated main guidelines and requirements for skin irritation/corrosion testing in vitro. The guidelines determined a panel of well-known chemicals as standards for validation of the models and approved tests for analysis. The validated tests include a viability assay, the evaluation of barrier integrity, and the examination morphology of each skin model after the exposure. Based on these guidelines, researchers are developing new assays applicable specifically in vitro. Thus, Saito et al. described the epidermal sensitization assay based on the microarray analysis of the expression of five genes related to cellular stress response (Saito et al. 2013), and Pfuhler et al. presented the Comet assay-based genotoxicity analysis on the reconstructed human epidermis (Pfuhler et al. 2021).

These in vitro models are poised to compete with in vivo models, not only for clinical studies, but also for the multitude of nonclinical trials, which rely on animal models to predict the effects of drugs, cosmetics, and chemicals. This is of particular importance considering not only the scientific, but also the ethical ramifications of using animals that may not precisely replicate the human condition. Organoid technology is closing the gap between 2D cell culture and the in vivo animal models, as an alternative, accurate in vitro model, and has proven its worth in developmental biology and personalized medicine. At present, organoid technology holds great potential for biomedical applications including disease modeling, drug screening, biobanks, regenerative therapy, genetic screening, and personalized medicine (Kim et al. 2020; Xu et al. 2018). The demand for a rapid, large-scale model suitable for in vitro toxicity and efficacy has also been growing from the pharmaceutical and cosmetic industries, further driving the commercialization of organoid technology.

2.6 Part 6: Summary and Conclusions

Skin tissue engineering is a powerful and highly versatile technology that can be applied for skin development research, dermatopathology, wound healing, and development of new topical drugs and vaccines. Original in vitro models, consisting of single-cell-type cultures, are now being replaced with organoids representing different human skin functions and broaden their scope in the industrial and clinical application. New trends in the fabrication of skin constructs have included the incorporation of stem cells along with pre-fabrication of skin appendages to generate self-supporting, highly functional systems. Parallel advancement of in situ and in vitro 3D skin bioprinting has resulted in promising technologies for on-site treatment of excessive wounds and the formulation patch-on-demand services for clinical applications. The rapid development of in vitro skin models is poised to transform the cosmetic industry, as 3D skin models facilitate the examination of cosmetic products and topical drug for efficiency and toxic influence, with their improvements in cost-efficacy and case-specific relevancy compared with animal models. Skin disease models, as showcased in this review, have a high utility not only for investigative pathology, but also as powerful tool for drug and vaccine development.

References

Abaci HE, Guo Z, Coffman A, Gillette B, Lee WH, Sia SK, Christiano AM (2016) Human skin constructs with spatially controlled vasculature using primary and iPSC-derived endothelial cells. Adv Healthc Mater 5(14):1800–1807. https://doi.org/10.1002/adhm.201500936

Abaci HE, Coffman A, Doucet Y, Chen J, Jacków J, Wang E et al (2018) Tissue engineering of human hair follicles using a biomimetic developmental approach. Nat Commun 9(1):1–11. https://doi.org/10.1038/s41467-018-07579-y

Albanna M, Binder KW, Murphy SV, Kim J, Qasem SA, Zhao W, Tan J, El-Amin IB, Dice DD, Marco J, Green J, Xu T, Skardal A, Holmes JH, Jackson JD, Atala A, Yoo JJ (2019) In situ bioprinting of autologous skin cells accelerates wound healing of extensive excisional full-thickness wounds. Sci Rep 9(1):1856. https://doi.org/10.1038/s41598-018-38366-w

Asselineau D, Bernard BA, Bailly C, Darmon M, Pruniéras M (1986) Human epidermis reconstructed by culture: is it "Normal"? J Investig Dermatol 86(2):181–186. https://doi.org/10.1111/1523-1747.ep12284237

Bates RC, Edwards NS, Yates JD (2000) Spheroids and cell survival. Crit Rev Oncol Hematol 36(2–3):61–74. https://doi.org/10.1016/s1040-8428(00)00077-9

Belkaid Y, Segre JA (2014) Dialogue between skin microbiota and immunity. Science 346(6212): 954–959. https://doi.org/10.1126/science.1260144

Bell E, Ehrlich HP, Buttle DJ, Nakatsuji T (1981) Living tissue formed in vitro and accepted as skin-equivalent tissue of full thickness. Science 211(4486):1052–1054. https://doi.org/10.1126/science.7008197

Binder KW, Zhao W, Aboushwareb T, Dice D, Atala A, Yoo JJ (2010) In situ bioprinting of the skin for burns. J Am Coll Surg 211(3):S76. https://doi.org/10.1016/j.jamcollsurg.2010.06.198

Boland T, Mironov V, Gutowska A, Roth EA, Markwald RR (2003) Cell and organ printing 2: fusion of cell aggregates in three-dimensional gels. Anat Rec A Discov Mol Cell Evol Biol 272(2):497–502. https://doi.org/10.1002/ar.a.10059

Buffoli B, Rinaldi F, Labanca M, Sorbellini E, Trink A, Guanziroli E, Rezzani R, Rodella LF (2014) The human hair: from anatomy to physiology. Int J Dermatol 53(3):331–341. https://doi.org/10.1111/ijd.12362

Camarena DEM, Matsuyama LSAS, Maria-Engler SS, Catalani LH (2020) Development of epidermal equivalent from electrospun synthetic polymers for *in vitro* irritation/corrosion testing. Nano 10(12):2528. https://doi.org/10.3390/nano10122528

Catarino CM, do Nascimento Pedrosa T, Pennacchi PC, de Assis SR, Gimenes F, Consolaro MEL, de Moraes Barros SB, Maria-Engler SS (2018) Skin corrosion test: a comparison between reconstructed human epidermis and full thickness skin models. Eur J Pharm Biopharm 125:51–57. https://doi.org/10.1016/j.ejpb.2018.01.002

Clevers H (2016) Modeling development and disease with organoids. Cell 165(7):1586–1597. https://doi.org/10.1016/j.cell.2016.05.082

Costello EK, Lauber CL, Hamady M, Fierer N, Gordon JI, Knight R (2009) Bacterial community variation in human body habitats across space and time. Science 326(5960):1694–1697. https://doi.org/10.1126/science.1177486

Cubo N, Garcia M, Del Canizo JF, Velasco D, Jorcano JL (2016) 3D bioprinting of functional human skin: production and in vivo analysis. Biofabrication 9(1):015006. https://doi.org/10.1088/1758-5090/9/1/015006

Dekio I, Hayashi H, Sakamoto M, Kitahara M, Nishikawa T, Suematsu M, Benno Y (2005) Detection of potentially novel bacterial components of the human skin microbiota using

culture-independent molecular profiling. J Med Microbiol 54(Pt 12):1231–1238. https://doi.org/10.1099/jmm.0.46075-0
Dieterich W, Neurath MF, Zopf Y (2020) Intestinal ex *vivo* organoid culture reveals altered programmed crypt stem cells in patients with celiac disease. Sci Rep 10(1):3535. https://doi.org/10.1038/s41598-020-60521-5
Dijkstra KK, van den Berg JG, Weeber F, van de Haar J, Velds A, Kaing S, Peters DDGC, Eskens FALM, de Groot DJA, Tesselaar MET, Voest EE (2021) Patient-derived organoid models of human neuroendocrine carcinoma. Front Endocrinol 12:627819. https://doi.org/10.3389/fendo.2021.627819
Dominijanni A, Mazzocchi A, Shelkey E, Forsythe S, Devarsetty M, Soker S (2020) Bioengineered tumor organoids. Curr Opin Biomed Eng 13:168–173. https://doi.org/10.1016/j.cobme.2020.03.005
Drost J, Clevers H (2018) Organoids in cancer research. Nat Rev Cancer 18(7):407–418. https://doi.org/10.1038/s41568-018-0007-6
Engevik MA, Engevik KA, Yacyshyn MB, Wang J, Hassett DJ, Darien B, Yacyshyn BR, Worrell RT (2015) Human Clostridium difficile infection: inhibition of NHE3 and microbiota profile. Am J Physiol Gastrointest Liver Physiol 308(6):G497–G509. https://doi.org/10.1152/ajpgi.00090.2014
Ettayebi K, Crawford SE, Murakami K, Broughman JR, Karandikar U, Tenge VR, Neill FH, Blutt SE, Zeng XL, Qu L, Kou B, Opekun AR, Burrin D, Graham DY, Ramani S, Atmar RL, Estes MK (2016) Replication of human noroviruses in stem cell-derived human enteroids. Science 353(6306):1387–1393. https://doi.org/10.1126/science.aaf5211
Finkbeiner SR, Zeng XL, Utama B, Atmar RL, Shroyer NF, Estes MK (2012) Stem cell-derived human intestinal organoids as an infection model for rotaviruses. MBio 3(4):e00159–e00112. https://doi.org/10.1128/mBio.00159-12
Firth AL, Menon T, Parker GS, Qualls SJ, Lewis BM, Ke E, Dargitz CT, Wright R, Khanna A, Gage FH, Verma IM (2015) Functional gene correction for cystic fibrosis in lung epithelial cells generated from patient iPSCs. Cell Rep 12(9):1385–1390. https://doi.org/10.1016/j.celrep.2015.07.062
Forsythe SD, Erali RA, Laney P, Sivakumar H, Li W, Skardal A, Soker S, Votanopoulos KI (2022) Application of immune enhanced organoids in modeling personalized Merkel cell carcinoma research. Sci Rep 12(1):13865. https://doi.org/10.1038/s41598-022-17921-6
Gaviria Agudelo C, Restrepo LM (2022) Human skin cancer: an overview of animal, ex vivo, and in vitro models. Curr Derm Rep 11:168–177. https://doi.org/10.1007/s13671-022-00361-w
Hartung T (2008) Thoughts on limitations of animal models. Parkinsonism Relat Disord 14(Suppl 2):S81–S83. https://doi.org/10.1016/j.parkreldis.2008.04.003
Heo I, Dutta D, Schaefer DA, Iakobachvili N, Artegiani B, Sachs N, Boonekamp KE, Bowden G, Hendrickx APA, Willems RJL, Peters PJ, Riggs MW, O'Connor R, Clevers H (2018) Modelling cryptosporidium infection in human small intestinal and lung organoids. Nat Microbiol 3(7):814–823. https://doi.org/10.1038/s41564-018-0177-8
Hoffmann M, Kleine-Weber H, Schroeder S, Krüger N, Herrler T, Erichsen S, Schiergens TS, Herrler G, Wu NH, Nitsche A, Müller MA, Drosten C, Pöhlmann S (2020) SARS-CoV-2 cell entry depends on ACE2 and TMPRSS2 and is blocked by a clinically proven protease inhibitor. Cell 181(2):271–280.e8. https://doi.org/10.1016/j.cell.2020.02.052
Jorgensen AM, Varkey M, Gorkun A, Clouse C, Xu L, Chou Z, Murphy SV, Molnar J, Lee SJ, Yoo JJ, Soker S, Atala A (2020) Bioprinted skin recapitulates Normal collagen remodeling in full-thickness wounds. Tissue Eng Part A 26(9–10):512–526. https://doi.org/10.1089/ten.TEA.2019.0319
Jung SY, You HJ, Kim MJ, Ko G, Lee S, Kang KS (2022) Wnt-activating human skin organoid model of atopic dermatitis induced by Staphylococcus aureus and its protective effects by Cutibacterium acnes. iScience 25(10):105150. https://doi.org/10.1016/j.isci.2022.105150
Kanitakis J (2002) Anatomy, histology and immunohistochemistry of normal human skin. Eur J Dermatol 12(4):390–399. Quiz 400–401

Kim BS, Lee JS, Gao G, Cho DW (2017) Direct 3D cell-printing of human skin with functional transwell system. Biofabrication 9(2):025034. https://doi.org/10.1088/1758-5090/aa71c8

Kim J, Kong JS, Han W, Kim BS, Cho DW (2020) 3D cell printing of tissue/organ-mimicking constructs for therapeutic and drug testing applications. Int J Mol Sci 21(20). https://doi.org/10.3390/ijms21207757

Lamers MM, Beumer J, van der Vaart J, Knoops K, Puschhof J, Breugem TI, Ravelli RBG, Paul van Schayck J, Mykytyn AZ, Duimel HQ, van Donselaar E, Riesebosch S, Kuijpers HJH, Schipper D, van de Wetering WJ, de Graaf M, Koopmans M, Cuppen E, Peters PJ et al (2020) SARS-CoV-2 productively infects human gut enterocytes. Science 369(6499):50–54. https://doi.org/10.1126/science.abc1669

Lee J, Koehler KR (2021) Skin organoids: a new human model for developmental and translational research. Exp Dermatol 30(4):613–620. https://doi.org/10.1111/exd.14292

Lee W, Debasitis JC, Lee VK, Lee JH, Fischer K, Edminster K, Park JK, Yoo SS (2009) Multi-layered culture of human skin fibroblasts and keratinocytes through three-dimensional freeform fabrication. Biomaterials 30(8):1587–1595. https://doi.org/10.1016/j.biomaterials.2008.12.009

Li M, Izpisua Belmonte JC (2019) Organoids—preclinical models of human disease. N Engl J Med 380(6):569–579. https://doi.org/10.1056/NEJMra1806175

Liao CC, Wu CY, Lin MH, Hsieh FK, Hsu LT, Chang SY, Chen KJ, Huang HT, Hsu HC, Lin CH, Lin PJ, Lai HM, Kojima H, Todo H, Lin SJ, Li JH, Chen W (2021) Validation study of a new reconstructed human epidermis model EPiTRI for *in vitro* skin irritation test according to OECD guidelines. Toxicol in Vitro 75:105197. https://doi.org/10.1016/j.tiv.2021.105197

Ma J, Liu J, Gao D, Li X, Zhang Q, Lv L, Wang Y, Li J, Zhu Y, Wu Z, Hu H, Li Y, Ma L, Liu Q, Hu Z, Zhang S, Zhou Y, Wang M, Leng L (2022) Establishment of human pluripotent stem cell-derived skin organoids enabled pathophysiological model of SARS-CoV-2 infection. Adv Sci 9(7):2270041. https://doi.org/10.1002/advs.202270041

Manita PG, Garcia-Orue I, Santos-Vizcaino E, Hernandez RM, Igartua M (2021) 3D bioprinting of functional skin substitutes: from current achievements to future goals. Pharmaceuticals 14(4):362. https://doi.org/10.3390/ph14040362

Marques AP, Reis R, Pirraco RP, Cerqueria M (eds) (2017) Skin tissue models. Academic, Cambridge

Mazzocchi A, Devarasetty M, Herberg S, Petty WJ, Marini F, Miller L, Kucera G, Dukes DK, Ruiz J, Skardal A, Soker S (2019) Pleural effusion aspirate for use in 3D lung cancer modeling and chemotherapy screening. ACS Biomater Sci Eng 5(4):1937–1943. https://doi.org/10.1021/acsbiomaterials.8b01356

Michael S, Sorg H, Peck C-T, Koch L, Deiwick A, Chichkov B, Vogt PM, Reimers K (2013) Tissue engineered skin substitutes created by laser-assisted bioprinting form skin-like structures in the dorsal skin fold chamber in mice. PLoS One 8(3):e57741. https://doi.org/10.1371/journal.pone.0057741

Mok BR, Shon SJ, Kim AR, Simard-Bisson C, Martel I, Germain L, Kim DH, Shin JU (2022) Structural and functional validation of a full-thickness self-assembled skin equivalent for disease modeling. Pharmaceutics 14(6):1211. https://doi.org/10.3390/pharmaceutics14061211

Monteil V, Kwon H, Prado P, Hagelkrüys A, Wimmer RA, Stahl M, Leopoldi A, Garreta E, del Pozo CH, Prosper F, Romero JP, Wirnsberger G, Zhang H, Slutsky AS, Conder R, Montserrat N, Mirazimi A, Penninger JM (2020) Inhibition of SARS-CoV-2 infections in engineered human tissues using clinical-grade soluble human ACE2. Cell 181(4):905–913.e7. https://doi.org/10.1016/j.cell.2020.04.004

Murphy SV, Atala A (2014) 3D bioprinting of tissues and organs. Nat Biotechnol 32(8):773–785. https://doi.org/10.1038/nbt.2958

Ng WL, Wang S, Yeong WY, Naing MW (2016) Skin bioprinting: impending reality or fantasy? Trends Biotechnol 34(9):689–699. https://doi.org/10.1016/j.tibtech.2016.04.006

Ng WL, Qi JTZ, Yeong WY, Naing MW (2018) Proof-of-concept: 3D bioprinting of pigmented human skin constructs. Biofabrication 10(2):025005. https://doi.org/10.1088/1758-5090/aa9e1e

OECD(2021) Test guideline no. 439: in vitro skin irritation: Reconstructed human epidermis test methods
OECD (n.d.) Test no. 431: in vitro skin corrosion: reconstructed human epidermis (RHE) test method. https://www.oecd.org/publications/test-no-431-in-vitro-skin-corrosion-reconstructed-human-epidermis-rhe-test-method-9789264242753-en.htm
Pfuhler S, Pirow R, Downs TR, Haase A, Hewitt N, Luch A et al (2021) Validation of the 3D reconstructed human skin comet assay, an animal-free alternative for following-up positive results from standard in vitro genotoxicity assays. Mutagenesis 36(1):19–35. https://doi.org/10.1093/mutage/geaa009
Pompaiah M, Bartfeld S (2017) Gastric organoids: an emerging model system to study helicobacter pylori pathogenesis. Curr Top Microbiol Immunol 400:149–168. https://doi.org/10.1007/978-3-319-50520-6_7
Rosdy M, Clauss LC (1990) Terminal epidermal differentiation of human keratinocytes grown in chemically defined medium on inert filter substrates at the air-liquid interface. J Invest Dermatol 95(4):409–414. https://doi.org/10.1111/1523-1747.ep12555510
Ruiz-Garcia H, Alvarado-Estrada K, Schiapparelli P, Quinones-Hinojosa A, Trifiletti DM (2020) Engineering three-dimensional tumor models to study glioma cancer stem cells and tumor microenvironment. Front Cell Neurosci 14:558381. https://doi.org/10.3389/fncel.2020.558381
Saito K, Nukada Y, Takenouchi O, Miyazawa M, Sakaguchi H, Nishiyama N (2013) Development of a new in vitro skin sensitization assay (Epidermal sensitization assay; EpiSensA) using reconstructed human epidermis. Toxicol in Vitro 27(8):2213–2224. https://doi.org/10.1016/j.tiv.2013.08.007
Scharschmidt TC, Fischbach MA (2013) What lives on our skin: ecology, genomics and therapeutic opportunities of the skin microbiome. Drug Discov Today Dis Mech 10(3–4):e83–e89. https://doi.org/10.1016/j.ddmec.2012.12.003
Schwank G, Koo BK, Sasselli V, Dekkers JF, Heo I, Demircan T, Sasaki N, Boymans S, Cuppen E, van der Ent CK, Nieuwenhuis EES, Beekman JM, Clevers H (2013) Functional repair of CFTR by CRISPR/Cas 9 in intestinal stem cell organoids of cystic fibrosis patients. Cell Stem Cell 13(6):653–658. https://doi.org/10.1016/j.stem.2013.11.002
Seol YJ, Lee H, Copus JS, Kang HW, Cho DW, Atala A, Lee SJ, Yoo JJ (2018) 3D bioprinted bio mask for facial skin reconstruction. Bioprinting 10:e00028. https://doi.org/10.1016/j.bprint.2018.e00028
Shanks N, Greek R, Greek J (2009) Are animal models predictive for humans? Philos Ethics Humanit Med 4:2. https://doi.org/10.1186/1747-5341-4-2
Shelkey E, Oommen D, Stirling ER, Soto-Pantoja DR, Cook KL, Lu Y, Votanopoulos KI, Soker S (2022) Immuno-reactive cancer organoid model to assess effects of the microbiome on cancer immunotherapy. Sci Rep 12(1):9983. https://doi.org/10.1038/s41598-022-13930-7
Skardal A, Devarasetty M, Rodman C, Atala A, Soker S (2015) Liver-tumor hybrid organoids for modeling tumor growth and drug response *in vitro*. Ann Biomed Eng 43(10):2361–2373. https://doi.org/10.1007/s10439-015-1298-3
Sun T, Jackson S, Haycock JW, MacNeil S (2006) Culture of skin cells in 3D rather than 2D improves their ability to survive exposure to cytotoxic agents. J Biotechnol 122(3):372–381. https://doi.org/10.1016/j.jbiotec.2005.12.021
Sutarjono B (2019) Can we better understand how Zika leads to microcephaly? A systematic review of the effects of the Zika virus on human brain organoids. J Infect Dis 219(5):734–745. https://doi.org/10.1093/infdis/jiy572
Thayer PS, Orrhult LS, Martínez H (2018) Bioprinting of cartilage and skin tissue analogs utilizing a novel passive mixing unit technique for bioink Precellularization. J Vis Exp 131:56372. https://doi.org/10.3791/56372
Wang Y, Jiang T, Qin Z, Jiang J, Wang Q, Yang S, Rivard C, Gao G, Ng TL, Tu MM, Yu H, Ji H, Zhou C, Ren S, Zhang J, Bunn P, Doebele RC, Camidge DR, Hirsch FR (2019) HER2 exon 20 insertions in non-small-cell lung cancer are sensitive to the irreversible pan-HER receptor tyrosine kinase inhibitor pyrotinib. Ann Oncol 30(3):447–455. https://doi.org/10.1093/annonc/mdy542

Wang X, Wang S, Guo B, Su Y, Tan Z, Chang M, Diao J, Zhao Y, Wang Y (2021) Human primary epidermal organoids enable modeling of dermatophyte infections. Cell Death Dis 12(1):35. https://doi.org/10.1038/s41419-020-03330-y

Weiswald LB, Bellet D, Dangles-Marie V (2015) Spherical cancer models in tumor biology. Neoplasia 17(1):1–15. https://doi.org/10.1016/j.neo.2014.12.004

Weng T, Zhang W, Xia Y, Wu P, Yang M, Jin R, Xia S, Wang J, You C, Han C, Wang X (2021) 3D bioprinting for skin tissue engineering: current status and perspectives. J Tissue Eng 12: 20417314211028576. https://doi.org/10.1177/20417314211028574

Wiener DJ, Basak O, Asra P, Boonekamp KE, Kretzschmar K, Papaspyropoulos A, Clevers H (2018) Establishment and characterization of a canine keratinocyte organoid culture system. Vet Dermatol 29(5):375–e126. https://doi.org/10.1111/vde.12541

Xu M, Lee EM, Wen Z, Cheng Y, Huang WK, Qian X, Tcw J, Kouznetsova J, Ogden SC, Hammack C, Jacob F, Nguyen HN, Itkin M, Hanna C, Shinn P, Allen C, Michael SG, Simeonov A, Huang W et al (2016) Identification of small-molecule inhibitors of Zika virus infection and induced neural cell death via a drug repurposing screen. Nat Med 22(10): 1101–1107. https://doi.org/10.1038/nm.4184

Xu H, Jiao Y, Qin S, Zhao W, Chu Q, Wu K (2018) Organoid technology in disease modelling, drug development, personalized treatment and regeneration medicine. Exp Hematol Oncol 7:30. https://doi.org/10.1186/s40164-018-0122-9

Yan WC, Davoodi P, Vijayavenkataraman S, Tian Y, Ng WC, Fuh JYH, Robinson KS, Wang C-H (2018) 3D bioprinting of skin tissue: from pre-processing to final product evaluation. Adv Drug Deliv Rev 132:270–295. https://doi.org/10.1016/j.addr.2018.07.016

Yang KC, Breitbart A, De Lange WJ, Hofsteen P, Futakuchi-Tsuchida A, Xu J, Schopf C, Razumova MV, Jiao A, Boucek R, Pabon L, Reinecke H, Kim DH, Ralphe JC, Regnier M, Murry CE (2018) Novel adult-onset systolic cardiomyopathy due to MYH7 E848G mutation in patient-derived induced pluripotent stem cells. JACC Basic Transl Sci 3(6):728–740. https://doi.org/10.1016/j.jacbts.2018.08.008

Yao B, Wang R, Wang Y, Zhang Y, Hu T, Song W et al (2020) Biochemical and structural cues of 3D-printed matrix synergistically direct MSC differentiation for functional sweat gland regeneration. Sci Adv 6(10):eaaz1094. https://doi.org/10.1126/sciadv.aaz1094

Human Skin Reconstructs Model Systems in Mechanistic Research, Safety, and Efficacy Studies of Cosmetics: Pros and Cons

3

Deepti Bhat, Shruti Goyal, and Shiv Poojan

Abstract

Toxicology testing has advanced significantly since the invention of in vitro testing procedures based on animal models, bridging the gap between the regulatory agencies that tested and approved new chemicals and the scientists that built and designed these in vitro testing models. A full review of the existing toxicity testing criteria for skin tissue models must be coupled with the optimum demands of the toxicological testing platform regarding production, testing, and screening technique (definition, testing models, concept, and limitations). Incorporating 3D bio-printing with microfluidics has caused a paradigm shift in toxicity testing. For the increased effectiveness of skin toxicological testing, this union would standardize fabrication processes, accuracy, and prompt deposition of test chemicals, real-time monitoring, and high-throughput screening.

Keywords

3D skin reconstruct · Efficacy studies · Safety studies · Toxicity studies · Cosmetics

3.1 Introduction

Toxicology testing is done to ascertain a chemical's potential dangers to a person and their surroundings (Krewski et al. 2010). These compounds include those found in cosmetics, household goods, and industrial activities. A total of 2000 new chemicals

D. Bhat · S. Goyal · S. Poojan (✉)
bioTox Solution Pvt. Ltd., Institute of Engineering and Technology, Lucknow, Uttar Pradesh, India

A. B. Pant et al. (eds.), *Skin 3-D Models and Cosmetics Toxicity*,
https://doi.org/10.1007/978-981-99-2804-0_3

are produced each year for varied uses, and every day, more and more new compounds are tested for safety by conventional toxicological procedures (Congress US 1995).

The in vitro toxicity testing market was predicted to increase from an anticipated USD 13 billion in 2016 to USD 20.8 billion by the end of 2021 (Markets and Markets 2016). Expecting studies to precisely determine a substance's toxicity to humans without using a substantial number of subjects who accurately reflect the diversity of the human population is unrealistic and immoral. Animal models are used as a result to give preliminary safety data to meet conservative regulatory standards. How accurately and consistently these animal models can foretell human responses is a crucial question. Using animal models has some drawbacks, including differences in how chemicals or substances are absorbed or dispersed, how they are metabolized, and animals' short lifespans (to monitor disease development accurately) 2019 (Van Norman 2019). As a result, practicing animal model experiment is still extremely contentious because of the vast differences in detrimental consequences. One of the most advanced in vitro-created structures frequently utilized by the cosmetics industry as an alternative testing tool to animal models is 3D skin tissue models (Suhail et al. 2019). As a result, the practice of animal models is still highly debatable due to the vast differences in detrimental consequences. One of the most advanced in vitro-created structures frequently utilized by the cosmetics industry as an alternative testing tool to animal models is 3D skin tissue models (Suhail et al. 2019). Human skin is a sophisticated organ comprised of various cell types, and the extracellular matrix (ECM) and cellular components are distributed anisotropically throughout the organ. Thanks to models of skin tissue that are relevant to humans, cosmetics testing is accurate and dependable (Frantz et al. 2010). One of the most sophisticated and fully comprehended in vitro constructs is the 3D skin tissue model. They have been widely employed by the cosmetics industry as an alternate testing approach to substitute animal models (Randall et al. 2018). The human skin is the largest and complex organ with a variety of cell types situated in relation to one another in a precise configuration. The extracellular matrix (ECM) and cellular components are dispersed in an anisotropic fashion (Kular et al. 2014; Tracy et al. 2016). Testing for cosmetics is made more reliable and accurate by using human-relevant skin tissue models. The Globally Harmonized System (GHS) of Classification and Labeling of Chemicals (UN), which is used by the Organization for Economic Cooperation and Development (OECD), tests guidelines (TG) that have thoroughly investigated and recorded a number of skin toxicity tests (OECD 2021a, b).

3.2 3D Skin Models in the Dermatological Studies

It has become easier to create various skin models for use in dermatological research or in the testing of cosmetics and pharmaceuticals. These skin models include organotypic, co-culture, and mono-cellular culture models, as well as in vivo and in vitro research. In vivo models, albeit the most physiological model system, have disadvantages including high costs, extensive development timeframes, and ethical and legal issues (Moniz et al. 2020). The effectiveness of medications has frequently been assessed using in vivo animal research (Arora et al. 2011). The current surge in ethical consciousness and legal concerns, which are driving alternatives to animal research, has, nevertheless, reduced the use of animal studies. Also, animal skin tissue has a different histology and physiology than human skin tissue, but there has not been a reliable replacement produced to replace the usage of animal models (Cho et al. 2013; Barré-Sinoussi and Montagutelli 2015).

3.3 Microfluidics: 3D Skin-on-a-Chip Platform

Microfluidics and 3D cell culture were combined for the first time in 2003. Since then, it has developed into the organ-on-a-chip technology of today (Coluccio et al. 2019). Due to the incorporation of programmed flow with pressure control, utilizing micro-engineered system that includes pumps and valves, in the use of microfluidics, facilitates the analysis of toxicological testing, high-throughput screening, and the evaluation of pharmacological modulation using well-characterized medicines (De Stefano et al. 2022). In a few recent research studies, full-thickness model of human skin equivalent prepared from human skin cells and different types of matrix was maintained above PET membranes using a microfluidic skin-on-a-chip technology. To give a physiologically appropriate blood residence for tissue maturation and terminal differentiation of HSEs, microfluidics chip design is distinctively built. This microfluidic device makes it possible to maintain HSEs for an extended period of time without a pump (up to 3 weeks with repeatable and extremely accurate skin permeation testing). Additionally, by adding more parallel microfluidic permeation arrays, high-throughput screening can be accomplished with ease (Ponmozhi et al. 2021; Risueño et al. 2021; Fernandez-Carro et al. 2022). This adaptability of combining a skin-on-a-chip method system with an effective organ mimics could lead to the development of a "human-on-a-chip," which could be used to investigate various drug delivery methods (oral, dermal, or aerosol) and assess the toxicity of various drug formulations using a multi-organ approach (Risueño et al. 2021).

3.4 Ex Vivo Skin Explant

Skin samples that were taken following a skin excision are used to create skin explants. The samples are cleaned of contaminants and subcutaneous fat tissue, and the remaining tissue is then grown. The epidermis and dermis of the skin explants can either be separated or the entire tissue can be employed (Cho et al. 2013; Corzo-León et al. 2019). Temperature, light, and humidity are important cultural variables. DMEM and DMEM/F12 Ham's medium are currently available culture media, whereas human scalp tissue is grown in William's E culture medium with fetal bovine serum added as a supplement (Cho et al. 2013).

3.5 Advantages of 3D Skin or Ex Vivo Skin Explants

Skin explants and reconstructed skins are examples of organotypic models, which have the advantage of showing three dimensions as opposed to mono-cellular culture and co-culture models, which only do so (Lebonvallet et al. 2010). Additionally, organotypic models incorporate intercellular connections, such as those between keratinocytes and fibroblasts, whereas mono-cellular cultures by definition only contain one kind of cell (Russo et al. 2020). In addition, compared to in vivo models, organotypic models are simpler to employ. However, drawbacks include the absence of desquamation, difficulties in manipulating sample conditions, and lack of circulation and nerve innervation, which are present in in vivo models. Skin explants can produce physiological results among these models because of their unique properties (Neil et al. 2020). All or most types of cells are present in skin explant models, which has an advantage over reconstructed skin, which only has fibroblasts, keratinocytes, and melanocytes (Henrot et al. 2020). Additionally, skin explants—as opposed to reconstructed skins—contain details about the subject's age and lifestyle, such as sun exposure, medical history, and allergies (Lebonvallet et al. 2010). Additionally, skin explants cost less to create and may be used right away, whereas reconstructive skin needs more time to prepare and the procedure itself can be difficult. Skin explants, as opposed to reconstructed skins, contain details about the subject's age and way of life. According to the culture conditions, skin explants should typically be employed within 10–14 days, although reconstructed skin can be used for a longer period of time during the maturation stage, which can run from a few days to a few weeks (Table 3.1) (Eberlin et al. 2020). Ex vivo can be utilized as an alternative

Table 3.1 Evaluation of reconstructed skin and ex vivo explant models

Model	Advantages	Limitations
3-D skin	Diverse circumstances	Missing appendages
	Altering the maturation period	A lack of circulation and innervation
	A lot of applications	Long and difficult to create
		High maintenance and creation costs
Skin explant	All common cell types are present	Unable to apply desired criteria
	Detailed skin information on several uses	Absence of maturational control
	Representative of personal behavior	Brief duration
	Simple to manufacture cheap to produce	Based on a human biopsy
	All common cell types are present	A lack of circulation and innervation
	Detailed skin information on several uses	

to animals for skin efficacy testing despite the aforementioned drawbacks. It might be possible to create more trustworthy models to test the anti-aging or whitening benefits of various treatments with more sophisticated skin explant systems. Because it is likely to portion a variety of useful needles of the skin to test reduction, wall effects, and wound healing, which are challenging to test in animals or in cell culture (Burger et al. 2016).

3.6 Limitations and Other Perspectives of 3D Skin or Ex Vivo Skin Explants

It seems that ex vivo skin experiments can replicate in vivo laser studies based on prior research (Park et al. 2015). Given the comparatively brief preservation and the lack of blood circulation in the ex vivo skin, it is anticipated that these advantages will be particularly noticeable in studies examining the early alterations following laser exposure. However, the dermal remodeling of collagen is assumed to be connected to the core skin alterations that follow fractional laser exposure (Borges et al. 2016). Skin explants cannot be kept for long enough to exhibit these histologic alterations. Additionally, the skin collagen is gradually deteriorating. These elements might make the dermal alterations seen in skin explants less reliable. In addition, the majority of the human skin tissues used in previous investigations were from the abdomen. However, the spreading of sebaceous glands and hair shafts in the skin tissue of the face and the belly may vary (Yousef et al. 2022). Additionally, because these skin samples are taken during surgery, there is a cap on how much can be taken during each excision. Another challenge is that the processed removed skin pieces need to be done quickly; thus, researchers need to carefully plan their investigations to allow for handling time (Table 3.2). Since human skin tissues are reported to resemble those of pigs and mini-pigs, utilizing in vivo pig models as an alternative to skin explants to study the effects of fractional laser therapy may also be advantageous (Hwang et al. 2021).

Table 3.2 Benefit and limits of skin explants in the other studies

Benefit	Limitations
Existence of sweat glands, hair follicles, and other skin-related structures	There is no blood flow
Cheaper	Evaluation of dermal alterations is challenging
Almost all of the skin's cellular components reflect the traits of the donors and originating parts.	It was difficult to plan the study timetable and collect enough skin samples from the different site
Reflects the three-dimensional structure of human skin effective for evaluating the early effects of laser exposure	No invasion of inflammatory cells
Existence of sweat glands, hair follicles, and other skin-related structures	Later stages of laser-induced alterations are challenging to see long-term maintenance challenges

References

Arora T et al (2011) Substitute of animals in drug research: an approach towards fulfillment of 4R's. Indian J Pharm Sci 73(1):1–6

Barré-Sinoussi F, Montagutelli X (2015) Animal models are essential to biological research: issues and perspectives. Future Sci OA 1(4):Fso63

Borges J, Manela-Azulay M, Cuzzi T (2016) Photoaging and the clinical utility of fractional laser. Clin Cosmet Investig Dermatol 9:107–114

Burger P et al (2016) Skin whitening cosmetics: feedback and challenges in the development of natural skin lighteners. Cosmetics 3(4):36

Cho HI et al (2013) Usefulness and limitations of skin explants to assess laser treatment. Medical Lasers 2(2):58–63

Coluccio ML et al (2019) Microfluidic platforms for cell cultures and investigations. Microelectron Eng 208:14–28

Congress US (1995) Screening and testing chemicals in commerce. OTA-BP-ENV-166. Office of Technology Assessment, Washington

Corzo-León DE, Munro CA, MacCallum DM (2019) An ex vivo human skin model to study superficial fungal infections. Front Microbiol 10:1172

De Stefano P, Bianchi E, Dubini G (2022) The impact of microfluidics in high-throughput drug-screening applications. Biomicrofluidics 16(3):031501

Eberlin S et al (2020) The ex vivo skin model as an alternative tool for the efficacy and safety evaluation of topical products. Altern Lab Anim 48(1):10–22

Fernandez-Carro E et al (2022) Modeling an optimal 3D skin-on-Chip within microfluidic devices for pharmacological studies. Pharmaceutics 14(7):1417

Frantz C, Stewart KM, Weaver VM (2010) The extracellular matrix at a glance. J Cell Sci 123 (Pt 24):4195–4200

Henrot P et al (2020) A method for isolating and culturing skin cells: application to endothelial cells, fibroblasts, keratinocytes, and melanocytes from punch biopsies in systemic sclerosis skin. Front Immunol 11:566607

Hwang JH et al (2021) Ex vivo live full-thickness porcine skin model as a versatile in vitro testing method for skin barrier research. Int J Mol Sci 22(2):657

Krewski D et al (2010) Toxicity testing in the 21st century: a vision and a strategy. J Toxicol Environ Health B Crit Rev 13(2–4):51–138

Kular JK, Basu S, Sharma RI (2014) The extracellular matrix: structure, composition, age-related differences, tools for analysis and applications for tissue engineering. J Tissue Eng 5: 2041731414557112

Lebonvallet N et al (2010) The evolution and use of skin explants: potential and limitations for dermatological research. Eur J Dermatol 20:671–684

Markets and Markets (2016) Global in vitro toxicology testing market by product, type (ADME) toxicity endpoints and tests (Carcinogenicity, Dermal Toxicity), technology (Genomics, Transcriptomics), method (Cellular Assays), industry (Pharmaceutical) forecast to y2021, Report 2016.

Moniz T, Costa Lima SA, Reis S (2020) Human skin models: from healthy to disease-mimetic systems; characteristics and applications. Br J Pharmacol 177(19):4314–4329

Neil JE, Brown MB, Williams AC (2020) Human skin explant model for the investigation of topical therapeutics. Sci Rep 10(1):21192

OECD (2021a) Test no. 439: in vitro skin irritation: reconstructed human epidermis test method.

OECD (2021b) Guideline no. 497: defined approaches on skin sensitisation.

Park GH et al (2015) Usefulness of skin explants for histologic analysis after fractional Photothermolysis. Ann Dermatol 27(3):283–290

Ponmozhi J et al (2021) Development of skin-on-A-Chip platforms for different utilizations: factors to be considered. Micromachines (Basel) 12(3):294

Randall MJ et al (2018) Advances in the biofabrication of 3D skin in vitro: healthy and pathological models. Front Bioeng Biotechnol 6:6

Risueño I et al (2021) Skin-on-a-chip models: general overview and future perspectives. APL Bioeng 5(3):030901

Russo B, Brembilla NC, Chizzolini C (2020) Interplay between keratinocytes and fibroblasts: a systematic review providing a new angle for understanding skin fibrotic disorders. Front Immunol 11:648

Suhail S et al (2019) Engineered skin tissue equivalents for product evaluation and therapeutic applications. Biotechnol J 14(7):e1900022

Tracy LE, Minasian RA, Caterson EJ (2016) Extracellular matrix and dermal fibroblast function in the healing wound. Adv Wound Care (New Rochelle) 5(3):119–136

Van Norman GA (2019) Limitations of animal studies for predicting toxicity in clinical trials: is it time to rethink our current approach? JACC Basic Transl Sci 4(7):845–854

Yousef H et al (2022) Histology, skin appendages. In: StatPearls. Stat Pearl, Treasure Island, FL

Skin-on-a-Chip Microfluidic Devices: Production, Verification, and Uses in Cosmetic Toxicology

4

Krishan Mistry and Michael H. Alexander

Abstract

It is increasingly recognized that the use of 2D cell culture assays to model keratinocyte and fibroblast responses to cosmetic products is being rapidly modified and updated. While these probes provide a time-efficient, simple, and cost-effective model of the skin, they have been repeatedly shown not to fully incorporate the in vivo cutaneous environment. The European Union has already implemented legislation that prevents the sale of cosmetic products developed through animal testing. This highlights the urgent need for novel, reproducible, cost-effective, and mass-producible model systems that can offer a comparable resource for cosmetic testing. While this need was initially attempted to be met through the development of 3D models of the skin, these systems have been found to lack the more complex biochemical and biophysical properties present in vivo. Thus, considerable interest has been shown in the development and optimization of "organ-on-a-chip" technology to produce a reproducible and ethically isolated model system of the skin. The following chapter will explain how to produce a skin-on-a-chip microfluidic device, the various factors that must be considered in its design and production, the methodology required to verify its reproduction of the in vivo cutaneous environment, and its applications in cosmetic toxicology.

K. Mistry
Translational and Clinical Research Institute, Newcastle University, Newcastle upon Tyne, UK

M. H. Alexander (✉)
Department of Dermatology and Cutaneous Biology, Thomas Jefferson University, Philadelphia, PA, USA
e-mail: Michael.Alexander2@jefferson.edu

A. B. Pant et al. (eds.), *Skin 3-D Models and Cosmetics Toxicity*,
https://doi.org/10.1007/978-981-99-2804-0_4

Keywords

Skin-on-a-chip · Microfluidic devices · Cosmetic toxicology and 3D bioprinting

4.1 Introduction

It is increasingly accepted that the use of 2D cell culture assays to model keratinocyte and fibroblast responses to cosmetic products needs to be revised and updated. While these assays provide a time-efficient, simple, and cost-effective model of the skin, they have been repeatedly shown to not fully encompass the full in vivo cutaneous environment (Duval et al. 2017). This issue of true representation has traditionally been resolved through the use of animal testing (Ngo and Maibach 2010). However, the growing movement in scientific and political fields to limit the use of animals in cosmetic product development due to ethical concerns may prevent the future use of animals as representative skin models. The European Union has already implemented legislation that prevents the sale of cosmetic products developed through animal testing (Adler et al. 2011; Hartung and Rovida 2009). This highlights the urgent need for novel, reproducible, cost-effective, and mass-producible model systems that can offer a comparable resource for cosmetic testing. As such, a significant effort has been made in recent decades to develop and improve in vitro systems to encompass all aspects of in vivo skin biology.

While this need was initially attempted to be met through the development of 3D models of the skin (Carlson et al. 2008; Antoni et al. 2015), which typically use commercially available biologically inert plastic scaffolds or ECM-derived hydrogels to produce a multicellular and multilayered model system of the skin, these have also been found to lack the more complex biochemical and biophysical properties present in vivo *(*Gangatirkar et al. 2007*;* El Ghalbzouri et al. 2009*)*. As such, considerable interest has been shown in the field for the development and optimization of "organ-on-a-chip" technology to produce a reproducible and ethically uncompromised model system of the skin.

Organ-on-a-chip technology originally emerged from material and computer research during the 1980s, which built upon advancements in micro-electric–mechanical systems (MEMS), a technology used in producing semiconductor chips and other electrical components that are only microns in size (Azizipour et al. 2020). Following this advancement, biomaterialists began to develop MEMS that incorporated biological material for potential use in multiple areas of life science research, including drug development and toxicology (Grayson et al. 2004). These biological MEMS (bioMEMS) are ultimately microfluidic devices, which are a type of device designed to allow for the precise control of extremely low volumes of liquid within microscopically small cell culture environments (Wu et al. 2020). This level of control allows for a new level of complexity to be introduced to tissue models previously lacking in all other systems. For example, these bioMEMS allow closer modeling of the in vivo tissue environment by mimicking the supply of nutrients and immune cells to tissues via a specifically designed and machined mock–vascular network (Vargas et al. 2021). They also allow for the establishment of biological molecule gradients within tissues that give broader context to the model and allow more significant distinction in tissue-to-tissue interfaces (Bhatia and Ingber 2014). This advancement in fluid control, combined with the previously developed tissue-specific cell culture techniques, allowed the first "organ-on-a-chip" devices to be produced.

The first reports of the successful implementation of organ-on-a-chip technology for skin modeling, or "skin-on-a-chip," occurred in 2013. Since this founding work was published, there has been a significant advancement in this technology (Ataç et al. 2013; Wagner et al. 2013).

The following chapter will not attempt to explain this history of skin-on-a-chip development but will instead aim to explain how to produce a skin-on-a-chip microfluidic device. This will be achieved in three separate sections. The first (Sect. 4.2) will explain the various factors that must be considered in the fabrication and production of the microfluidic device itself. The second (Sect. 4.3) will explain the methodology that needs to be employed to verify that the device is able to fully capture the in vivo cutaneous environment, and the final section (Sect. 4.4) will explain how these skin-on-a-chip devices can actually be utilized in the field of cosmetic toxicology.

4.2 Production of Skin-on-a-Chip Devices

4.2.1 Sourcing and Production of Skin Models for Skin-on-a-Chip Devices

When developing a skin-on-a-chip microfluidic device, for the purpose of toxicity and cosmetic testing, there are many factors that need to be considered. However, the central and most vital factor to consider is how to source or produce the physiologically relevant skin model housed within the device itself. When considering the best skin model system for a given device, its intended use, access to patient tissue or cells, and the need for incorporating other vital components, such as those needed for its maintenance and monitoring, all must be considered (Zoio and Oliva 2022).

Typically, the skin models housed within skin-on-a-chip devices have two separate sources. The first is human skin biopsies, which are extracted from clinical patients and transferred directly into the housing units of the device (Risueño et al. 2021). Access to this source of material does pose a challenge, given the clinical training required to remove the sample and the patient/ethical permissions that need to be attained prior to removal. As such, humanized full-thickness skin equivalents have been increasingly seen as a viable source of the material. These skin equivalents are produced using a specific cell culture technique that can accurately replicate the three-dimensional structures of skin (Hill et al. 2015). These humanized skin equivalents can either be produced externally, then biopsied, and inserted into the microfluidic device, or they can be generated de novo within the skin-on-a-device itself. To date, multiple tissue culture protocols have been developed and optimized to incorporate multiple cell types from various biological sources into humanized skin equivalent models.

Conventional full-thickness skin equivalent models are generally constructed using primary human cutaneous cells, such as keratinocytes and dermal fibroblasts, isolated from healthy human skin biopsies, as the use of these cells will capture an in vivo phenotype exceptionally well (Zoio and Oliva 2022; van den Broek et al. 2017). Previous studies have successfully demonstrated that this protocol can also be used to generate 3D psoriatic skin models using patient-derived cells (Rioux et al. 2021). Additional work has also shown that by incorporating activated T cells, isolated from whole human blood, into the skin model, they were able to create an immunocompetent model that can reflect the psoriatic inflammatory environment more precisely, thereby creating a suitable model that could be used for both fundamental and translational research studies (Rioux et al. 2021). Overall, using primary cells presents a unique set of advantages, such as capturing the in vivo phenotype to ensure cell–cell communication within the skin-on-a-chip device is consistent with the in vivo environment; however, there are still disadvantages associated with using primary cells. Some main disadvantages include the limited availability of donor skin and donor variation, which may affect experimental reproducibility. Also, given that the extraction, growth, and maintenance of primary cells require more specialized culture techniques than immortalized cell culture, laboratories with little experience in primary cell culture may find it harder to integrate primary cells into the design of their skin-on-a-chip devices.

An alternative to using primary cells in the construction of humanized skin equivalents is immortalized cell lines. Cell lines with validated purity and viability have significant advantages due to their high availability and reliability for cell population expansion (Zoio and Oliva 2022). However, cell lines are only approximations of primary cell function and can deviate from the original phenotype. This can be observed in the widely used keratinocyte cell line, HaCaT. While HaCaT cells can be used to form epidermal tissue, they have a low differentiation potential compared to primary keratinocytes, making generating a functional *stratum corneum* challenging. This highlights the potential drawbacks of incorporating immortalized cell lines into the skin model retained within the skin-on-a-chip device, which may be used to assess the specific toxicological effects of a given compound on skin barrier function (Brohem et al. 2011; Jung et al. 2016).

Another cell line that could be considered is human telomerase reverse transcriptase (hTERT)-immortalized keratinocyte and dermal fibroblast cell lines. Previous studies have demonstrated that hTERT-immortalized keratinocytes and dermal fibroblasts could form a high-quality full-thickness skin equivalent with a fully differentiated epidermis, comparable to that formed when using primary keratinocytes (Reijnders et al. 2015). As these hTERT-immortalized cell lines are both reliably expansive and able to generate a fully stratified epidermis, 3D skin models generated using these cells should be considered when designing skin-on-a-chip devices needed for higher throughput screening of new drugs and cosmetics. However, depending on the intended use of the skin-on-a-chip model, donor variation can be an essential feature to accurately resemble the population (Zoio and Oliva

2022). Cell lines lack patient specificity, which is particularly important for disease modeling. Furthermore, previous studies have suggested that some proteins overexpressed in cell lines are associated with toxicity-related pathways and, therefore, limit the use of these cell lines in 3D skin models for toxicity testing (Astashkina et al. 2012).

Another promising cell source for skin-on-a-chip is induced pluripotent stem cells (iPSCs). These cells are derived from adult somatic cells via reprogramming with octamer-binding transcription factor 3/4 (Oct 3/4), SRY (sex-determining region Y)-box 2 (Sox2), c-Myc, and Kruppel-like factor 4 (Klf4) expression factors (Rowe and Daley 2019). By expressing these factors, genes responsible for cell differentiation are suppressed, and the cells revert to a pluripotent state. Using iPSCs to construct 3D skin models could overcome the limitations of full-thickness skin models that are only comprised of keratinocytes, fibroblasts, and endothelial cells, as they have unlimited growth potential and the ability to differentiate into multiple cell types. Previous studies have demonstrated that iPSC-derived keratinocytes, dermal fibroblasts, and melanocytes generated a full-thickness skin equivalent that showed similar morphology and physiology to normal human skin (Gledhill et al. 2015). This model also reported efficient melanin production and transfer within epidermal–melanin units of the iPSC-derived skin equivalent, thereby demonstrating the potential to generate increased complexity in the model system, thus better mimicking normal skin function and physiology (Gledhill et al. 2015). Furthermore, cells that have been differentiated from iPSCs retain characteristics of the original donor, such as disease phenotypes and therapeutic response, thereby offering an alternative source of cells for modeling cutaneous diseases (Khurana et al. 2021). However, many factors still need to be considered before iPSC-derived skin-on-a-chip models can be viable, such as high cost, retention of epigenetic memory, and genomic instability (Zoio and Oliva 2022).

Overall, each cell type and skin model has its own set of advantages and disadvantages that need to be carefully considered when producing a physiologically relevant skin-on-a-chip model, especially when designing a skin-on-a-chip model for a specific downstream application such as toxicology and cosmetic testing.

4.2.2 Fabrication Methodologies of the Housing Units of Skin-on-a-Chip Devices

A central consideration when producing skin-on-a-chip devices for cosmetic toxicology is the design and fabrication of the housing units that hold all the components, producing the microfluidic system and device.

While these housing units can be produced through a number of different methods, all must contain several critical elements to ensure primary function. The first element is a central space capable of housing a functional skin model. As explained previously, this skin model will typically be either an extracted skin biopsy from either native skin or an externally produced humanized skin equivalent or a de novo generated skin equivalent produced within the device. The second required element is a complete microfluidic system, which must either mimic or allow for the production of a vasculature that supplies the skin model with all required nutrition through the perfusion of cell culture medium. This system must also be accessible to replace the cell culture medium after nutrition depletion.

In addition to these required elements, the housing unit may also contain space for integrating a wide range of other potential components. This may include space for the installation of pumps to allow for powered perfusion of the cell culture medium, biosensors to capture an output of interest, or additional organ housing spaces, to allow for multi-organ testing within the same microfluidic device. This section will briefly describe the most common production methods used in building skin-on-a-chip microfluidic devices and examine some of the more complex and advanced methods.

The most common production method in fabricating microfluidic housing units is lithography, specifically a combination of two separate lithographic processes named "photolithography" and "soft lithography." In short, these processes begin with pouring a light-sensitive material onto a silicon-based support block. When this light-sensitive material, also named a "photoresist," is exposed to UV light, it will liquefy, allowing it to be removed. This property allows the photoresist material to be shaped into the desired design. Once this photoresist has been poured onto the support block and allowed to solidify, this material is overlapped with the housing unit design. This design is printed or etched onto a "photomask," which either blocks or allows light penetration into the photoresist material. Areas exposed to UV light by the photomask will be liquefied and removed, while any area blocked from UV penetration will remain solid. This allows the two-dimensional design on the photomask to become a three-dimensional solid object made by the photoresist. The solid remains of the photoresist material can either act as a mold, casting the desired shape when a desired material is poured into it, or it can be used as a stamp to imprint the design on other materials (Duffy et al. 1998; Lee et al. 2017). While this is a standard production method for the housing units of skin-on-a-chip devices, the number, complexity, design, and arrangement of these housing units themselves can produce unique devices with different uses and different levels of in vivo representation. It is important to note that the descriptions in this section are generalized and can be deviated from based on the required use.

The most common use of this lithography-based technique in skin-on-a-chip production is named membrane-based soft lithography. In this technique, photolithography and soft lithography are initially used to produce two separate housing units, an apical and a basal plate. While many materials can be used, which will be discussed later, they are typically cast using the elastomer material polydimethylsiloxane (PDMS) (Lee et al. 2017; Maschmeyer et al. 2015; Song et al. 2017a, b).

The apical housing plate is typically designed to have a hole in the center that spans its entire thickness. This central hole allows for the placement and support of either the externally sourced skin model or will act as the cell culture chamber itself for de novo skin equivalent generation. In addition to this structure, the upper plate will commonly have additional holes that allow access to the microfluidic system, which will be housed in the basal plate. These can either be large holes at opposing ends of the plate, allowing access to cell culture medium reservoirs, which are also present in the basal plate, or smaller holes, which act as inlets and outlets for an external perfusion system. The cell culture medium reservoirs store the cell culture medium, which ultimately perfuses the microfluidic device and supports the skin model. This access in the upper plate has to be present to allow for the easy replacement of the medium after sufficient culture time and nutrient use.

The basal plate is typically more complex, as it must contain the entire microfluidic system within it. As previously mentioned, this system is typically comprised of two cell culture medium reservoirs, placed at opposite ends of the basal plate, which are connected to a central hole, which is aligned with the apical plate hole, and connected through either a single or a series of microfluidic channels. This allows cell culture medium to be perfused through the microfluidic system through either the use of pumps or gravity. In some designs, the medium reservoirs are only present in the apical plate, and the basal plate connects these to the microfluidic system (; Song et al. 2017a, b).

The entire housing unit can be assembled when both plates have been produced. Initially, the bottom plate is typically adhered to a glass slide through low-pressure plasma oxidization, which irreversibly seals these plates together, producing a fluid-tight seal. Following this, a porous membrane is fused to the top of the basal plate, using either glue or low-pressure plasma oxidation. This membrane must span the central cell culture hole present in both plates but be prevented from spanning the cell culture medium reservoirs. This porous membrane supports the external skin model or acts as a base for the construction of de novo skin equivalents. The membrane also allows for the transmission of nutrients to whichever functional skin model is used in the upper central cavity. It also prevents cell exposure to the direct flow of the medium solution, which can be damaging. Finally, the upper plate can be fixed to this structure to complete the arrangement.

This form of two-plate membrane-based soft lithography has been expanded on by adding a third PDMS plate into the structure (Wufuer et al. 2016; Jeon et al. 2020a). This additional middle plate, inserted between the apical and basal plates, typically has a central hole that spans its entire thickness. This ensures that when all the plates are fused, with each one separated by a porous membrane, there is a central cavity, which is not present when using a two-plate design. The presence of this central cavity serves a modeling function, as it allows for the epidermis of the skin model to be directly seeded onto the top of the upper membrane, with the dermis to be seeded in the central cavity. The bottom of the lower membrane can then be populated with endothelial cells, allowing for direct endothelial/dermal/epithelial cell contact, which better mimics the in vivo blood supply of the skin than using a single porous membrane to separate the dermis from the mock vasculature. While membrane-based soft lithography has been repeatedly used in the generation of skin-on-a-chip devices and can be argued to be the most popular form of production, there has been increased interest in the development of other production methods that can address some of its shortcomings, such as improving the reproducibility and speed of production of skin-on-a-chip devices. One of these emergent production methods is micromilling.

Micromilling is an ultra-precise production process that uses a cutting edge less than 1000 μm in size to machine complex 3D structures on a chip material (O'Toole et al. 2021; Câmara et al. 2012). The design of these complex 3D structures mirrors those produced by membrane-based soft lithography, such as a full microfluidic system and inlet/outlet ports. The cutting process is often guided through computer numerical control (CNC) and will often use poly methyl methacrylate (PMMA) as a base material rather than PDMS. Once multiple micromilled plates are produced, they can be assembled to form a full microfluidic device. This method ultimately produces a similar device to membrane-based soft lithography, with an apical plate used to house the skin construct and a basal plate with microfluidic channels to guide the medium solution, which is separated through the use of a porous membrane (Sriram et al. 2018). The best example of this production method is by Sriram et al. (Sriram et al. 2018), who utilized micromilling to produce this form of a device, with the addition of multiple other plates that are layered above the apical plate to allow for additional elements to be inserted into it, such as diffusion chambers, which allows for compound absorption testing.

Another method of production that does not employ any form of lithography is named the "layer-by-layer" production method, which has been previously used in many forms of organ-on-a-chip production. This process typically breaks down the aforementioned apical and basal plates into multiple smaller layers, which are then assembled to produce a fully functional microfluidic device. While this process can produce full devices, it can also produce the master molds, which allow for the repeated casting of PDMS plates for skin-on-a-chip production (Sasaki et al. 2019).

When directly producing microfluidic devices via layer-by-layer assembly, multiple vinyl UC or acrylic layers are cut from a single starting sheet. As this cutting process involves passing a single sheet through the narrow opening of the cutting tool, these layers must be much thinner than those plates produced via soft lithography. The multiple separate layers are then assembled to form the apical and basal plates of the skin-on-a-chip device. Some descriptions of this production method suggest that as many as seven separate sheets are required to produce a single plate, and assembly holes must be included in the design. These assembly holes allow for the correct alignment of the layers, which ensures the various microfluidic channels and housing units form correctly and provide a fluid-tight seal (Risueño et al. 2021; Valencia et al. 2021). These assembly holes prevent more complex structures from being included in the design of devices, as they occupy some of the limited available space. Still, cutting and assembly speed allows for a high-throughput approach to testing.

These production methods, layer-by-layer, micromilling, and membrane-based soft lithography have produced significant advancements in the modeling of skin and the production of skin-on-a-chip devices. However, these production methods have limited design potential compared to the housing units that can be produced through 3D printing technologies and have less in vivo representation compared to what can be achieved by incorporating 3D bioprinting. As such, multiple studies have investigated and developed viable production methodologies that employ 3D printing/bioprinting. This revolutionary technology can be utilized to either improve the microfluidic device housing units produced by these other methods or directly fabricate a more representative model of in vivo skin for insertion into the device.

One of the simplest forms of the incorporation of 3D printing into skin-on-a-chip housing units is through the printing of accessory elements that either support more standard production methods or adapt an already established device to allow for de novo skin equivalent generation within it.

An example of 3D printing to produce support structures is described by Bajza et al. While the actual microfluidic device was produced using standard PDMS polymer techniques, the surrounding support structures of PMMA and PLC were printed using a variety of commercially available 3D printers. These printed support structures secure the entire microfluidic device together and allow for the placement of a heating element in close contact with the skin model (Bajza et al. 2020).

3D printing has also been shown to be able to alter the housing unit in an already established skin-on-a-chip device, allowing for a different skin model to be utilized. This relatively simple 3D-printed apparatus is made of three separate elements. The first is a "sample holder," a short cylindrical tube with a faired base produced through filament-based 3D printing. This holds a collagen-based gel within it, forming a viable dermal compartment to support different cell cultures. The second is a circular mesh structure produced through the electrospinning of polycaprolactone (PCL). After coating with an ECM protein, it is placed on top of the collagen gel's sample holder. This mesh further supports cell attachment, as keratinocytes are seeded directly onto its surface. The final element of the 3D-printed apparatus is a "ring holder," which holds the other two components together. This sample holder device, with functioning epidermal and dermal compartments, can be inserted directly into the established skin-on-a-chip system, allowing a full-thickness humanized skin equivalent model to be utilized rather than a human skin biopsy (Tárnoki-Zách et al. 2021).

3D printing in skin-on-a-chip devices has a broader application than these listed alterations of already established production methods. It can fully produce the entire housing unit itself, with more accurate and complex components.

One of the earliest incorporations of broader 3D printing into the design of skin-on-a-chip housing units was the work by Abaci et al., who utilized this method to integrate a more complex vasculature into a humanized skin equivalent model. While not a complete skin-on-a-chip device, this methodology could be readily incorporated with little difficulty.

Through computer-aided design (CAD), several separate components are designed, printed, and cured. The first part of this novel structure was a ring-shaped support, which attaches directly to the top of the transwell insert and provides connection points for the other components. The second set of components are the inlet and outlet pipes, which are connected directly to the ring support structure and reach the base of the transwell insert. Finally, separate molds were used to cast sacrificial alginate channels, which will be used to guide vasculature formation. These sacrificial channels are laid on the insert's base and connected to the inlet and outlet pipes. Once the 3D-printed structures and alginate channels are assembled in the transwell, a dermal collagen matrix was seeded over the sacrificial alginate channels to form a functional dermal compartment and support base for the seeding of an epithelial layer. Once this matrix has solidified and cross-linked, the precast alginate channels are removed by passing a solution of sodium citrate through the 3D-printed inlet and outlet pipes. This dissolves the alginate, leaving continuous hollow channels in the dermal compartment. These channels can then be perfused with a solution containing a single endothelial cell suspension, which allows endothelial cells to adhere to the hollow channels' walls and form a de novo vasculature embedded in the dermal compartment, which is more representative of in vivo skin physiology.

Another methodology that employs 3D printing and de novo vasculature generation in the production of a skin-on-a-chip microfluidic device is the work conducted by Mori et al. (Mori et al. 2017). In this methodology, 3D printing technology initially produces a base plate that will go on to support the rest of skin-on-a-chip device. Following this, a hollow four-walled compartment was produced using the same extrusion-based printing technique. Each of these walls contained three hollow anchoring structures that spanned its length and projected in and out of the walls. These components are treated with perylene and subjected to O_2 plasma etching to increase ECM adhesion and general cell biocompatibility. To provide further support, nano-strings are fed between the anchoring structures in the walls to create an overlain grid of wire that provides a solid support on which a dermal compartment can be cast. A collagen solution containing fibroblasts is then added to the central cavity of the skin equivalent holder and allowed to solidify to form a dermal compartment. The anchoring structure and nanofibers continue to support the dermis as it contracts and detaches from the support structure walls. The nanowires are then slowly removed from the dermal compartment to create hollow channels in the dermis, which can be seeded with endothelial cells to form a network of de novo vasculature vessels that closely mimic those present in the in vivo dermal compartment. To form a full-thickness skin model, keratinocytes are seeded onto the dermal compartment within a silicone ring placed on its surface, which helps to promote keratinocyte adhesion and cornification directly on the dermis. The entire structure is then placed in a cell culture dish, with a series of silicone tubes connected to a pump to allow the perfusion of culture media (Mori et al. 2017).

This work by Abaci et al. and Mori et al. demonstrates that the use of 3D printing technology has advanced the production of skin-on-a-chip devices substantially, allowing for more complex incorporations of vascular networks into skin equivalents, the alteration of existing microfluidic devices to hold a wider variety of skin models, and the production of more specific support structures. These methodologies, however, only used standard plastic-based 3D printing. Further work has shown that 3D bioprinting, which utilizes bioink to produce a more comprehensive humanized skin equivalent model, has the potential to improve skin-on-a-chip devices further and replace animal-based model systems.

Studies by Kim et al. used 3D bioprinting to produce a total thickness skin equivalent and standard 3D printing to produce the surrounding housing unit (Kim et al. 2019). Again, while this skin equivalent production method is not a complete skin-on-a-chip microfluidic device, it could be readily expanded upon or incorporated into existing devices.

Initially, 3D printing produces a PCL mesh base plate, which, once filled with gelatin, acts as the support plate for the entire housing unit and the skin model itself. Unlike other 3D-printed housing units, which often fully produce the housing unit first, with the skin model produced after, this methodology allows for support walls to be printed in tandem with the various skin compartments. The first compartment of the skin is the hypodermal compartment, which is printed directly onto the PCL/gelatin mesh base plate using an adipose-fibrinogen bioink containing preadipocytes. This is the first skin model that includes a hypodermal compartment of the examples in this section. Following this, a thrombin–gelatin hydrogel containing endothelial cells was printed directly onto the hypodermis in a continuous cylinder that connects to inlet and outlet channels printed during the production of the support structure. This vascular bioink is surrounded by a 3D-printed dermal compartment, produced by the extrusion of a fibrinogen-based bioink containing fibroblasts directly onto both the vascular and hypodermal compartments. With these different compartments printed in the correct architecture, a series of temperature changes are then conducted to cross-link and mature the printed bioink. Initially, the device is incubated at 30 °C, which induces partial cross-linking of the collagen proteins in the hypodermal and dermal bioink, increasing its structural integrity. The temperature is then further increased to 37 °C, which completes the cross-linking of ECM components in the hypodermal and dermal compartments while completely liquefying the vascular bioink. This liquidation allows the endothelial cells present in the vascular bioink to sink under the influence of gravity and attach to the solid walls of the dermal compartment. This means that as the device is rotated, endothelial cells will adhere to all the free surfaces of the dermal compartment, forming a complete vasculature. The entire device is then submerged in fibroblast supporting medium, and a peristaltic pump is used to circulate endothelial supporting medium through the vasculature, to further help the maturation of these tissue compartments. The final step in this production method is that keratinocytes are printed directly onto the dermis surface using an in-house printing method and allowed to form a functional epidermis. This completes a comprehensible skin equivalent model containing all three primary skin layers and a representative vasculature present in the dermis.

Overall, all the numerous techniques listed in this section produce viable skin-a-on-chip microfluidic devices that are well suited as apparatus for cosmetic toxicology.

4.2.3 Material Selection for Fabrication of Skin-on-a-Chip Housing Plates

As explained in the previous section, there are numerous methods for producing housing plates for skin-on-a-chip devices. Still, even with these differences, the materials used to manufacture the housing plates have a level of consistency.

Excluding those that use 3D printing and 3D bioprinting, which use extrusion of specific plastic and bioink material to produce the device design, the materials selected in the other listed methodologies are chosen due to their specific characteristics.

The most common material used in skin-on-a-chip fabrication is polydimethylsiloxane (PDMS), which is characterized as an elastomer (Ding et al. 2020). Elastomers are a class of polymers that are noted for their viscoelastic properties. Viscoelasticity is defined by viscosity, the solidity of the material, and elasticity, the ability of a material to stretch and return to its original shape and structure upon removal of the force (Gogoi et al. 2022; Touchet and Cosgriff-Hernandez 2016). The preference for PDMS of all the potential elastomers is well documented in organ-on-a-chip designs due to numerous factors (Nge et al. 2013). The first is its biocompatibility, being well tolerated by numerous body tissues and cell types (Miranda et al. 2021; Bélanger and Marois 2001; Guo and Liu 2017; Hassler et al. 2011). The second is its gaseous permeability, allowing oxygen and carbon dioxide to permeate throughout the device (Markov et al. 2014). The third is its ability to be used in rapid production processes, reducing manufacturing time compared to processes that use silicon and glass as base materials and being easily removable from molds (McDonald and Whitesides 2002; Friend and Yeo 2010). The final is its optical transparency, allowing for real-time analysis via microscopy, which is critical to the verification and implementation of organ-on-a-chip devices (Liu et al. 2020).

While the elasticity of elastomers makes them desirable materials for lithographic production methods, this quality makes them unsuitable for other methods, such as micromilling (Sriram et al. 2018). These types of production methods tend to employ plastics as a base material. Plastics are defined as synthetic or semisynthetic polymers that have thermoplastic and thermosetting properties, meaning they are easily shaped through the use of heat (Halden 2010). The specific plastic, poly (methyl methacrylate) (PMMA), has been demonstrated to be a suitable material in the micromilling production of skin-on-a-chip housing plates. This is chiefly due to the increased rigidity of PMMA compared to PDMS, which allows the micromilling cutting edge to shape the plastic without causing deformities that may occur in more elastic materials. PMMA specifically has this improved structural integrity while retaining the transparent and biocompatible properties of PDMS, with the additional benefit of reduced autofluorescence (Ding et al. 2020; Piruska et al. 2005). The use of PMMA as a fabrication material has been shown to have a reduced permeability to oxygen, which should be accounted for when designing the apical plate of the skin-on-a-chip device (Zoio and Oliva 2022; Zahorodny-Burke et al. 2011).

The production method, "layer-by-layer," has been shown to chiefly employ acrylic and vinyl UC as a base material for skin-on-a-chip devices. While these materials have been demonstrated to produce viable skin-on-a-chip devices able to support the extensive viability of the housed skin model, the complete and comparative biocompatibility of these materials is yet to be elucidated and requires further analysis (Risueño et al. 2021; Valencia et al. 2021).

While the materials listed are the most frequently used in non-3D-printed housing plate production techniques for skin-on-a-chip devices, other materials have been investigated for their potential use in other organ-on-a-chip device contexts. These include paper, a variety of hydrogels, and combinations of different materials (Ding et al. 2020; Seliktar 2012; Sapp et al. 2015). While not explicitly shown to produce working skin-on-a-chip devices, all these should be considered potential base materials.

4.2.4 Design of the Nutrient Support System Within the Skin-on-a-Chip Device

Nutrient support is an essential factor when designing a skin-on-a-chip system. The supply of nutrients, such as glucose, calcium, and hormonal growth factors to the skin model, retained within the device are important to improve the longevity of the housed skin model, the induction of differentiation and stratification of the epidermis, and its downstream applications (Bikle et al. 2012; Salameh et al. 2021).

The vasculature of in vivo human skin is located within the dermal layer and acts to deliver nutrients to cells while also removing unwanted metabolic waste products. The vasculature also functions as a conduit for immune system components and helps regulate temperature (Low et al. 2020). While important for normal cutaneous function, the cutaneous vasculature also plays a role in different pathological conditions such as inflammatory conditions, cancer metastasis, and wound healing (Kashani-Sabet et al. 2001; Huggenberger and Detmar 2011). Furthermore, previous studies have shown that the cutaneous vasculature can impact the transdermal diffusion of substances. Therefore, cutaneous vasculature and its distribution should be considered in the generation of 3D skin models, especially those used for topical or transdermal drug delivery studies (Cevc and Vierl 2007).

Conventional humanized 3D skin models do not incorporate a vasculature, limiting their ability to fully replicate the function of human skin properly. Furthermore, the lack of vascularization within a skin model reduces the perfusion of nutrients and oxygen within the model system, reducing cell viability (Magliaro et al. 2019). In particular, the diffusion limit of oxygen within cell-rich tissues is approximately 200 μm. This value can be used to determine the smallest cubic volume of cells that can function and survive without a vasculature (Magliaro et al. 2019; Ehsan and George 2013). Therefore, the culture of 3D skin models thicker than 200 μm is more likely to undergo hypoxia-induced apoptosis if they do not include a complete vascular system.

There have been many approaches to inducing vasculogenesis in vitro, such as endothelial cell seeding onto support matrixes, such as hydrogels or scaffolds, and cell encapsulation (Shafiee et al. 2021; Zhang et al. 2018; Heo et al. 2019). However, these approaches are complex and slow, and are thus unsuitable for high-throughput applications (Shafiee et al. 2021; Sorrell et al. 2007). Additionally, the vascular channels formed have a random formation, meaning that the vascular channels within the 3D skin models cannot achieve complete perfusion, thereby limiting their applicability.

Advancements in biotechnology, which led to the development of microfluidic devices and 3D bioprinting, have allowed the development of perfusable vascular networks within full-thickness skin models (Salameh et al. 2021; Dellaquila et al. 2021). Various approaches have also been explored for perfusing culture medium. The two most frequent methods are either pump or gravity-driven approach (Kaarj and Yoon 2019).

The most common method for perfusing culture media is to use external pumps, such as a syringe or peristaltic pumps, as these systems can deliver an accurate, fine-tuned fluid flow (Mori et al. 2017; Salameh et al. 2021; Ramadan and Ting 2016). However, using these pumps can be time-consuming, and the need for external tubing and multiple connections can increase the risk of contamination (Zoio and Oliva 2022).

An alternative method is gravity-driven approach (Wufuer et al. 2016; Wang and Shuler 2018; Abaci et al. 2015). These approaches usually involve using custom-built rocking platforms to recirculate culture media through the microfluidic device. However, these gravity-driven approaches typically need more refined control as they do not perfuse fresh culture media and remove waste products.

The development of 3D bioprinting technology has allowed further advancements in mimicking the cutaneous vasculature within reconstructed skin models. Abaci et al. (Abaci et al. 2016) used 3D-bioprinted molds to micropattern sacrificial alginate channels that were used to cast a simulated vasculature inside a collagen matrix of the dermal compartment of a 3D skin model. Following stratification of the epidermis, the alginate was removed, leaving behind hollow tubes for perfusion. Additionally, Abaci et al. (Abaci et al. 2016) seeded either iPSC-derived endothelial cells or human umbilical vein endothelial cells (HUVECs) onto the inner surface of these channels. Adding these cells decreased the permeability and diffusivity compared to unseeded channels and demonstrated values similar to a real microvasculature. Similar 3D skin models were utilized by Mori et al. (Mori et al. 2017), with nylon threads used to create hollow channels before seeding with HUVECs. This perfusable skin model was used to study the percutaneous absorption of caffeine and isosorbide dinitrate solutions applied topically by measuring the amount of these molecules in the perfused medium flowing through the microvasculature and in the medium at the bottom of the culture device. Mori et al. (Mori et al. 2018) further improved the design of their perfusable skin model by introducing a motor to the system. This approach allowed them to apply mechanical force to the tissue, recreating skin stretching and thus demonstrating enhanced epidermal differentiation and stratification (Mori et al. 2018).

These advances in 3D bioprinting and microfluidics have allowed for better tissue formation and maturation ex vivo. Additionally, the incorporation of perfusable vascular channels demonstrates good vascular permeability properties, making skin-on-a-chip a promising platform for drug and cosmetic testing.

4.2.5 Biosensor Integration into the Skin-on-a-Chip Device

When analyzing 3D tissue models cultured on scaffolds or membranes, conventional microscopy techniques cannot be used as 3D cultures may be too thick and have high scattering effects, thereby limiting light penetration (Graf and Boppart 2010). Conventional 3D tissue morphology and function analyses mainly rely on endpoint assessment techniques such as immunohistochemistry of tissue sections. This usually requires the removal of the tissue from its original housing, chemical fixation, and labeling. Alternatively, assays using tracer compounds such as fluorescein isothiocyanate-labeled dextran can also be used. Furthermore, these compounds may affect tissue integrity and are not sensitive enough to detect subtle changes in tissue function (Arık et al. 2018). However, this limitation can be circumvented by integrating microsensors into organ-on-a-chip systems to measure physical or chemical parameters in situ.

Integrating sensors into organ-on-a-chip systems can help characterize engineered tissue models while also giving prompt insights into tissue interactions with different stimulants. Zhang et al. (2017) demonstrated that integrating an array of on-chip sensors, such as optical and pH, oxygen, and temperature monitors, allowed them to monitor and capture real-time changes in organoid behavior to obtain more information about what was happening within the organ-on-a-chip system. In the skin-on-a-chip platforms, integrating various sensors would be extremely beneficial as establishing a full-thickness skin model within the platform can be a long process ranging from around 2–6 weeks. Conventional endpoint assays provide no information about the period during skin formation and, therefore, may result in low experimental reproducibility. Ideally, skin-on-a-chip devices should incorporate physical sensors to monitor cell culture parameters, e.g., pH, oxygen, and temperature; electrochemical sensors to measure soluble protein biomarkers; and transepithelial electrical resistance (TEER) sensors to measure skin barrier integrity and function.

Currently, most skin-on-a-chip devices mainly use TEER sensors as it enables non-destructive real-time quantification of barrier integrity and function by quantifying alterations in the transcellular and paracellular permeability of epithelial and epidermal cell cultures (Zoio et al. 2021b; Petrova et al. 2014). TEER values will gradually increase during skin culture, correlating with epidermal differentiation and formation of the *stratum corneum,* thus allowing researchers to use these results as a quality control to ensure barrier integrity of skin models prior to their use in downstream assays (Gorzelanny et al. 2020; Zoio et al. 2022a). Additionally, TEER sensors can also be used for drug testing purposes, with TEER measurements used as a testing parameter for the Organization for Economic Cooperation and Development (OCED) test guidelines 430 (In vitro skin corrosion: Transcutaneous electrical resistance test method) (Zoio and Oliva 2022; Zoio et al. 2022a). Subsequently, various studies have investigated using TEER to assess the potential skin irritation of different compounds. Wei et al. (2020) demonstrated that TEER measurements could be used to evaluate the skin irritation potential of 46 compounds tested on 3D-bioprinted skin models. Groeber et al. (2015) used TEER as a complimentary endpoint in cutaneous toxicity analysis to distinguish between the effects of solid irritants and non-irritants while concluding that TEER could be an instrumental measurement to identify sub-irritative effects such as burning and itching sensations in the skin.

TEER measurements can be performed on skin-on-a-chip devices by incorporating electrodes on either side of the cellular barrier; however, this can be difficult because of the micrometer-sized channels within the device (Henry et al. 2017). Electrodes can also be inserted into the chip's inlets and outlets. This method does not affect access to the cells; however, it has low reliability due to variations in the placement of the electrodes and geometry of the small channels (van der Helm et al. 2016). Alternative methods involve integrating the electrodes into the chip closer to the cell culture chamber, to decrease resistance from the cell culture medium and noise generated by the motion of the electrode (Arık et al. 2018). However, to ensure the successful integration of electrode systems into on-chip devices, researchers must consider the electrode size, geometry of the chip, and placement. Studies by Odijk et al. (2015) reported problems when integrating electrodes into their organ-on-a-chip device, resulting in overstimulation of TEER values. Other potential sources of error for TEER measurements include the presence of air bubbles within the microchannels and inadequate cell coverage. Previous studies have reported that even a small gap in tissue coverage (0.4%) can cause TEER values to drop by approximately 80% (Odijk et al. 2015). This problem is particularly relevant for full-thickness skin models generated using animal-derived collagen, as this matrix source is prone to shrinkage during culture over time.

The idea of a modular microfluidic device with interchangeable organ-on-a-chips and sensors has been investigated; however, a skin-specific system has yet to be achieved (Zhang et al. 2017). The ability to alter the circuits to suit the requirements of a skin-on-a-chip device will enhance the functionality of the skin models while allowing for automated analysis procedures and real-time monitoring of tissue health.

4.3 Verification of Skin Structure and Function Within the Skin-on-a-Chip Device

As explained earlier in this chapter, the design and production of skin-on-a-chip devices is a complicated process that requires the consideration of multiple factors, which must be chosen carefully to produce a microfluidic device that is best suited for its intended purpose. However, of all the elements considered in its design, the skin model housed within it is the most critical. As this skin model aims to reproduce in vivo skin, this skin model system must be verified to ensure that it retains the same structure, function, and physiology of in vivo human skin over the entire course of its time in culture within the microfluidic device. Without this verification, any toxicological information gathered about a given compound from its testing using this model would not be truly representative of the expected response in consumers.

During the course of the development of skin-on-a-chip devices, numerous verification methods have been established as standard assessments of the skin model. This section will aim to explain what those methods are and how best to apply them for verification of skin-on-a-chip performance.

4.3.1 Histological and Immunohistological Examination of Skin Model Structure

The use of classical histopathology stains is one of the main techniques that can be used to examine the physical structure of a housed skin model and determine whether it is representative of in vivo human skin. This has been achieved using three basic histopathology stains: hematoxylin and eosin (H&E), Masson's trichrome, and Sirius Red.

H&E staining can be applied to formalin-fixed paraffin-embedded (FFPE) or optimal cutting temperature (OCT) frozen sections of the skin model isolated at various stages of culture in the microfluidic device and exploits the different preferential binding of these stains to examine cell morphology (Song et al. 2017b; Abaci et al. 2016; Abaci et al. 2015). Hematoxylin has a high binding potential for nucleotide-based molecules, such as DNA, and strongly stains the cells' nucleus. In contrast, eosin has a solid binding affinity for proteins and, as such, binds nonspecifically to the cytoplasm. This staining technique allows all cells within the skin model to be visible, where noticeable phenotypic differences in cell morphology can be observed (Feldman and Wolfe 2014; Fischer et al. 2008).

Clinical histopathologists routinely use this stain to access tissue structure. While it does require a level of knowledge of typical skin structure to use, it gives the broadest indication of the structure and maintenance of the epidermal and dermal compartments of the skin.

While H&E staining broadly labels all skin compartments, other stains can be used to specifically examine the dermal compartment's structure (Lim et al. 2018). Two of these are Masson's trichrome and Sirius Red (Rieppo et al. 2019; Rittié 2017), which only stain the collagenous fibers and other connective tissue in the dermis. These fibers are secreted by dermal fibroblasts and are critical for forming an extracellular matrix required for normal skin function and resistance. The density and intensity of these stains correlate to the level of these fibers, which, when compared to the level in in vivo skin, can be used to ensure consistency between the two pieces of tissue.

While these histological stains are helpful in the verification of skin model structure, many studies use immunofluorescence/immunohistochemistry of FFPE or OCT skin model sections to label cell type-specific markers to achieve a closer examination of its structure. Numerous markers are available that identify cells at various stages of differentiation in various skin compartments. While examining all is not required, examining some in all the compartments is recommended.

One of the critical skin structures to investigate is the stratum corneum, the outermost layer of the epidermis. This stratum comprises fully differentiated keratinocytes, or corneocytes, and provides the initial and strongest barrier to the external environment. Two markers are commonly used to judge these cell level of terminal differentiation, involucrin, and loricrin (Lee et al. 2017; Abaci et al. 2016). These two proteins, upon terminal differentiation, cross-link and help form the cornified envelope, which aids in providing the skin's barrier function. When staining skin models, there should be a continuous line of expression of both of these proteins in this stratum, which indicates a sufficient level of terminal differentiation (Eckert et al. 1993; Kalinin et al. 2001). If the expression is absent, the barrier function of the skin model should be questioned.

Before reaching the stratum corneum and undergoing terminal differentiation, keratinocytes progress through the multiple other squamous stratums in the epidermis, including the stratum spinosum and stratum granulosum. The cells in these layers can be identified as the keratinocytes begin to undergo differentiation, which causes visible phenotypic changes in their morphology. This change in keratinocytes' phenotype is accompanied by an alteration to their protein expression, allowing their demarcation from the cells in the more columnar basal stratum. One of the most commonly examined protein families, which marks this difference, is the keratin family (Maschmeyer et al. 2015; Abaci et al. 2016; Plaza et al. 2021). Keratins are essential structural fibrous proteins that aid in forming the skin barrier, and the expression of specific keratin proteins is specific to a given skin stratum. Keratin 1 and keratin 10 ensure normal differentiation of the epidermis, as they should only be detectable in differentiating keratinocytes (Totsuka et al. 2017). As such, no expression should be detected in the stratum basale. Keratin 19 may also be examined, as some have reported that this protein has weak staining throughout the epidermis and can be readily observed in normal skin, making it suitable for comparison between de novo and native skin (Kim et al. 2019).

To ensure that the epidermis is normally proliferating, keratin 16 can be examined, as this protein only appears in inflamed or hyperproliferating epidermises, such as those actively undergoing wound repair (McGowan and Coulombe 1998). An absence of this marker ensures that the epidermis is forming correctly and not increasing at a rate where it will be unable to maintain itself (Sriram et al. 2018).

Keratins are not the only proteins induced through keratinocyte differentiation (Jusoh et al. 2019). To ensure normal epidermal function, the adhesion and tight junctional proteins desmoglein-1 and claudin-1 can also be examined. The expression of these markers increases in line with differentiation; as such, a gradient of expression should be observed, increasing toward the stratum corneum (Tsukita et al. 2001; Hammers and Stanley 2013).

The combined expression of these markers indicates that the differentiation of keratinocytes is proceeding as expected and not unsustainably.

The differentiation of keratinocytes present in the upper stratum layers of the epidermis is critical to skin function. Still, the ability of the keratinocytes present in the basal layer to continuously proliferate to provide the replacement cells lost to shedding is equally as important. Without this activity, the epidermis will fully differentiate into corneocytes and ultimately be lost, reducing the longevity of the skin model. The structural proteins keratin 14 and keratin 15 have mostly restricted expression to the stratum basale, thus acting as good markers for the preservation of this stratum (Abaci et al. 2016; Wei et al. 2016; Bose et al. 2013). Additionally, the expression of Ki-67 and p63 should also be examined (Abaci et al. 2016; Kim et al. 2019), as these proteins are involved in the initiation of proliferation, and as such, their expression is vital to maintain both the basal population and to replace the corneocytes in the stratum corneum (Pellegrini et al. 2001; Chang et al. 2010). Occasional identification of these proteins in cells within the stratum basale is a good indicator of a functional epidermis that can continuously replenish itself, mirroring in vivo behavior and suggesting the skin model has the potential for long-term cell culture.

The expression of these markers demonstrates the presence of differentiated, differentiating, and basal keratinocytes within the epidermis. However, this epidermis needs to be securely attached to the dermal compartment of the skin to capture in vivo biology fully. Observing that the epidermis and dermis are tightly attached can be achieved by staining the extracellular matrix (ECM) proteins that form the epidermal–dermal junction, which are a series of structural proteins that attach basal keratinocytes to the dermis. Markers of the epidermal–dermal junction include collagen VII, collagen XVII, integrin-β1, and laminin-5, which can be identified when using immunofluorescence or immunohistochemistry, as the expression of these proteins form a solid line of expression directly beneath the columnar basal keratinocytes at the base of the epidermis (Sriram et al. 2018; Chung and Uitto 2010; Gatalica et al. 1997; Nishiyama et al. 2000; Liu and Leask 2013).

The formation and structure of the dermal compartment also need to be examined to ensure sufficient secretion and cross-linking of the ECM proteins that infer the tensile strength and elasticity of human skin. Numerous ECM proteins ensure this integrity and rigidity, but the most common ECM fibrous proteins examined are collagen I, collagen IV, and fibronectin (Pankov and Yamada 2002; Matsuura-Hachiya et al. 2018; Hwang et al. 2021). When stained, these proteins should be visible throughout the dermis, and the staining levels can be compared to that seen in native skin to ensure a sufficient level of similarity between the two structures (Sriram et al. 2018; Lim et al. 2018).

As the previous sections explain, some skin-on-a-chip models have been constructed to contain a full vascular network produced using endothelial cells (Abaci et al. 2016; Mori et al. 2017; Kim et al. 2019). As these networks are commonly formed through the perfusion of hollow chambers with single endothelial cell suspensions, work must be undertaken to ensure they form a complete and tight endothelial cell sheet. This can be achieved by staining for ZO-1, which is an occludin protein that contributes to the formation of tight functions between endothelial cells, which should be readily observable in the membrane of these cells (Kim and Kim 2017).

This is not the only marker used; however, several "cluster of differentiation" (CD) proteins are also used to indicate the level of differentiation in endothelial cells (Abaci et al. 2016). Most skin models are commonly stained with CD31 and CD14, which helps identify cells as either endothelial progenitor cells or as less-proliferative mature endothelial cells, respectively (Krenning et al. 2009).

Functional markers can also determine the level of function within the cells. One of them is the expression of endothelial nitric oxidase synthase, an enzyme that maintains cardiovascular endothelial homeostasis (Shimokawa and Tsutsui 2010).

Finally, while the inclusion of a hypodermis is unusual within a skin model present in a skin-on-a-chip model, advancements in 3D printing have allowed this often neglected compartment to be utilized in skin modeling (Kim et al. 2019). A standard fluorescent stain that examines the formation of this compartment is boron dipyrromethene (BODIPY), which can determine the maturity of the adipocytes present in this compartment (Nicu et al. 2018). BODIPY is a probe that emits a fluorescent signal in the presence of lipids. As adipose cells accumulate lipids following maturation, the fluorescent intensity level indicates the level of adipocyte maturation (Sarantopoulos et al. 2018).

4.3.2 Determining Cell Viability of the Retained Skin Model

While using immunofluorescence to examine critical markers in the skin is vital, it can only demonstrate the maintenance of the skin's structure. Equally important is determining the viability of the cells in the model for the length of time required. Multiple direct and indirect measurements have been developed and utilized to assess the level of viability during the production of different skin-on-a-chip models.

Cell viability assays such as the use of calcein and ethidium homodimer-1 solution, more commonly known as the LIVE/DEAD viability assay, can directly measure the housed skin model viability. The housed model is harvested at several time points during culture, with the epidermal and dermal compartments separated before treatment with the LIVE/DEAD assay (Lee et al. 2017). Calcein is a derivative of fluorescein, which is converted to green fluorescent calcein after it passes the cell membrane of live cells. At the same time, ethidium homodimer-1 can only cross severely damaged cell membranes, where it binds to nucleic acids and increases in fluorescent intensity, identifying dead cells (Decherchi et al. 1997; Bratosin et al. 2005). Confocal microscopy can be used to capture fluorescent images, which can be quantified using standard image processing software. The proportion of alive cells can then be determined by calculating the live cell fraction (calcein positive) over the total cell fraction (calcein and ethidium homodimer-1 positive).

While this is a valuable method, using LIVE/DEAD viability assays is problematic as it is destructive and does not allow for continuous assessment of the skin model's viability. As such, other methods have been established for continual modeling, which often examines either metabolic byproducts or the secretion of damage-related markers.

One method that examines damage-related markers utilizes the measurement of lactate dehydrogenase (Maschmeyer et al. 2015). Lactate dehydrogenase is only released by cells that have suffered some damage. As such, the relative concentration in the circulating cell culture medium over time within the microfluidic device indicates the level of damage within the whole model system. To assess the levels of lactate dehydrogenase, cell culture samples can be extracted at any time during the skin model culture and incubated with lactate. Lactate dehydrogenase converts lactate to pyruvate through the reduction of NAD+ to NADH. The addition of diaphorase then utilizes the available NADH formed by this reaction to convert endogenously added tetrazolium salt to red formazan (Kumar et al. 2019). The level of red formazan present in the culture medium, determined by standard optical density, acts as a reporter for the lactate dehydrogenase level, indicating the level of damage in the skin model.

Lactate dehydrogenase is not the only damage-related marker, and optical density is not the only measurement system used in determining cell viability. Standard experimental assays can also be utilized to examine the level of damage-related markers in the circulating medium, such as multiplex assays and ELISA that determine the concentration of a given inflammatory or damage-related marker, such as interleukin-1β, interleukin-6, interleukin-8, or TNF-α (Wufuer et al. 2016; Kim et al. 2020).

Another possible indicator of declining cell health is through the monitoring of the metabolic activity of the skin model. Regular sampling of the cell culture medium can be conducted, with the level of glucose and lactate analyzed through detection assays, such as GLU 142 and LAC 142, which allow the concentration of these two metabolites to be quantified with a photometer (Maschmeyer et al. 2015).

A similar method for determining metabolic activity is assessing extracellular acidification rate (EAR) (Alexander et al. 2018). This methodology exploits the charged nature of H+ ions, a natural byproduct of cellular metabolic processes. The skin model is submerged in a fresh medium in the presence of metal oxide sensors. The metabolites present in this fresh medium are actively utilized, producing an increase in H+ ions that alter the charge of the solution, which is detected and plotted as pH (mV) (Alexander et al. 2017). While a helpful assay, the requirement of submerging the whole skin model in a cell culture medium and the need for incorporating metal oxide sensors into the design of microfluidic devices, prevent its use in all microfluidic devices.

4.3.3 Accessing the Flow Rate and Perfusion of the Retained Skin Model by the Vascular/Microfluidic System

The validation methods explained so far act to verify that the skin's structure retained within the model reflects in vivo physiology and that the cells within the model remain viable during its use. However, given that the significant advantage of skin-on-a-chip devices is the mimicking or generation of a complex vascularization able to supply nutrition to the retained skin, this system also requires verification. This is to ensure that the fluid flow rate within the system closely aligns with that observed in the vasculature of native skin and that it can adequately perfuse the skin model.

The first of these verifications is achieved by analyzing microparticle image velocity. This can only be performed on machined microfluidic channels, and not endothelial-derived channels, as the channel must be transparent to allow for observation (Maschmeyer et al. 2015). This methodology uses either polymeric microparticles or isolated red blood cells suspended in PBS. As these particles are circulated through the manufactured chip, several points of interest are defined within the microfluidic system. As the particle/PBS solution passes through these points of interest, a high-speed CMOS (complementary metal oxide semiconductor) camera connected to a microscope takes serial images at 4 μs exposures. This series of images can be used to track the distance moved by these polymer/cells, which in turn can be used in the calculation of the mean velocity of the solution using standard image analysis software (Stamhuis and Thielicke 2014; Schimek et al. 2013). With the mean velocity calculated, the flow rate of the solution can then be determined with the following equation:

$$Q = w \times h \times v_{av} = w \times h \times k \times v_{\max} \tag{4.1}$$

Q	Flow rate
w	Microfluidic channel width
h	Microfluidic channel height
k	Flow coefficient
v_{av}	Average velocity
v_{max}	Maximum velocity

Equation (4.1): Equation for calculating the flow rate within a microfluidic channel. Adapted from Maschmeyer et al. (2015), Schimek et al. (2013)

While the flow rate must match that of standard in vivo vasculature to prevent the buffeting of the cells present in the skin model, the ability of the chosen vasculature system to sufficiently perfuse a skin model must also be assessed.

This ability is most commonly determined using fluorescein isothiocyanate (FITC)-dextran, a fluorescent probe that can cross vascular cell barriers (Natarajan et al. 2017). The use of FITC-dextran, however, is skin-on-a-chip design dependent. For skin-on-a-chip devices that partially mimic endothelial/dermal transport by culturing endothelial and fibroblasts cells on opposing sides of a membrane, the central cavity in the microfluidic device is sampled at regular periods to determine the level of perfusion. The level of fluorescence in these samples, caused by the perfusion of FITC-dextran, plotted over time, indicates the level of perfusion and, thus, the performance of the microfluidic channels (Wufuer et al. 2016).

Skin models generated with de novo endothelial cell vasculatures are more complex to assess. Most commonly, FITC-dextran is added to the circulating medium solution and allowed to pass through the skin model for a given time. The permeability level is then assessed by tracking the fluorescent intensity in the dermal area surrounding the vascular structures. The changes in intensity are measured via fluorescent microscopy, with a time series of fluorescent images taken at a given dermal location. The level of fluorescence in these areas can be quantified with standard image processing software. The permeability of the vascular network can then be calculated using the following equation (Lee et al. 2017; Abaci et al. 2016; Kim et al. 2019):

$$Pd = \frac{1}{l_1 - l_b} \times \left(\frac{l_2 - l_1}{t}\right) \times \frac{d}{4} \tag{4.2}$$

P_d	Diffusion permeability coefficient
l_1	Initial average fluorescent intensity
l_2	Average fluorescent intensity after a given time (t)
l_b	Background fluorescent intensity
t	Time
t	Time
d	Microfluidic channel diameter

Equation (4.2): Equation for calculating vascular channel permeability using FITC-Dextran. Adapted from Kim et al. (2019).

4.3.4 Determining Barrier Performance of the Retained Skin Model

The verification methods explained so far provide a good indication of the health and structure of the housed skin model; however, none of those listed examine the skin's natural functions. One of the most critical in vivo functions of the skin is to prevent the penetration of foreign bodies and liquids into the body. The barrier performance needs to be assessed in the housed skin model.

The simplest form of assessment, which can be conducted on almost any skin-on-a-chip device, is a water-wicking test. A small volume of PBS can be applied to the skin model's surface, and the stratum corneum's ability to repel this solution into a single droplet can be observed by the eye, indicating a certain level of barrier function (Mori et al. 2017). As the easiest and simplest method, water-wicking does not provide quantifiable measurements, making the continuous monitoring of barrier function complex.

As such, the measurement of transepithelial resistance (TEER) is more frequently used as an assessment methodology (Alexander et al. 2018). TEER does not directly measure the permeability of the skin model. Instead, it is inferred from the level of resistance observed in a current passing through the skin model. This resistance is generated from the tight junctions that connect the cells in the stratum corneum, with a higher resistance equating to more tight junctions equating to a lower level of permeability (Benson et al. 2012). TEER is measured using two electrodes on opposing areas of the visible skin model. A defined DC voltage can then be applied between the two electrodes, with the outputted current produced measured. The ohmic resistance of the stratum corneum can then be calculated through Ohm's law.

$$V = I \times R \quad (4.3)$$

V	Voltage
I	Current
R	Resistance

Equation (4.3): Ohm's law

The level of resistance can be tracked over time in culture, indicating if the barrier function is maintaining its function or losing it over time.

This method has been improved upon due to the potentially detrimental effects of the direct application of current. One example is the epithelial voltohmmeter (EVOM), which uses alternating current at a fixed voltage rather than a direct current, to avoid the direct application of electricity to the cells (Srinivasan et al. 2015). As is the case in the measurement of the EAR, the potential need to incorporate electrodes into the design of the microfluidic device may be a limiting factor in its use.

A permeation methodology can be employed to avoid this potential limiting factor. The methodology utilizes oligonucleotides labeled with fluorescein amidites (FAM), which are applied to the apical side of the skin model at a given concentration, and sufficient time is given to allow their penetration into the skin model (Abaci et al. 2016). The circulating medium solution within the microfluidic device can then be sampled at regular time points and assessed for their level of fluorescence using a fluorescence spectrometer, which indicates the concentration of FAM-labeled oligonucleotides. The difference in concentration between that applied to the apical surface and that present in the circulating medium can be used to determine transepithelial and transdermal permeation, utilizing Fick's law:

$$J_{ss} = \frac{\left(\frac{C_s}{C_m}\right) D_s}{\delta_s} \times A_s \times (C_d - C_b) \quad (4.4)$$

J_{ss}	Steady-state mass transfer of compound
C_s	Skin concentration of compound
C_m	Medium concentration of compound
D_s	Diffusion coefficient of compound
δ_s	Skin model thickness
A_s	Skin model surface area
C_d	Compound concentration in vehicle
C_b	Compound concentration in bottom chamber

Equation (4.4): Equation to find the transepithelial/transdermal permeation of the FAM-labeled oligonucleotides. Adapted from Abaci et al. (2015).

This methodology is advantageous as the application, collection, and determination of the concentration of FAM nucleotides can be performed at any point during culture. It is not damaging and is suitable for all microfluidic devices.

Another potential method of barrier function assessment is confocal Raman spectroscopy. This method is destructive, as the skin model must be removed from the culture and placed directly on the instrument, making it unsuitable for continual assessment (Sriram et al. 2018). Confocal Raman spectroscopy is a novel method that exploits the physical phenomenon of the Raman effect, which observes that the vibrations that occur in the chemical bonds of molecular structures can cause light to either gain or lose energy predictably. This change in light energy can then be measured at specific wavelengths, allowing for the identification of specific types of bonds (Butler et al. 2016; Caspers et al. 2001). This phenomenon can be used to measure the presence of structural proteins, such as keratin, and base molecules, such as water. The water-integrated intensity, determined by the OH stretching vibrations in the range of 3350–3550 cm^{-1}, and the keratin-integrated content, determined by the CH stretching vibrations in the range of 2910–2960 cm^{-1}, can then be used to find the water content of the outer layer of skin using the following equation:

$$WC = 100\% \times \frac{\left(\frac{W}{P}\right)}{\left(\frac{W}{P} + R\right)} \tag{4.5}$$

WC	Water content
W	Water-integrated intensities
P	Protein-integrated intensities
R	Water-to-protein signal proportionality constant

Equation (4.5): Equation to find the water content of a skin model following confocal Raman spectroscopy. Adapted from Sriram et al. (2018).

The determined water content of the stratum corneum can be used to determine its thickness, and when confocal Raman spectroscopy is performed on native skin, a direct comparison between the model and in vivo skin can be made, inferring barrier function (Caspers et al. 2003; Mahrhauser et al. 2015).

4.3.5 Confirming Cell Position within the Retained Skin Model

The final verification that can be employed is an examination of the position of the cells within the housed skin model. As with other verification methods, this methodology depends on the skin model's form in the skin-on-a-chip device.

A cell tracker system is best for skin models that aim to mimic the cell–cell contacts present in the skin through the seeding of monolayer sheets on membranes within the skin-on-a-chip device (Wufuer et al. 2016). Prior to seeding, keratinocytes, fibroblasts, and endothelial cells are incubated with three different cell tracker solutions with three different colored fluorescent probes. The cells are then seeded onto the microfluidic device and allowed to reach confluence. The microfluidic device can then be imaged using a standard fluorescent microscope, and by examining the z-axis, the various cell types and their 3D position can be easily visualized. This can be used to confirm the confluence of the cell layers and that the required cells are making contact.

For more complex skin models, such as those that utilize de novo skin equivalents or bioprinted models, a more complex microscopy method has to be used, two-photon excited fluorescence microscopy (Sriram et al. 2018). Due to the thickness of these retained skin models, standard microscopy techniques would be unable to visualize targets in the deeper compartments of the skin model without causing severe photodamage. Two-photon microscopy prevents this by visualizing targets using two photons aimed at converging on a single target area. Each photon has half the required energy to excite the required fluorescence, so converging on a single molecule supplies enough energy to visualize the target without causing photodamage (So et al. 2000; Denk et al. 1990). This form of microscopy can be used to visualize cell positions within any skin model without fluorescent labels, as nicotinamide adenine dinucleotide phosphate (NADPH) and flavin adenine dinucleotide (FAD) have endogenous fluorescence. The fluorescence of these molecules can help examine the confluence of keratinocytes and fibroblasts in the epidermal and dermal compartments.

4.4 Advancements and Applications of Skin-on-a-Chip Technology

Cutaneous and subcutaneous disorders affect approximately one-third of the global population. Given the high prevalence of skin disorders, there has been an increase in new technological advancements for cutaneous drug development. Skin-targeted drug delivery includes topical (e.g., creams and repellents), dermal (e.g., corticosteroids and antibiotics), and transdermal approaches (Zoio and Oliva 2022). Traditional preclinical drug testing relies on in vitro 2D cell culture or animal models. While 2D cell culture systems are rapid and reproducible, they are unable to mimic complex interactions observed in vivo. Also, while animal models provide insight into systemic effects, they cannot replicate human skin anatomy and physiology. Furthermore, animal models have low throughput and interspecies variability, limiting their accuracy for human cutaneous drug testing. Additionally, from an ethical perspective, the replacement and reduction in the use of animal models fulfill the growing societal concern surrounding animal testing, with ethical guidelines dictating animal testing should be replaced, reduced, or refined (3R principle) (Zoio and Oliva 2022). Since 2009, the European Commission has been authorizing regulations on cosmetic testing, establishing a prohibition that prevents testing finished cosmetic products or ingredients on animals and commercializing any cosmetic product or ingredient tested on animals within the European Union (Taylor and Rego 2020). These restrictions on animal testing have thereby led to advancements in the development of more physiologically relevant skin models that can replace current inefficient methods. The need for more physiologically relevant and functional tissue models has led to the development of skin-on-a-chip technology. Even though skin-on-a-chip technology is still in its infancy, these devices show promise to improve upon the current limitations of 3D-based cell culture platforms and increase the ability to determine the toxicity and efficacy of new drugs.

The most common application of skin-on-a-chip has been the maintenance of skin tissue under dynamic perfusion to increase longevity or to establish a co-culture of different tissues (Ataç et al. 2013; Abaci et al. 2015; Tavares et al. 2020). These studies provide valuable insight into the potential use of skin-on-a-chip for clinical applications such as multi-organ crosstalk and assessing drug sensitivity and toxicity. By culturing monolayers of cells on a chip with each layer separated by a porous membrane, researchers can co-culture and mimic the different skin compartments to observe and analyze interlayer communication (epidermal, dermal, and vascular) when testing new drugs or cosmetics (Wufuer et al. 2016; Jeon et al. 2020b). The measurement of dual parameters such as cell viability and tight junction has allowed researchers to use this skin-on-a-chip device to assess skin irritation and distinguish between irritants and non-irritants with 80% more sensitivity, specificity, and accuracy compared to in vivo data (Jeon et al. 2020b). Other studies have further adapted this design and used a porous membrane sandwiched between branched microchannels to an on-chip device to culture a HaCaT monolayer that can be used in permeation assays to test skin irritants (Sasaki et al. 2019). Studies by Ren et al. used a microfluidic skin-on-a-chip device to study transendothelial and transepithelial migration of T cells in a mimicked skin inflammatory microenvironment (Ren et al. 2021). This microfluidic device allowed Ren et al. to quantitatively study the effect of cutaneous inflammatory mediators on T-cell transmigration at a single cell level and identify potential anti-inflammatory drugs for treating skin diseases such as psoriasis (Ren et al. 2021). While these 2D skin-on-a-chip devices were able to culture cells directly within the microfluidic device and have shown the ability to stimulate cell responses to drug treatment, they do not fully mimic the complex 3D architecture of native human skin as the cells are cultured in 2D monolayers.

Various groups have developed skin-on-a-chip devices to overcome this limitation that transfer skin models to the device or the in situ formation of 3D skin equivalent models within the device. Lukács et al. developed a microfluidic diffusion chamber to monitor the transdermal delivery of topical drugs (Lukács et al. 2019). The device comprised three functional units: a top compartment where the drug of interest is applied, a middle compartment that houses an integrated skin sample, and a bottom compartment that houses the receptors. This device was capable of producing similar reproducible results when compared to traditional drug penetration assays such as the Franz diffusion cell system, as well as demonstrating other advantages such as small drug and skin consumption, low sample volume, and a dynamic arrangement with a continuous flow to mimic blood circulation through the dermal compartment (Lukács et al. 2019). Further studies by Bajza et al. used the same skin-on-a-chip device to study the role of P-glycoprotein in dermal drug delivery using two P-glycoprotein substrate model drugs: quinidine and erythromycin to further demonstrate the suitability of the skin-on-a-chip device as a tool to investigate dermal drug delivery (Bajza et al. 2020).

While skin-on-a-chip can be used to assess new cosmetics and drug toxicity within the cutaneous microenvironment, these skin-on-a-chip models can be further modified to determine if any toxic effects are exhibited in other tissues and organs and observe any crosstalk between the skin and these other tissues and organs. Lee et al. developed a skin–nerve hybrid on-chip model by incorporating differentiated neural stem cells in a collagen matrix adjacent to and below an epidermal layer, thus enabling real-time quantification of skin sensitization by measuring alterations in neuronal activity following chemical treatment (Lee et al. 2022).

In addition, Lee et al. also developed a skin–liver hybrid on-chip model by incorporating hepatic cells derived from pluripotent stem cells in a matrix distant from the skin model (Lee et al. 2022). This model evaluated potential hepatotoxicity from topically applied chemicals to the cutaneous layer by quantifying glutathione and reactive oxygen species.

To further increase the complexity of skin-on-a-chip models to mimic the dermal microvasculature for systemic applications, researchers have used various techniques such as 3D bioprinting, templating, and sacrificial molding to generate full-thickness skin models with perfusable lumens (Abaci et al. 2016; Mori et al. 2017; Salameh et al. 2021). Abaci et al. used 3D printing to print sacrificial channels of cross-linked alginate embedded in a collagen I gel (Abaci et al. 2016). These microchannels were removed using sodium citrate following epidermal differentiation leaving behind hollow channels. Endothelial cells derived from HUVECs or iPSCs were then used to coat the inner surface of the channels, allowing the researchers to recapitulate endothelial barrier function. Other studies have used nylon wires to create perfusable vascular channels (Mori et al. 2017). However, this technique lacked a microvascular network, resulting in only one microchannel. Alternative approaches for generating perfusable vascularized skin models involve 3D bioprinting. Kim et al. used a bioink of gelatin, glycerol, and thrombin embedded with endothelial cells to print vascular channels (Kim et al. 2019). While proper tissue formation and good vascular permeability properties were reported, this model was still limited to one microchannel. More recently, studies have used 3D templating techniques to develop a vascularized full-thickness skin model (Salameh et al. 2021). This technique produces hollow channels similar to the work by Mori et al. (Mori et al. 2017); however, to induce vasculogenesis, the hollow channels were seeded with HUVECs, and perfusion was achieved using a peristaltic pump. This model generated a differentiated epidermis, a perfusable vascular network with angiogenic sprouts, and an adjacent microvascular network. Furthermore, the potential of this model for downstream topical and systemic applications was validated using various compounds such as caffeine, minoxidil, and benzo[a]pyrene pollutant.

Aside from skin-on-a-chip devices with perfusable lumens, some researchers have used microfluidic-based techniques to develop full-thickness skin models with a basal perfusion system for testing new drugs and cosmetics. These skin-on-a-chip devices consist of two layers of PDMS assembled on top of a glass base, with the bottom PDMS layer containing a fluidic chamber, while the top layer houses a central chamber for skin model formation (Lee et al. 2017). After generating the reconstructed skin models, they were transferred to the on-chip device. They were cultured for an additional 6 days at an air–liquid interface before assessing skin barrier function using testosterone and caffeine as reference substances. Studies have also used these skin-on-a-chip models with perfusion platforms.

Applying mechanical forces and shear stress to the skin models to better mimic the in vivo cutaneous microenvironment (Strüver et al. 2017).

Recently, skin-on-a-chip models have been used for studying skin aging and testing the efficacy of new anti-aging cosmetics. Studies conducted by Lim et al. incorporated magnets into a dedicated cavity within a PDMS layer of the skin-on-a-chip device and applied an external electromagnetic field to uniaxially stretch the membrane and create a wrinkled skin-on-a-chip (WSOC) (Lim et al. 2018). This WSOC demonstrated reduced collagen production and decreased fibronectin and keratin 10 expression similar to that observed in aged skin, thus highlighting the potential of this tool to evaluate the efficacy of new anti-aging cosmetics and treatments without having to rely on ex vivo human skin biopsies.

Despite advancements in the development and application of skin-on-a-chip for cosmetic toxicology, most studies use animal-derived ECM proteins such as rat rail collagen to produce the dermal compartment. Consequently, this results in the formation of dermal compartments with inadequate biomechanical properties due to the contraction and degradation of the matrix by fibroblasts, thereby limiting the lifespan and reproducibility of the skin-on-a-chip models. To overcome this issue, chemical and physical modifications of the ECM matrix through the addition of synthetic polymers, natural polymers, or peptide motifs were considered. Due to the poor mechanical stability of collagen and fibrin, Sriram et al. combined fibrinogen with PEG polymers before pipetting into a device comprised of a multi-chamber microfluidic chip that contained two fluidic compartments separated by a permeable microporous membrane (Sriram et al. 2018). Using this technique, Sriram et al. produced a stratified epidermis with an enhanced basement membrane, demonstrated by increased deposition of collagens IV, VII, and XVII (Sriram et al. 2018). Studies by Zoio et al. used rapid prototyping techniques to develop a modular device integrated with electrodes for TEER measurements. This method combined the production of a fibroblast-derived matrix with an inert polystyrene porous scaffold integrated on-chip, thereby excluding the need for exogenous hydrogels and membranes (Zoio et al. 2021a, 2022b). The integration of electrodes allowed for TEER measurements to be obtained in situ during skin culture and also allowed the analysis of irritants on skin barrier function.

Overall, skin-on-a-chip technology shows promise to surpass current conventional drug testing assays and provide an alternative and more representative model than animal testing, especially given the ethical guidelines surrounding the use of animals for testing new cosmetics. Despite being in its infancy, skin-on-a-chip will continue to evolve, thus allowing its successful translation into the field of cosmetic toxicology and use as the new gold standard over conventional 2D assays and animals when testing new cosmetics and drugs.

References

Abaci HE, Gledhill K, Guo Z, Christiano AM, Shuler ML (2015) Pumpless microfluidic platform for drug testing on human skin equivalents. Lab Chip 15(3):882–888

Abaci HE, Guo Z, Coffman A, Gillette B, Lee WH, Sia SK et al (2016) Human skin constructs with spatially controlled vasculature using primary and iPSC-derived endothelial cells. Adv Healthc Mater 5(14):1800–1807

Adler S, Basketter D, Creton S, Pelkonen O, van Benthem J, Zuang V et al (2011) Alternative (non-animal) methods for cosmetics testing: current status and future prospects-2010. Arch Toxicol 85(5):367–485

Alexander FA, Eggert S, Wiest J (2017) A novel lab-on-a-chip platform for spheroid metabolism monitoring. Cytotechnology 70:375–386

Alexander FA, Eggert S, Wiest J (2018) Skin-on-a-chip: transepithelial electrical resistance and extracellular acidification measurements through an automated air-liquid Interface. Genes (Basel) 9(2):114

Antoni D, Burckel H, Josset E, Noel G (2015) Three-dimensional cell culture: a breakthrough in vivo. Int J Mol Sci 16(3):5517–5527

Arık YB, van der Helm MW, Odijk M, Segerink LI, Passier R, van den Berg A et al (2018) Barriers-on-chips: measurement of barrier function of tissues in organs-on-chips. Biomicrofluidics 12(4): 042218

Astashkina A, Mann B, Grainger DW (2012) A critical evaluation of in vitro cell culture models for high-throughput drug screening and toxicity. Pharmacol Ther 134(1):82–106

Ataç B, Wagner I, Horland R, Lauster R, Marx U, Tonevitsky AG et al (2013) Skin and hair on-a-chip: in vitro skin models versus ex vivo tissue maintenance with dynamic perfusion. Lab Chip 13(18):3555–3561

Azizipour N, Avazpour R, Rosenzweig DH, Sawan M, Ajji A (2020) Evolution of biochip technology: a review from lab-on-a-chip to organ-on-a-chip. Micromachines (Basel) 11(6):599

Bajza Á, Kocsis D, Berezvai O, Laki AJ, Lukács B, Imre T et al (2020) Verification of P-glycoprotein function at the dermal barrier in diffusion cells and dynamic "skin-on-a-chip" microfluidic device. Pharmaceutics 12(9):804

Bélanger MC, Marois Y (2001) Hemocompatibility, biocompatibility, inflammatory and in vivo studies of primary reference materials low-density polyethylene and polydimethylsiloxane: a review. J Biomed Mater Res 58(5):467–477

Benson K, Cramer S, Galla HJ (2012) Impedance-based cell monitoring: barrier properties and beyond. Fluids Barriers CNS 10:5

Bhatia SN, Ingber DE (2014) Microfluidic organs-on-chips. Nat Biotechnol 32(8):760–772

Bikle DD, Xie Z, Tu CL (2012) Calcium regulation of keratinocyte differentiation. Expert Rev Endocrinol Metab 7(4):461–472

Bose A, Teh MT, Mackenzie IC, Waseem A (2013) Keratin k15 as a biomarker of epidermal stem cells. Int J Mol Sci 14(10):19385–19398

Bratosin D, Mitrofan L, Palii C, Estaquier J, Montreuil J (2005) Novel fluorescence assay using calcein-AM for the determination of human erythrocyte viability and aging. Cytometry A 66(1): 78–84

Brohem CA, Cardeal LB, Tiago M, Soengas MS, Barros SB, Maria-Engler SS (2011) Artificial skin in perspective: concepts and applications. Pigment Cell Melanoma Res 24(1):35–50

Butler HJ, Ashton L, Bird B, Cinque G, Curtis KM, Dorney J et al (2016) Using Raman spectroscopy to characterize biological materials. Nat Protoc 11:664–687

Câmara MA, Rubio JCC, Abrão AM, Davim JP (2012) State of the art on micromilling of materials, a review. J Mater Sci Technol 28(8):673–685

Carlson MW, Alt-Holland A, Egles C, Garlick JA (2008) Three-dimensional tissue models of normal and diseased skin. Curr Protoc Cell Biol Chapter 19:Unit 19.9

Caspers PJ, Bruining HA, Puppels GJ, Lucassen GW, Carter EA (2001) In vivo confocal Raman microspectroscopy of the skin: noninvasive determination of molecular concentration profiles. J Investig Dermatol 116(3):434–442
Caspers PJ, Lucassen GW, Puppels GJ (2003) Combined in vivo confocal Raman spectroscopy and confocal microscopy of human skin. Biophys J 85(1):572–580
Cevc G, Vierl U (2007) Spatial distribution of cutaneous microvasculature and local drug clearance after drug application on the skin. J Control Release 118(1):18–26
Chang SL, Hu S, Hung SI, Huang YL, Hsiao WC, Chung WH (2010) A comparison of Ki-67 antigen presentation in acute generalized exanthematous pustulosis and pustular psoriasis. Arch Dermatol Res 302(7):525–529
Chung HJ, Uitto J (2010) Type VII collagen: the anchoring fibril protein at fault in dystrophic epidermolysis bullosa. Dermatol Clin 28(1):93–105
Decherchi P, Cochard P, Gauthier P (1997) Dual staining assessment of Schwann cell viability within whole peripheral nerves using calcein-AM and ethidium homodimer. J Neurosci Methods 71(2):205–213
Dellaquila A, Le Bao C, Letourneur D, Simon-Yarza T (2021) In vitro strategies to vascularize 3D physiologically relevant models. Adv Sci (Weinh) 8(19):e2100798
Denk W, Strickler JH, Webb WW (1990) Two-photon laser scanning fluorescence microscopy. Science 248(4951):73–76
Ding C, Chen X, Kang Q, Yan X (2020) Biomedical application of functional materials in organ-on-a-chip. Front Bioeng Biotechnol 8:823
Duffy DC, McDonald JC, Schueller OJA, Whitesides GM (1998) Rapid prototyping of microfluidic Systems in Poly(dimethylsiloxane). Anal Chem 70(23):4974–4984
Duval K, Grover H, Han L-H, Mou Y, Pegoraro AF, Fredberg J et al (2017) Modeling physiological events in 2D vs. 3D cell culture. Physiology (Bethesda) 32(4):266–277
Eckert RL, Yaffe MB, Crish JF, Murthy S, Rorke EA, Welter JF (1993) Involucrin–structure and role in envelope assembly. J Invest Dermatol 100(5):613–617
Ehsan SM, George SC (2013) Nonsteady state oxygen transport in engineered tissue: implications for design. Tissue Eng Part A 19(11–12):1433–1442
El Ghalbzouri A, Commandeur S, Rietveld MH, Mulder AA, Willemze R (2009) Replacement of animal-derived collagen matrix by human fibroblast-derived dermal matrix for human skin equivalent products. Biomaterials 30(1):71–78
Feldman AT, Wolfe D (2014) Tissue processing and hematoxylin and eosin staining. Methods Mol Biol 1180:31–43
Fischer AH, Jacobson KA, Rose J, Zeller R (2008) Hematoxylin and eosin staining of tissue and cell sections. CSH Protoc 2008:pdb.prot4986
Friend J, Yeo L (2010) Fabrication of microfluidic devices using polydimethylsiloxane. Biomicrofluidics 4(2):026502
Gangatirkar P, Paquet-Fifield S, Li A, Rossi R, Kaur P (2007) Establishment of 3D organotypic cultures using human neonatal epidermal cells. Nat Protoc 2(1):178–186
Gatalica B, Pulkkinen L, Li K, Kuokkanen K, Ryynänen M, McGrath JA et al (1997) Cloning of the human type XVII collagen gene (COL17A1), and detection of novel mutations in generalized atrophic benign epidermolysis bullosa. Am J Hum Genet 60(2):352–365
Gledhill K, Guo Z, Umegaki-Arao N, Higgins CA, Itoh M, Christiano AM (2015) Melanin transfer in human 3D skin equivalents generated exclusively from induced pluripotent stem cells. PLoS One 10(8):e0136713
Gogoi R, Manik G, Sahoo SK (2022) 10–viscoelastic behavior of elastomer blends and composites. In: Rangappa SM, Parameswaranpillai J, Siengchin S, Ozbakkaloglu T (eds) Elastomer blends and composites. Elsevier, Amsterdam, pp 171–194
Gorzelanny C, Mess C, Schneider SW, Huck V, Brandner JM (2020) Skin barriers in dermal drug delivery: which barriers have to be overcome and how can we measure them? Pharmaceutics 12(7):684

Graf BW, Boppart SA (2010) Imaging and analysis of three-dimensional cell culture models. Methods Mol Biol 591:211–227

Grayson ACR, Shawgo RS, Johnson AM, Flynn NT, Yawen LI, Cima MJ et al (2004) A BioMEMS review: MEMS technology for physiologically integrated devices. Proc IEEE 92(1):6–21

Groeber F, Engelhardt L, Egger S, Werthmann H, Monaghan M, Walles H et al (2015) Impedance spectroscopy for the non-destructive evaluation of in vitro epidermal models. Pharm Res 32(5): 1845–1854

Guo R, Liu J (2017) Implantable liquid metal-based flexible neural microelectrode array and its application in recovering animal locomotion functions. J Micromech Microeng 27(10):104002

Halden RU (2010) Plastics and health risks. Annu Rev Public Health 31:179–194

Hammers CM, Stanley JR (2013) Desmoglein-1, differentiation, and disease. J Clin Invest 123(4): 1419–1422

Hartung T, Rovida C (2009) Chemical regulators have overreached. Nature 460(7259):1080–1081

Hassler C, Boretius T, Stieglitz T (2011) Polymers for neural implants. J Polym Sci B Polym Phys 49(1):18–33

Henry OYF, Villenave R, Cronce MJ, Leineweber WD, Benz MA, Ingber DE (2017) Organs-on-chips with integrated electrodes for trans-epithelial electrical resistance (TEER) measurements of human epithelial barrier function. Lab Chip 17(13):2264–2271

Heo DN, Hospodiuk M, Ozbolat IT (2019) Synergistic interplay between human MSCs and HUVECs in 3D spheroids laden in collagen/fibrin hydrogels for bone tissue engineering. Acta Biomater 95:348–356

Hill DS, Robinson ND, Caley MP, Chen M, O'Toole EA, Armstrong JL et al (2015) A novel fully humanized 3D skin equivalent to model early melanoma invasion. Mol Cancer Ther 14(11): 2665–2673

Huggenberger R, Detmar M (2011) The cutaneous vascular system in chronic skin inflammation. J Investig Dermatol Symp Proc 15(1):24–32

Hwang SJ, Kim SH, Seo WY, Jeong Y, Shin MC, Ryu D et al (2021) Effects of human collagen α-1 type I-derived proteins on collagen synthesis and elastin production in human dermal fibroblasts. BMB Rep 54(6):329–334

Jeon B, Lee G, Wufuer M, Huang Y, Choi Y, Kim S et al (2020a) Enhanced predictive capacity using dual-parameter chip model that simulates physiological skin irritation. Toxicol in Vitro 104955:104955

Jeon B, Lee G, Wufuer M, Huang Y, Choi Y, Kim S et al (2020b) Enhanced predictive capacity using dual-parameter chip model that simulates physiological skin irritation. Toxicol in Vitro 68:104955

Jung MH, Jung SM, Shin HS (2016) Co-stimulation of HaCaT keratinization with mechanical stress and air-exposure using a novel 3D culture device. Sci Rep 6:33889

Jusoh N, Ko J, Jeon NL (2019) Microfluidics-based skin irritation test using in vitro 3D angiogenesis platform. APL Bioeng 3(3):036101

Kaarj K, Yoon JY (2019) Methods of delivering mechanical stimuli to organ-on-a-chip. Micromachines (Basel) 10(10):700

Kalinin A, Marekov LN, Steinert PM (2001) Assembly of the epidermal cornified cell envelope. J Cell Sci 114(Pt 17):3069–3070

Kashani-Sabet M, Sagebiel RW, Ferreira CM, Nosrati M, Miller JR 3rd. (2001) Vascular involvement in the prognosis of primary cutaneous melanoma. Arch Dermatol 137(9):1169–1173

Khurana P, Kolundzic N, Flohr C, Ilic D (2021) Human pluripotent stem cells: An alternative for 3D in vitro modelling of skin disease. Exp Dermatol 30(11):1572–1587

Kim S, Kim GH (2017) Roles of claudin-2, ZO-1 and occludin in leaky HK-2 cells. PLoS One 12(12):e0189221

Kim BS, Gao G, Kim JY, Cho DW (2019) 3D cell printing of Perfusable vascularized human skin equivalent composed of epidermis, dermis, and hypodermis for better structural recapitulation of native skin. Adv Healthc Mater 8(7):e1801019

Kim J, Kim K, Sung GY (2020) Coenzyme Q10 efficacy test for human skin equivalents using a pumpless skin-on-a-chip system. Int J Mol Sci 21:8475
Krenning G, van der Strate BWA, Schipper M, van Seijen XJGY, Fernandes BCA, van Luyn MJA et al (2009) CD34+ cells augment endothelial cell differentiation of CD14+ endothelial progenitor cells in vitro. J Cell Mol Med 13:2521
Kumar G, Ertel A, Naranjo N, Shahid S, Languino L, Fortina P (2019) Hematopoietic cell cluster identifier (HCCI): A knowledgebase computational tool to facilitate the characterization of hematopoietic cell populations in single cell RNA-sequencing. Blood 134:3587
Lee S, Jin SP, Kim YK, Sung GY, Chung JH, Sung JH (2017) Construction of 3D multicellular microfluidic chip for an in vitro skin model. Biomed Microdevices 19(2):22
Lee JS, Kim J, Cui B, Kim SK, Cho S-A, An S et al (2022) Hybrid skin chips for toxicological evaluation of chemical drugs and cosmetic compounds. Lab Chip 22(2):343–353
Lim HY, Kim J, Song HJ, Kim K, Choi KC, Park S et al (2018) Development of wrinkled skin-on-a-chip (WSOC) by cyclic uniaxial stretching. J Ind Eng Chem 68:238–245
Liu S, Leask A (2013) Integrin β1 is required for dermal homeostasis. J Invest Dermatol 133(4): 899–906
Liu J, Zheng H, Dai X, Poh PSP, Machens HG, Schilling AF (2020) Transparent PDMS bioreactors for the fabrication and analysis of multi-layer pre-vascularized hydrogels under continuous perfusion. Front Bioeng Biotechnol 8:568934
Low DA, Jones H, Cable NT, Alexander LM, Kenney WL (2020) Historical reviews of the assessment of human cardiovascular function: interrogation and understanding of the control of skin blood flow. Eur J Appl Physiol 120(1):1–16
Lukács B, Bajza Á, Kocsis D, Csorba A, Antal I, Iván K et al (2019) Skin-on-a-Chip device for ex vivo monitoring of transdermal delivery of drugs-design, fabrication, and testing. Pharmaceutics 11(9):445
Magliaro C, Mattei G, Iacoangeli F, Corti A, Piemonte V, Ahluwalia A (2019) Oxygen consumption characteristics in 3D constructs depend on cell density. Front Bioeng Biotechnol 7:251
Mahrhauser DS, Nagelreiter C, Gehrig S, Geyer A, Ogris M, Kwizda K et al (2015) Assessment of Raman spectroscopy as a fast and non-invasive method for total stratum corneum thickness determination of pig skin. Int J Pharm 495(1):482–484
Markov DA, Lillie EM, Garbett SP, McCawley LJ (2014) Variation in diffusion of gases through PDMS due to plasma surface treatment and storage conditions. Biomed Microdevices 16(1): 91–96
Maschmeyer I, Lorenz AK, Schimek K, Hasenberg T, Ramme AP, Hübner J et al (2015) A four-organ-chip for interconnected long-term co-culture of human intestine, liver, skin and kidney equivalents. Lab Chip 15(12):2688–2699
Matsuura-Hachiya Y, Arai KY, Muraguchi T, Sasaki T, Nishiyama T (2018) Type IV collagen aggregates promote keratinocyte proliferation and formation of epidermal layer in human skin equivalents. Exp Dermatol 27(5):443–448
McDonald JC, Whitesides GM (2002) Poly(dimethylsiloxane) as a material for fabricating microfluidic devices. Acc Chem Res 35(7):491–499
McGowan K, Coulombe PA (1998) The wound repair-associated keratins 6, 16, and 17. Insights into the role of intermediate filaments in specifying keratinocyte cytoarchitecture. Subcell Biochem 31:173–204
Miranda I, Souza A, Sousa P, Ribeiro J, Castanheira EMS, Lima R et al (2021) Properties and applications of PDMS for biomedical engineering: a review. J Funct Biomater 13(1)
Mori N, Morimoto Y, Takeuchi S (2017) Skin integrated with perfusable vascular channels on a chip. Biomaterials 116:48–56
Mori N, Morimoto Y, Takeuchi S (2018) Perfusable and stretchable 3D culture system for skin-equivalent. Biofabrication 11(1):011001
Natarajan R, Northrop N, Yamamoto B (2017) Fluorescein Isothiocyanate (FITC)-dextran extravasation as a measure of blood-brain barrier permeability. Curr Protoc Neurosci 79:9.58.1–9.58.15

Nge PN, Rogers CI, Woolley AT (2013) Advances in microfluidic materials, functions, integration, and applications. Chem Rev 113(4):2550–2583
Ngo MA, Maibach HI (2010) Dermatotoxicology: historical perspective and advances. Toxicol Appl Pharmacol 243(2):225–238
Nicu C, Pople J, Bonsell L, Bhogal R, Ansell DM, Paus R (2018) A guide to studying human dermal adipocytes in situ. Exp Dermatol 27:589–602
Nishiyama T, Amano S, Tsunenaga M, Kadoya K, Takeda A, Adachi E et al (2000) The importance of laminin 5 in the dermal-epidermal basement membrane. J Dermatol Sci 24(Suppl 1):S51–S59
Odijk M, van der Meer AD, Levner D, Kim HJ, van der Helm MW, Segerink LI et al (2015) Measuring direct current trans-epithelial electrical resistance in organ-on-a-chip microsystems. Lab Chip 15(3):745–752
O'Toole L, Kang CW, Fang FZ (2021) Precision micro-milling process: state of the art. Adv Manuf 9(2):173–205
Pankov R, Yamada KM (2002) Fibronectin at a glance. J Cell Sci 115(Pt 20):3861–3863
Pellegrini G, Dellambra E, Golisano O, Martinelli E, Fantozzi I, Bondanza S et al (2001) P 63 identifies keratinocyte stem cells. Proc Natl Acad Sci USA 98(6):3156–3161
Petrova A, Celli A, Jacquet L, Dafou D, Crumrine D, Hupe M et al (2014) 3D in vitro model of a functional epidermal permeability barrier from human embryonic stem cells and induced pluripotent stem cells. Stem Cell Rep 2(5):675–689
Piruska A, Nikcevic I, Lee SH, Ahn C, Heineman WR, Limbach PA et al (2005) The autofluorescence of plastic materials and chips measured under laser irradiation. Lab Chip 5(12):1348–1354
Plaza C, Meyrignac C, Botto JM, Capallere C (2021) Characterization of a new full-thickness in vitro skin model. Tissue Eng Part C Methods 27(7):411–420
Ramadan Q, Ting FC (2016) In vitro micro-physiological immune-competent model of the human skin. Lab Chip 16(10):1899–1908
Reijnders CM, van Lier A, Roffel S, Kramer D, Scheper RJ, Gibbs S (2015) Development of a full-thickness human skin equivalent in vitro model derived from TERT-immortalized keratinocytes and fibroblasts. Tissue Eng Part A 21(17–18):2448–2459
Ren X, Getschman AE, Hwang S, Volkman BF, Klonisch T, Levin D et al (2021) Investigations on T cell transmigration in a human skin-on-chip (SoC) model. Lab Chip 21(8):1527–1539
Rieppo L, Janssen L, Rahunen K, Lehenkari P, Finnilä MAJ, Saarakkala S (2019) Histochemical quantification of collagen content in articular cartilage. PLoS One 14(11):e0224839
Rioux G, Simard M, Morin S, Lorthois I, Guérin SL, Pouliot R (2021) Development of a 3D psoriatic skin model optimized for infiltration of IL-17A producing T cells: focus on the crosstalk between T cells and psoriatic keratinocytes. Acta Biomater 136:210–222
Risueño I, Valencia L, Holgado M, Jorcano JL, Velasco D (2021) Generation of a simplified three-dimensional skin-on-a-chip model in a micromachined microfluidic platform. J Vis Exp (171)
Rittié L (2017) Method for Picrosirius red-polarization detection of collagen fibers in tissue sections. Methods Mol Biol 1627:395–407
Rowe RG, Daley GQ (2019) Induced pluripotent stem cells in disease modelling and drug discovery. Nat Rev Genet 20(7):377–388
Salameh S, Tissot N, Cache K, Lima J, Suzuki I, Marinho PA et al (2021) A perfusable vascularized full-thickness skin model for potential topical and systemic applications. Biofabrication 13(3): 035042
Sapp MC, Fares HJ, Estrada AC, Grande-Allen KJ (2015) Multilayer three-dimensional filter paper constructs for the culture and analysis of aortic valvular interstitial cells. Acta Biomater 13:199–206
Sarantopoulos CN, Banyard DA, Ziegler ME, Sun B, Shaterian A, Widgerow AD (2018) Elucidating the Preadipocyte and its role in adipocyte formation: a comprehensive review. Stem Cell Rev Rep 14(1):27–42
Sasaki N, Tsuchiya K, Kobayashi H (2019) Photolithography-free skin-on-a-Chip for parallel permeation assays. Sens Mater 31:107

Schimek K, Busek M, Brincker S, Groth B, Hoffmann S, Lauster R et al (2013) Integrating biological vasculature into a multi-organ-chip microsystem. Lab Chip 13(18):3588–3598

Seliktar D (2012) Designing cell-compatible hydrogels for biomedical applications. Science 336(6085):1124–1128

Shafiee S, Shariatzadeh S, Zafari A, Majd A, Niknejad H (2021) Recent advances on cell-based co-culture strategies for Prevascularization in tissue engineering. Front Bioeng Biotechnol 9: 745314

Shimokawa H, Tsutsui M (2010) Nitric oxide synthases in the pathogenesis of cardiovascular disease. Pflugers Arch 459:959–967

So PT, Dong CY, Masters BR, Berland KM (2000) Two-photon excitation fluorescence microscopy. Annu Rev Biomed Eng 2:399–429

Song HJ, Lim HY, Chun W, Choi KC, Lee TY, Sung JH et al (2017a) Development of 3D skin-equivalent in a pump-less microfluidic chip. J Ind Eng Chem 60:355–359

Song HJ, Lim HY, Chun W, Choi KC, Sung JH, Sung GY (2017b) Fabrication of a pumpless, microfluidic skin chip from different collagen sources. J Ind Eng Chem 56:375–381

Sorrell JM, Baber MA, Caplan AI (2007) A self-assembled fibroblast-endothelial cell co-culture system that supports in vitro vasculogenesis by both human umbilical vein endothelial cells and human dermal microvascular endothelial cells. Cells Tissues Organs 186(3):157–168

Srinivasan B, Kolli AR, Esch MB, Abaci HE, Shuler ML, Hickman JJ (2015) TEER measurement techniques for in vitro barrier model systems. J Lab Autom 20(2):107–126

Sriram G, Alberti M, Dancik Y, Wu B, Wu R, Feng Z et al (2018) Full-thickness human skin-on-chip with enhanced epidermal morphogenesis and barrier function. Mater Today 21(4):326–340

Stamhuis E, Thielicke W (2014) PIVlab–towards user-friendly, affordable and accurate digital particle image velocimetry in MATLAB. J Open Res Softw 2(1):e30

Strüver K, Friess W, Hedtrich S (2017) Development of a perfusion platform for dynamic cultivation of in vitro skin models. Skin Pharmacol Physiol 30(4):180–189

Tárnoki-Zách J, Mehes E, Varga-Medveczky Z, Isai DG, Barany N, Bugyik E et al (2021) Development and evaluation of a human skin equivalent in a semiautomatic microfluidic diffusion chamber. Pharmaceutics 13(6)

Tavares RSN, Tao TP, Maschmeyer I, Maria-Engler SS, Schäfer-Korting M, Winter A et al (2020) Toxicity of topically applied drugs beyond skin irritation: static skin model vs. two organs-on-a-chip. Int J Pharm 589:119788

Taylor K, Rego AL (2020) Regulatory drivers in the last 20 years towards the use of in silico techniques as replacements to animal testing for cosmetic-related substances. Comput Toxicol 13:100112

Totsuka A, Omori-Miyake M, Kawashima M, Yagi J, Tsunemi Y (2017) Expression of keratin 1, keratin 10, desmoglein 1 and desmocollin 1 in the epidermis: possible downregulation by interleukin-4 and interleukin-13 in atopic dermatitis. Eur J Dermatol 27(3):247–253

Touchet TJ, Cosgriff-Hernandez EM (2016) 1–hierarchal structure–property relationships of segmented polyurethanes. In: Cooper SL, Guan J (eds) Advances in polyurethane biomaterials. Woodhead Publishing, Cambridge, pp 3–22

Tsukita S, Furuse M, Itoh M (2001) Multifunctional strands in tight junctions. Nat Rev Mol Cell Biol 2(4):285–293

Valencia L, Canalejas-Tejero V, Clemente M, Fernaud I, Holgado M, Jorcano JL et al (2021) A new microfluidic method enabling the generation of multi-layered tissues-on-chips using skin cells as a proof of concept. Sci Rep 11(1):13160

van den Broek LJ, Bergers L, Reijnders CMA, Gibbs S (2017) Progress and future prospectives in skin-on-chip development with emphasis on the use of different cell types and technical challenges. Stem Cell Rev Rep 13(3):418–429

van der Helm MW, Odijk M, Frimat JP, van der Meer AD, Eijkel JCT, van den Berg A et al (2016) Direct quantification of transendothelial electrical resistance in organs-on-chips. Biosens Bioelectron 85:924–929

Vargas R, Egurbide-Sifre A, Medina L (2021) Organ-on-a-chip systems for new drugs development. ADMET DMPK 9(2):111–141

Wagner I, Materne EM, Brincker S, Süssbier U, Frädrich C, Busek M et al (2013) A dynamic multi-organ-chip for long-term cultivation and substance testing proven by 3D human liver and skin tissue co-culture. Lab Chip 13(18):3538–3547

Wang YI, Shuler ML (2018) UniChip enables long-term recirculating unidirectional perfusion with gravity-driven flow for microphysiological systems. Lab Chip 18(17):2563–2574

Wei X, Fricker K, Enk AH, Hadaschik EN (2016) Altered expression of keratin 14 in lesional epidermis of autoimmune skin diseases. Int J Dermatol 55:620

Wei Z, Liu X, Ooka M, Zhang L, Song MJ, Huang R et al (2020) Two-dimensional cellular and three-dimensional bio-printed skin models to screen topical-use compounds for irritation potential. Front Bioeng Biotechnol 8:109

Wu Q, Liu J, Wang X, Feng L, Wu J, Zhu X et al (2020) Organ-on-a-chip: recent breakthroughs and future prospects. Biomed Eng Online 19(1):9

Wufuer M, Lee G, Hur W, Jeon B, Kim BJ, Choi TH et al (2016) Skin-on-a-chip model simulating inflammation, edema and drug-based treatment. Sci Rep 6:37471

Zahorodny-Burke M, Nearingburg B, Elias AL (2011) Finite element analysis of oxygen transport in microfluidic cell culture devices with varying channel architectures, perfusion rates, and materials. Chem Eng Sci 66(23):6244–6253

Zhang YS, Aleman J, Shin SR, Kilic T, Kim D, Mousavi Shaegh SA et al (2017) Multisensor-integrated organs-on-chips platform for automated and continual in situ monitoring of organoid behaviors. Proc Natl Acad Sci USA 114(12):E2293–Ee302

Zhang L, Qian Z, Tahtinen M, Qi S, Zhao F (2018) Prevascularization of natural nanofibrous extracellular matrix for engineering completely biological three-dimensional prevascularized tissues for diverse applications. J Tissue Eng Regen Med 12(3):e1325–e1e36

Zoio P, Oliva A (2022) Skin-on-a-chip technology: microengineering physiologically relevant in vitro skin models. Pharmaceutics 14(3):682

Zoio P, Lopes-Ventura S, Oliva A (2021a) Barrier-on-a-Chip with a modular architecture and integrated sensors for real-time measurement of biological barrier function. Micromachines (Basel) 12(7):816

Zoio P, Ventura S, Leite M, Oliva A (2021b) Pigmented full-thickness human skin model based on a fibroblast-derived matrix for long-term studies. Tissue Eng Part C Methods 27(7):433–443

Zoio P, Lopes-Ventura S, Marto J, Oliva A (2022a) Open-source human skin model with an in vivo-like barrier for drug testing. ALTEX 39(3):405–418

Zoio P, Lopes-Ventura S, Oliva A (2022b) Biomimetic full-thickness skin-on-a-Chip based on a fibroblast-derived matrix. Micro 2(1):191–211

3D Bioprinting of Skin Tissue Model

5

Ashis Kumar Bera and Falguni Pati

Abstract

The challenges and potential of using tissue engineering and regenerative medicine to create functional skin tissue constructs. While traditional skin grafts have limitations, advancements in biofabrication technology and natural polymers such as collagen offer promising avenues for creating skin tissue that mimics the cellular architecture of native tissue and organs. Laser-assisted bioprinting is a high-resolution printing method that can print delicate substrates with high precision, but it is not well-suited for printing large-scale tissue constructs and has a slower printing speed compared to other bioprinting methods. Creating ideal biomaterials and including skin appendages in the fabrication process are still challenges that need to be addressed, but the potential applications for tissue-engineered constructs in congenital defect surgery, surgical reconstruction, and epidermolysis bullosa treatment could improve patient outcomes and reduce overall costs. Overall, the development of functional skin tissue constructs with a prevascularized network and an effective barrier function has significant potential to impact patient care and advance the field of regenerative medicine.

Keywords

3D skin · 3D bioprinting · Cellular architecture · Skin model

A. K. Bera · F. Pati (✉)
Department of Biomedical Engineering, Indian Institute of Technology Hyderabad, Kandi, Telangana, India
e-mail: falguni@bme.iith.ac.in

A. B. Pant et al. (eds.), *Skin 3-D Models and Cosmetics Toxicity*,
https://doi.org/10.1007/978-981-99-2804-0_5

5.1 Introduction

Skin envelops the entire body and is the identification of individual human beings. It also reflects the body's internal condition. Its perfect conformity to the body's contour maintains dynamic equilibrium while continuously replenishing the outer surface by underneath stem cells. As the skin is the outer part of the body, damage by chemicals, burns, mechanical injury is common and different skin diseases due to polluted environment occur in daily life. Although the regeneration capacity of skin is relatively higher than other organs, it is limited by the degree and frequency of injury and the appendages can hardly regenerate under natural conditions. In the case of large-scale skin defects caused by accident or burn, if the effective treatment is not taken timely, it can threaten patient's lives due to loss of excess tissue fluid (Vaezi and Yang 2015). The current effective treatment for skin tissue is mainly autograft as it has no immune response, but it is limited by donor site morbidity and number of donors. Allograft and xenograft can be the ultimate strategies to treat large-scale skin damage, though immune rejection is the primary cause of its failure (Wang et al. 2018).

3D bioprinting, also called bio-additive manufacturing, is a layer-by-layer manufacturing process that fabricates tissue analogous to predefined architectural design. The fundamental idea of 3D bioprinting is the deposition of cell-laden bioink with a highly bionic structure like native tissue with computer-aided design (CAD) modeling. The advancement of bioprinting has the characteristics of user-defined controllability, short production time, and patient-specific customization that open a tremendous possibility for manufacturing complex tissue structures.

Human organs are not just a collection of cells and extracellular matrix; cells reside in a specific niche and manner embedded in the ECM matrix. The spatial context of cellular arrangement enables cells to respond to specific chemical cues, leading to the activation of a specific pathway and the inhibition of others. In 3D cell culture, the biomaterials support a provisional extracellular matrix to create a 3D microenvironment for imparting cellular and tissue level activity for a specific function. The cell-encapsulated biomaterial formulation (referred to as "bioink") supports the seeded cells and can be used for bioprinting to fabricate tissue analogs. The advancement of materials science in tissue engineering unlocks the possibility of using different biomaterials that mimic the 3D microenvironment at the cellular level and the native ECM and support the seeded cells, inducing proliferation, differentiation, migration, and ultimate tissue remodeling.

Bioprinting allows the deposition of bioink on receiving substrate in a predefined position in the x-, y-, and z-axes with the potential to mimic the cellular architecture like native tissue and organs. The flexibility of design and manufacturing enables the production of various forms of tissue construction, including tubes, patches, sheets, and organoid structures with high shape and size fidelity. With respect to skin tissue engineering, the dermis layer contains a more elevated amount of extracellular matrix (mainly type 1 collagen) (Montagna et al. 1992). In the epidermis, the keratinocyte cell layer is present without much extracellular matrix with several types of appendages (Fig. 5.1). The basement membrane presents between the

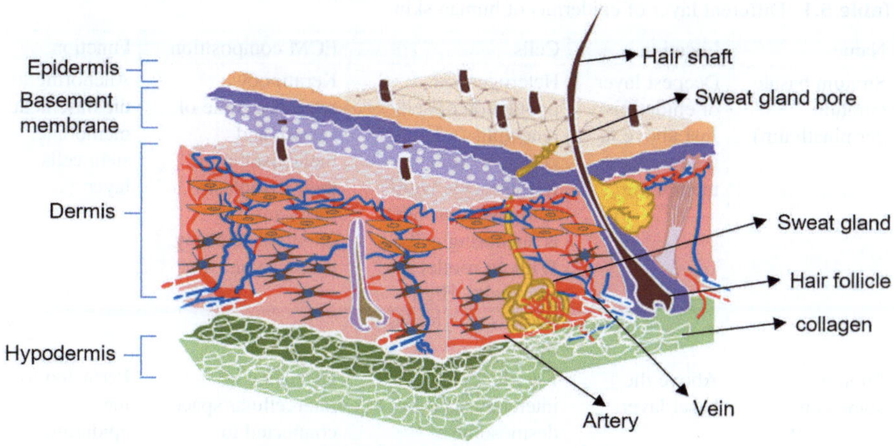

Fig. 5.1 Anatomical structure of human skin

dermis and epidermis, bonded tightly with each side, and prevents vascularization of the epidermis. Skin tissue has very high strength and elastic properties due to the presence and orientation of collagen fiber in the dermis that response to external stress and strain. The innermost layer, called hypodermis, comprises adipocytes, lipocytes, and macrophages. The function of the hypodermis is to provide insulation, vascularize the dermis, and padding that acts as a shock absorber.

In this chapter, we briefly review different bioprinting strategies for skin tissue fabrication, outlining the importance of structural requirements, highlighting the limitation of available engineered skin tissue, and pointing to the need for bioprinting of human skin tissue. Finally, the future challenges and future scope are also discussed.

5.2 Ultrastructure and ECM Composition

Skin is the largest and heaviest organ of our body. Though the thickness of skin up to a few millimeters only, its weighs from 3.5 to 10 kg depending upon height and body mass index. The surface area of 1.5–2 m^2 makes it about one-seventh of our body weight. The skin mirrors the internal body's condition and provides primary information about a person's age and health conditions. Changes in skin color and structure can signify medical conditions and display initial signals for chronic diseases.

5.2.1 Epidermis

Epidermis, the outer layer of the body, is maintained in a dynamic equilibrium state with continuously renewing by underneath cells. Depending on the site, the

Table 5.1 Different layer of epidermis of human skin

Name	Location	Cells	ECM composition	Function
Stratum basale (stratum germinativum)	Deepest layer of epidermis. Just above to the basement membrane	Heterogeneous population of self-renewing keratinocyte. Keratinocyte with cytoplasmic rootlets is anchoring cells Without rootlet are stem cells. High nucleus to cytoplasm ratio	Keratinocyte content bundle of keratin and melanosome in their cytoplasm Cells are interconnected by desmosome	Anchoring to the basement membrane, stem cells layer
Stratum spinosum	Above the basal layer	Keratinocyte interconnected by desmosome	Extension of intercellular space connected to dermis for perfusion and keratin deposition	Perfusion to the epidermis, anchoring plates for keratin filament
Stratum granulosum	Between spinosum and lucidum layer	Viable granular cells with keratohyalin granule	Filaggrin made from keratohyalin granule, keratin	Transition of viable
Stratum lucidum	Non-viable cell layer above the granular layer	Thin flatter keratinocyte cell layer	Keratin, involucrin	Flattened horny keratinocyte function as a major barrier
Stratum corneum	Top layer of epidermis	Strong, dead keratinocyte called corneocyte cell layer	Keratin	It protects from light, heat, pathogen, and chemicals

epidermis varies in thickness, and it is only 0.3 mm thick on elbows and in soles of our feet and palms of our hands up to 4 mm. The predominant cells in the epidermis are keratinocytes with different layers (Table 5.1) containing viable inner cells called stratum basale or stratum Malpighi and outer anucleated horny cells layer are called stratum corneum (Fig. 5.2). The keratinocytes in basale are continuously renewing and pushing up the upper layer, where they harden and eventually die off. The hardened keratinocyte (corneum) is closely packed and sealed from the outside environment. This continuously renewing keratinocyte replaces the cells, which shed off as tiny flakes. The cells in the epidermis grow faster, become thicker, hardened, and develop a callus to withstand pressure or rubbing to protect itself (Montagna et al. 1992).

The epidermis also contains other cell types with special functions:

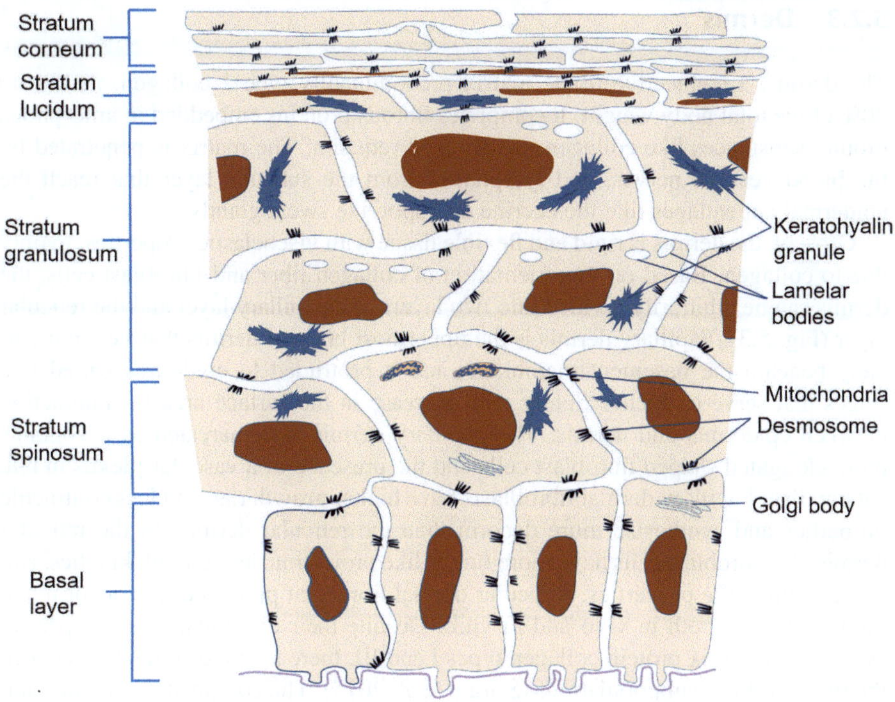

Fig. 5.2 Schematic diagram of different layer of epidermis

- Melanocytes: Melanocytes secrete melanin, the color pigment of skin in response to sunlight. This protects the skin from harmful UV rays.
- Lymphocytes and Langerhans Cells: The predominant immune cells present in epidermis.
- Merkel Cells: Merkel cells are special nerve cells in the skin that get activated in response to pressure and send signal to the central nervous system.

5.2.2 Basement Membrane

A periodic acid–Schiff (PAS) reactive membrane separates epidermis from underneath dermis by physically and functionally at the dermal–epidermal junction. In the epidermal side, basal keratinocyte cells produce hemidesmosomes, a junction between basal keratinocytes and lamina densa that attached basal keratinocytes to the basement membrane matrix. The keratin filament forms a cytoskeleton within the keratinocyte that attaches to the hemi desmosome, and anchoring fibrils bind to the lamina densa to the dermis collagen matrix. In the lamina densa, type IV collagen is predominant to anchor with the dermis and epidermis side, and the anchoring plaque in the dermis matrix is composed of collagen type VII.

5.2.3 Dermis

The dermis, a connective tissue matrix predominantly type 1 collagen, makes up 20% of the total body weight. It comprises fibrous proteins embedded in amorphous ground substances like collagen, elastin, and reticulin. The matrix is penetrated by the blood vessels, nerves, and lymphatics from the subcutis layer that reach the epidermal appendages like the eccrine and apocrine sweat glands.

Grossly, the dermis is rigid and flexible tissue with viscoelastic properties, mainly due to collagen. Based on the orientation of collagen fiber and fibroblast cells, the dermis can be arbitrarily divided into two layers: the papillary layer and the reticular layer (Fig. 5.3). Papillary dermis is the uppermost layer of dermis that lies immediately beneath the basement membrane, and it protruded to epidermis called rete ridges that serve to anchor tightly with increase in the surface area for interaction between epidermis and dermis. The papillary dermis is demarcated as it contains more elongated shaped fibroblast cells and the presence of a vascular plexus in rete subpapillae. Papillary dermal fibroblasts have higher growth rate and less contractile properties and synthesize more decorin than the reticular dermis. In the reticular dermis, the fibroblast cells have more finger-like projection, less growth kinetics, and higher contractile properties. Reticular dermal fibroblast produced less decorin and higher versican both in vivo and in vitro. Despite their dissimilarity, in respect to synthesis of matrix protein collagen types I and III, there are no differences between the two fibroblast subpopulations (Sriram et al. 2015). The composition and orientation of ECM also vary in papillary and reticular dermis based upon their secretion of ECM and functionality. In the papillary layer, the collagen fibers are thin with random orientation, whereas the reticular layer composes thick collagen fibers with orientation along the dermis–epidermis junction (DEJ). As the papillary layer

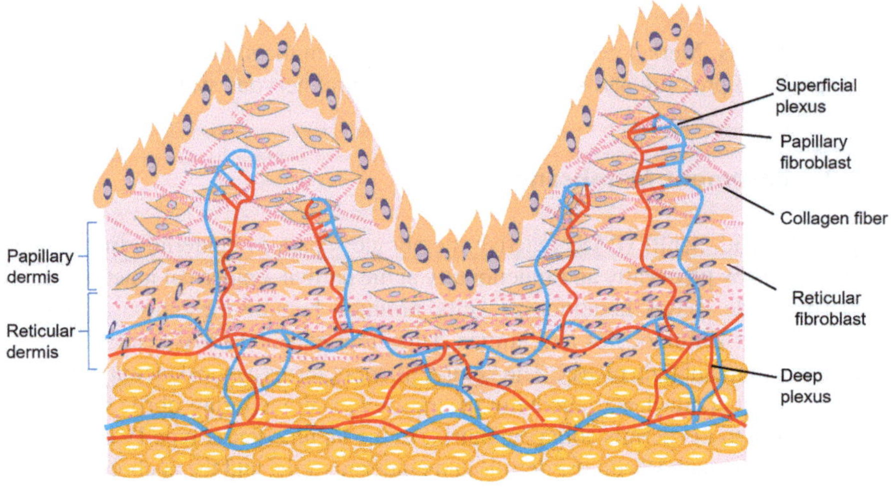

Fig. 5.3 Schematic diagram of different layer of dermis and hypodermis

contains more delicate collagen fiber, the inter-fibrillar space has a greater ground substance composed of proteoglycan.

5.2.4 Hypodermis

The hypodermis, innermost layer of skin containing fatty layer, differs in thickness with sex, race, hormonal, and individual nutritional status. The primary functions of the hypodermis layer are thermoregulation, contouring the body shape, cushioning from the outside mechanical force, filling space, and serving as an instantly available source of energy (Fig. 5.3). The hypodermis consists of three fatty layers separated from each other by connective tissue sheaths. The upper first layer is thicker and has a smaller number of connective tissue septa in women than in men.

5.3 3D Bioprinting Methods Applied in Skin Bioprinting

Bioprinting is a promising technology for the commercial fabrication of tissue constructed by layer-by-layer deposition of cells and biomaterials with computer-controlled precision. The bioink is mainly composed of live cells and biomaterials. Additional biological substances like growth factors and cytokines are often mixed in the bioink for better functionality of seeded cells. This technique enables the creation of a construct with patient-specific architecture using CT-scan data. The bioprinted construct aims to support and promote the seeded cells for maturation, proliferation, and differentiation, ensuring tissue remodeling. With the advances in additive manufacturing, researchers have developed different 3D bioprinting methods to fabricate a complex tissue construct. Three methods commonly used for the bioprinting of skin tissue are extrusion-based bioprinting, inkjet bioprinting, and laser-assisted bioprinting (Fig. 5.4).

5.3.1 Extrusion-Based Bioprinting

Extrusion-based bioprinting is the most popular and well-established fabrication process for the fabrication of tissue construct. The bioink is extruded out from the nozzle tip by using pneumatic pressure pushed by a plunger or mechanical screw. Though hydrogel-based biomaterials are generally used as a bioink for extrusion-based bioprinting, thermoplastic materials also can be used for printing scaffolds using a temperature-controlled extruder. As the melting temperatures of thermoplastic materials are much higher than the physiological temperature, the cells can only be seeded on the scaffolds after printing and used for various tissue engineering applications. Extrusion-based 3D bioprinters can print different hydrogels with shear thinning behavior and natural polymers like collagen, decellularized extracellular matrix, and synthetic biopolymers like poly (lactic-co-glycolic acid).

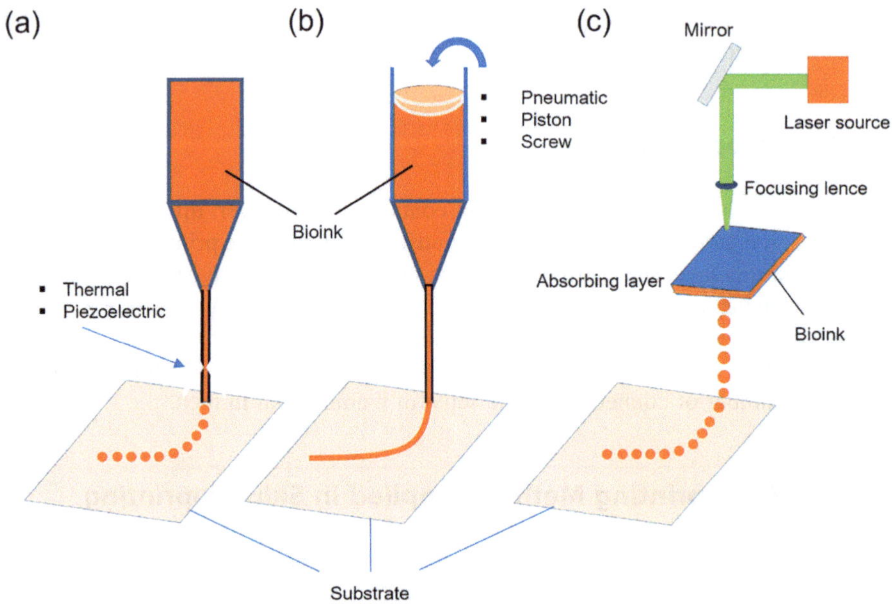

Fig. 5.4 Schematic representation of different bioprinting strategies applied for skin tissue fabrication. (**a**) Inkjet bioprinting, (**b**) extrusion-based bioprinting, and (**c**) laser-assisted bioprinting

The resolution of extrusion-based bioprinting is fundamentally governed by the diameter of nozzle diameter, mechanical and physical properties of hydrogels, and the printing parameters (Fig. 5.4). The extrusion pressure, either pneumatic or screw-based, applied on hydrogel, generates shear stress that in turn causes reduction in viscosity of bioink intended to dispense from the nozzle. The volumetric flow rate of bioink increases with increasing the applied pressure, which needs to be optimized. The printed line width correlates with the inner diameter of the nozzle. Further, it directly correlates with the applied pressure and inversely correlates with the printing speed. Although the final resolution (in X-, Y-, and Z-directions) depends on the gelation kinetics of the bioink and the viscosity (how quickly the material spread before it got gelled), the possible way to increase the printing resolution for hydrogel is to choose a minimum diameter of the nozzle and low applied pressure with high printing speed. The applied shear stress on bioink and cells is directly related to the applied pressure, printing speed, and nozzle diameter, which limits the further printing resolution as higher shear stress directly damages the cells.

Bioinks often used in extrusion-based bioprinting are not robust enough to support themselves and hold the weight of the additional layer printed on the top of it. Recently, researchers developed a support bath to counter this by providing support to the printed structure. This support bath, made up of protein or polysaccharide with very small size of particles, behaves as a solid substrate under zero or low shear. As the nozzle passes through the media, it generates shear strain to decrease storage modulus (G') and increase loss modulus (G"), resulting in it behaving like a liquid while allowing dispensing of bioink. The technology has expanded the printability of low viscous materials for fabricating complex tissue.

5.3.2 Inkjet-Based Bioprinting

Inkjet-based bioprinting techniques have been developed simultaneously along with extrusion bioprinting. The bioink forms into picometer-level droplets and is dispensed by the nozzle. The pressure pulse, generated by microvalve, thermal, piezoelectric, or acoustic actuators, is the driving force to generate bioink droplets (Fig. 5.4). The bioinks needed for inkjet-based bioprinting typically have low viscosity with a high gelation rate as they become gel before deposition on the receiving substrate. Low viscosity and a high surface area-to-volume ratio are needed for extrusion, and fast gelation is required to prevent shape deformation, but these limit the number of materials used in droplet-based bioprinting. Ideally, the bioink droplets should form a gel after they are ejected from the nozzle to prevent nozzle clogging. Methods have also been developed to print bioink droplets into other liquids that induce gelation containing cross-linking agents to preserve the shape and volume. This printing strategy is beneficial for controlling the porosity of printed scaffold by allowing micelle formation of sacrificial materials with precise size and shape. This high controllable porosity is being applied for bioprinting vascular tissue fabrication.

The bioinks used in inkjet printing typically have low viscosity compared to extrusion-based bioprinting. They also need less pressure to extrude from the nozzle, resulting in high cell viability compared to extrusion-based bioprinting. Inkjet printing also benefits from the small size of bioink droplet formation that offers excellent printing resolution.

5.3.3 Laser-Assisted Bioprinting

Laser-assisted bioprinting (LAB) is an emerging technique to engineer tissue mimics by direct writing based on laser-induced forward transfer. The three important major parts of a laser-assisted bio printer include a ribbon, a pulse laser, and a receiving substrate (Fig. 5.4). The multilayered ribbon comprises a transparent glass, a fine layer of gold or titanium (laser-absorbing), and a bioink layer. A pulsed laser beam focuses on the ribbon to heat the thin metal layer, which in turn creates a high-pressure bubble to push the bioink on the receiving substrate. Culture media present on the receiving substrate supports the newly formed droplets that are transferred from the ribbon. The resolution of LAB varies from picometer to micrometer depending upon several factors like thickness of bioink coating on the ribbon, viscosity, surface tension of bioink, laser intensity, and the gap between the ribbon and substrate. LAB has high precision and resolution and can print viscous bioink-containing cells without imparting mechanical stress to the cells.

5.4 Bioprinted Skin Tissue

Standard 3D skin models are significantly advanced from the traditional 2D skin cell culture. Still, their regulation does not reflect like in vivo situation. Production of automated functional engineered skin tissue is possible by the advancements in current 3D tissue development technology. Commercially available skin-like tissue constructs for in vivo applications are Apligraf® (Eaglstein and Falanga 1998), TheraSkin®, Dermagraft (Hart et al. 2012), and OrCelR and in Vitro applications such as EpiSkin (L'Oréal) (Roguet et al. 1994), EpiDerm™ (MatTek corp.) (Cannon et al. 1994), Leiden epidermal skin model and total thickness human skin model (biomimetic Aeon Astron Europe), EpiCSR RHE (CellAystemsR GmbH and Atera SAS), and LabCyte EPI-MODEL (Japan Tissue Engineering Co., Ltd). The developed model underlines the progress of tissue models toward more physiologically relevant tissue models through collaboration between industry and academia (Fig. 5.5).

5.4.1 Toward Well-Defined Physiological Matrix

The microenvironment of living tissue is a mechano-physiological space provided to tissue, maintaining its structural and functional integrity. Owing to the versatility and heterogeneity of multicellular organisms, it is difficult to define the native microenvironment of specific tissue and cellular activity is lost when we modify their microenvironment during in vitro cellular study. Mimicking the physiological microenvironments is the primary target for fabricating functional tissue analogs in vitro. Natural polymers like collagen, hyaluronic acid, and alginate; natural polymer blends like gelatin/chitosan, collagen/alginate, gelatin/silk fibroin; and composites like PCL/collagen are commonly used biomaterials for 3D bioprinting skin constructs.

Collagen is the structural and functional unit of skin ECM; in particular, type 1 and type III collagens are the significant elements of skin ECM. Content and ratio of these two types of collagens vary with respect to age and injury. In fetal skin, type III collagen accounts for 34–65%, while in adults, type I collagen is 80–85% (Sriram et al. 2015). The changes in ECM content with time infer that it evolves to support the residing cells and architecture that is highly dynamic and heterogeneous. Collagen is most extensively explored in tissue engineering and regenerative medicine due to its abundance in target organs and in nature and because of its cells' supporting behavior. Although pure collagen has low mechanical strength and slow gelation properties, 5% collagen solution is printable with extrusion-based bioprinting, but 10% collagen was used rarely because in higher collagen concentration, the dense fibrous architecture limits cell migration and viability (Hospodiuk et al. 2017; Yoon et al. 2016; Cross et al. 2010). The possible approach to increase the mechanical strength is to either increase NaCl concentration, temperature, and collagen concentration or mix with other higher modulus biomaterials (Duan et al.

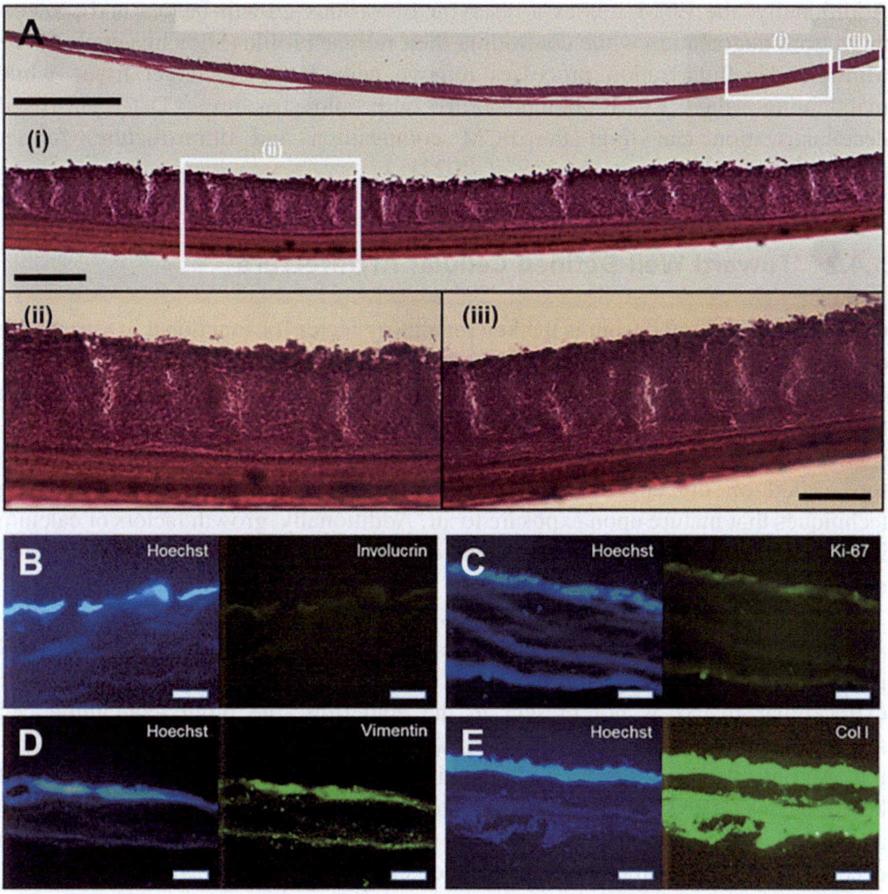

Fig. 5.5 Histological and immunohistochemical assessment of 3D printed skin equivalent. (**a**) H&E-stained cross-sectional image of cultured multilayered skin. Scale bar (i) 1 mm, (ii) 200 μm, and (iii) 50 μm. (**b–e**) Expression of skin-specific markers, keratinocyte differentiation: anti-involucrin (**b**), cell proliferation: anti-Ki-67 (**c**), anti-vimentin for viable epidermal/dermal layer (**d**), and anti-collagen 1 for type 1 collagen production (**e**) (scale bar: 20 μm). Reproduced from reference Lee et al. (2021) with permission from ELSEVIER

2013; Lai et al. 2008; Rhee et al. 2016; Ng et al. 2016; Kim et al. 2011; Xiong et al. 2017).

Designing ideal biomaterials for tissue engineering application is extremely challenging as the tissue-specific microenvironment undergoes an evolutionary development from embryo to maturity. Additionally, the composition and ultrastructure of ECM are still yet to be fully characterized using the most sophisticated approach; the synthetic approach cannot generate tissue construct in vitro. Decellularized extracellular matrices could be better materials to mimic the tissue-specific microenvironment with respect to ECM composition and ultrastructure.

Additionally, the ECM acts as a reservoir of various growth factors and protects them from degradation while controlling their release profile (Shpichka et al. 2019). Multiple decellularization processes remove cells from the target tissue while maintaining the ECM composition and ultrastructure. Detergent-based decellularization can limit the ECM composition and ultrastructure, further restricting cellular activity (Bera et al. 2022).

5.4.2 Toward Well-Defined Cellular Architecture

Cellular architectural design is the key regulatory factor for functional in vitro tissue construction. Skin is a multicellular and multilayered complex organ, and it is challenging to build a skin tissue model with high reproducibility. The current technology focuses on emulating skin anatomy by co-culturing major cell types such as fibroblasts, keratinocytes, and melanocytes. In general, keratinocyte cells are top seeded on the engineered dermis scaffold by manually or by bioprinting techniques that mature upon exposure to air. Additionally, growth factors or calcium ions can be used to promote keratinocyte cells to differentiation and stratification to mature stratum corneum (Pruniéras et al. 1983; Bikle 2014; Bikle et al. 2012).

To fabricate a more stable and reliable skin model, the high-resolution positioning and arrangement of keratinocyte cells are crucial that mimic the epidermal–dermal junction and further helps in the formation of mature corneocyte. Although the extracellular matrix is not present in the epidermis, the bioink-containing cell suspension can be printed with inkjet or extrusion-based printing for precise patterning. A controlled volume of cell-laden bioink can be used. The human epidermal keratinocytes (HEKs) suspended in a culture medium have been extruded uniformly as mono-layered basal keratinocytes on top of the fibroblast cells encapsulated decellularized dermal matrix. Notably, the keratinocyte-suspended culture medium needs to extrude with a small diameter nozzle as it has very low viscosity and requires very low pressure for dispensing. Typically, the epidermis is very thin and bioprinting of dense epithelium on the bulky dermis layer limits reproducibility and is the limitation of extrusion-based technology. The piezoelectric inkjet printing technology may be appropriate for depositing very thin layers of keratinocyte cells (Lee et al. 2021). Collagen has been used for laser-assisted bioprinting as a bioink for printing NIH 3 T3 fibroblast cells on top of the supportive MatriDerm scaffold. The keratinocyte cells (HaCaT) were printed on top of the NIH 3 T3 layer to generate a bi-layer dermis and epidermis. After 10 days of culture, the epidermis becomes intact and the presence of Connexin 43 confirms the formation of gap junction (Koch et al. 2012).

In connective tissue, the predominant cells are fibroblasts that secrete extracellular matrix (ECM) and maintain their architectural design. The epidermis layer is underpinned by its connection to the dermis, which contains the dermal fibroblast, blood vessels, immune cells, nerve fiber, hair follicles, and secretory gland. The spindle-shaped fibroblasts cells have capability to attach to tissue culture plastics and express vimentin and collagen 1. Traditionally, fibroblasts maintain a static population that maintains and supports the skin by secretion and degradation of ECM. Still, it plays crucial roles in almost every process throughout life: embryogenesis and pathology like psoriasis, aging, fibrosis healing, and skin cancer. The fibroblast in the skin resides as morphologically and functionally heterogeneous subpopulations in different dermis compartments. Despite their typical phenotype, dermal fibroblasts in the superficial layer (papillary dermis) exhibit different gene expression patterns than deeper layer (reticular dermis) fibroblasts (Chang et al. 2002; Fries et al. 1994; Sorrell et al. 2004; Rinn et al. 2006; Sorrell and Caplan 2004). In the papillary dermis, the collagen fibers are thin and poorly organized, while in the reticular dermis it is well-organized and thick (Sorrell and Caplan 2004).

Additionally, the papillary dermis has a higher collagen III to collagen I ratio and a higher level of decorin than the reticular dermis and undergoes better remodeling during wound healing (Sorrell et al. 2004; Sorrell and Caplan 2004; Jahoda 2003; Taylor et al. 2000). The differences in composition and organization of the ECM in different sublayers play an important role in functionalizing the fibroblast cells in response to wound healing and scar formation. Despite the heterogeneous fibroblast subpopulation in papillary and reticular subtypes, the fibroblasts in association with hair follicles are also subdivided into follicular dermal papilla (FDP) and dermal sheath (DS) fibroblast. Fibroblasts associated with hair follicles are the regulatory factor to maintain homeostasis and generation of epithelial including epidermis and hair (Jahoda 2003; Taylor et al. 2000).

Various cells have been used for 3D bioprinting for skin tissue engineering. The primary choice to fabricate dermis is human neonatal dermal fibroblasts (hNDFs) due to their high expansion rate, relatively low immunogenicity, secretion of bioactive molecules like basic fibroblast growth factor (bFGF), vascular endothelial growth factor (VEGF), and keratinocyte growth factor (Shpichka et al. 2019). Some studies also use other types of cells for bioprinting dermal substitutes listed in Table 5.2.

5.4.3 Toward Well-Defined Skin Appendages

The advancement of tissue engineering and regenerative medicine has made a breakthrough in the fabrication of engineered skin but still it needs to incorporate many functional units like sweat glands, blood vessels, sensory neurons, hair follicles, and pigmentation. The ideal tissue-engineered skin must have all kinds of skin appendages and could develop pigmentation followed by transplantation. Creating a microenvironment with matrix and stem cells could regenerate appendages like sweat glands and hair follicles in a biofabricated skin tissue graft. The embryonic ectoderm develops all the skin appendages in developmental stages, and the epidermis acts as a reservoir of stem cell niches. However, compared to relatively well-known skin appendages like hair follicles, cell inductive response about sweat glands is less known. Recreating the microenvironment with adult epidermal progenitor cells through bioprinting technology is suitable for regenerating and restoring the sweat gland function (Todd 2015). In another study,

Table 5.2 Representative biomaterials, cells, and bioprinting approaches used for bioprinting skin substitutes

Type of biomaterial	Polymer	Cells	Printing technique	Primary characterization	Reference
Natural	Collagen	NIH 3 T3, HaCaT	Laser-assisted bioprinting	Keratinocyte are proliferated and form gap junction	Koch et al. (2012)
		HFF-1, HaCaT	Extrusion	Air–liquid interface culture for maturation of keratinocyte	Lee et al. (2014)
		HFB, HKC	Extrusion	Collagen micro-standards frozen immediately after extrusion to control pore size and shrinkage	Kim et al. (2009)
	Alginate	NIH 3 T3, HaCaT, hMSC	Laser-assisted bioprinting	Alginate hydrogels enable well-defined cell structure and function	Koch et al. (2010)
	Chitosan	HDFs and HaCaTs	Extrusion	3D printed patch showed superior healing	Intini et al. (2018)
Natural composite	Chitosan and genipin	HKC, HDF	Extrusion	Cells viable >85% in day 7	Hafezi et al. (2020)
	Alginate and gelatin	HDF	Extrusion	Combining different cross-linking process to create dermal scaffold	Diani and Gall (2006)
	Gelatin, collagen–fibrinogen, and elastin	Human neonatal dermal fibroblast and keratinocyte	Extrusion inkjet	The hydrogel for dermis and basement membrane suitable for high-throughput bioprinting	Derr et al. (2019)
	GelMA and ulvan polysaccharide	Human dermal fibroblast	Extrusion	The bioink provides mechanical support to hold the printed structure and stability	Chen et al. (2021)

	GelMA and alginate	Human dermal fibroblast, keratinocyte, and endothelial cells	Extrusion	7.5% GelMA and 2% alginate support endothelial. Cells and modulate stiffness by produced pro-collagen I	Barros et al. (2021)
	GelMA and collagen with tyrosinase	HaCaTs, HDFs, and human melanocyte	Extrusion	The novel bioink GelMA and collagen doped with tyrosinase can help to form dermis and epidermis	Yoon et al. (2018)
	Gelatin and sulfonated silk	Child foreskin fibroblast (CFFs)	Extrusion	The composite (gelatin and silk) scaffold promotes vascularization and keratinocyte maturation	Xiong et al. (2017)
	Nano-fibrillated cellulose (NFC), carboxymethyl cellulose (CMC) and alginate	Human-derived skin fibroblast (hSF)	Extrusion	The composite bioink supports cellular activity	Zidarič et al. (2020)
Composite	PCL/collagen	HDF, HEK	Extrusion and inkjet	They print collagen-based construct with polycaprolactone(PCL) mesh by extrusion-based bioprinting and by inkjet-based bioprinting top seed keratinocyte on it	Kim et al. (2017)

a bioprinted matrix with different pore sizes has been shown to induce epidermal progenitor cells to differentiate into sweat glands (Liu et al. 2016).

Like other skin appendages, skin pigmentation is essential in protecting from damage by harmful UV rays. The melanocyte in the epidermis secretes melanin, and the pigment determines the skin color. The biofabrication of pigmented skin construct remains a bottleneck in skin tissue engineering. However, incorporating melanocytes into the 3D bioprinted skin construct has been used for developing pigmented skin models for cosmetics and toxicological studies. The sequential printing of keratinocyte and melanocyte on top of the collagen layer induces pigmentation when it is exposed to air–liquid interface culture (Min et al. 2018).

Hair follicles are another important skin appendage that plays an important role in thermoregulation, barrier function, secretion, and wound healing (Schneider et al. 2009). Compared to other skin appendages, regeneration of hair follicles from specialized dermal papilla cells (DPCs) is relatively slow. Although there have been some proof of concepts established for inducing hair follicles in mice when the DPCs are transplanted intracutaneously, developing therapeutic strategies of integrating hair follicles within artificial skin graft is challenging (Toyoshima et al. 2012). 3D bioprinting has been used for recreating specialized 3D microenvironments to develop hair follicles from dermal papilla cells (DPCs). Seeding DPCs in the microwell formed aggregation and developed hair follicles, creating micro-fabricated plastics containing hair follicles shaped extensions in the collagen I matrix (Abaci et al. 2018).

5.4.4 Toward Well-Defined Vascular Bed

Vascularization is an important indicator of engineered tissue to reach application level in the field of clinical and in tissue engineering and regenerative medicine. The bioengineered tissue constructs need a well-connected capillary network to develop as a drug testing model and to use as a graft. Additionally, when the bioengineered construct is used as a graft, it needs a prevascularized capillary network for maturation and functionalization. The diffusion limit of oxygen and nutrients is approximately 100–200 μm, also called Krogh length. When the fabricated tissue construct is relatively large, the cells in core are deprived of oxygen and nutrients and die.

Moreover, as the tissue matures upon time the encapsulated cells proliferate and fill the pores in the internal area of construct, which causes block of media transport channels. Therefore, a continuous vascular network is significant for the viability and maturation of fabricated tissue constructs. Sacrificial materials or coaxial nozzle-based 3D printing strategies may be the solution to develop vascular networks in skin tissue grafts (Abaci et al. 2016). The major limitation of printing with sacrificial materials is the channel diameter, which is larger than the actual diameter in native tissue. Furthermore, the endothelial lining is often not uniform and intact like the tunica intima of native blood vessels. Cell-based strategies could be the best approach to stimulate seeded endothelial cells to form the vascular network in the fabricated scaffold. The best strategy would be to create an optimum microenvironment for endothelial cells with supporting cells or growth factors, including a lumen, and further maturing into a vascular network. Zhu e al. (Zhu et al. 2017) introduced the prevascularization of skin tissue by digital light processing (DLP)-based bioprinting method. The endothelial and 10 T1/2 cell suspension with a photopolymer is exposed to hexagonally patterned UV light to produce cell-laden vascular channels. Then, the scaffold was transplanted to a dorsal region of immune-deficient mice that formed a lumen-like structure by integrating host blood vessels.

5.5 Validation of Bioprinted Functional Artificial Skin

The technology leading to a 3D reconstructed skin model was first introduced by James Rheinwald and Howard Green almost 40 years ago (Green et al. 1979; Connor et al. 1981). They cultivate human keratinocytes on a dermis tissue and expose it to an air–liquid interface culture to mature a fully differentiated epidermis to treat burn injury. Besides skin grafting in burn injuries, reconstructed skin tissue models are mainly used for research purposes. However, tissue-engineered skin models are being used for broader purposes, including the cosmetic industry, assessment of hazards by chemicals and pesticides, and preclinical testing of novel drugs. Following the fabrication of skin tissue grafts by researchers or by commercial developers, there is a need for international validation and acceptance for its use globally in both industrial and academic research (Fig. 5.6).

An essential feature of tissue-engineered skin is an effective barrier function that can resist the penetration of cytotoxic materials through skin. A well-established test to address skin's barrier function is the assessment of ET-50 value (effective time at which fixed concentration of toxic materials penetrates tissue to 50%) or IC-50 (inhibition concentration of a toxic substance causes 50% reduction of cell viability). Usually, 1% Triton X-100 is used for ET-50, and a minimum of three concentrations of SDS is used for IC-50 for skin validation tests (Kandarova and Hayden 2021). As a structural validation of bioengineered skin products, the presence of different cells in a precise location with the secretion of specific extracellular matrix proteins is assessed by histology and immunohistochemistry techniques for further use. Immunofluorescent localization of epidermal markers like keratins 5, 10, and 14, involucrin, loricrin, filaggrin, and transglutaminase is required to confirm the maturation of keratinocytes. In the dermal–epidermal junction, there is a need to develop a functional basement membrane that could hold and separate epidermis from the dermis with the presence of expression markers such as laminin 5, collagen IV, and collagen VII. In the dermis, fibroblasts cell maturation is required, and papillary and reticular compartmentalization is required for the elasticity and toughness of skin tissue. The expression of different markers such as decorin, versican, alpha-SMA, and collagen I to III ratios is also essential to validate the skin model for dermal maturation. A further critical parameter for barrier function and evaporation is the lipid profile of the available skin model, which needs to be assessed with standard quality control (QC).

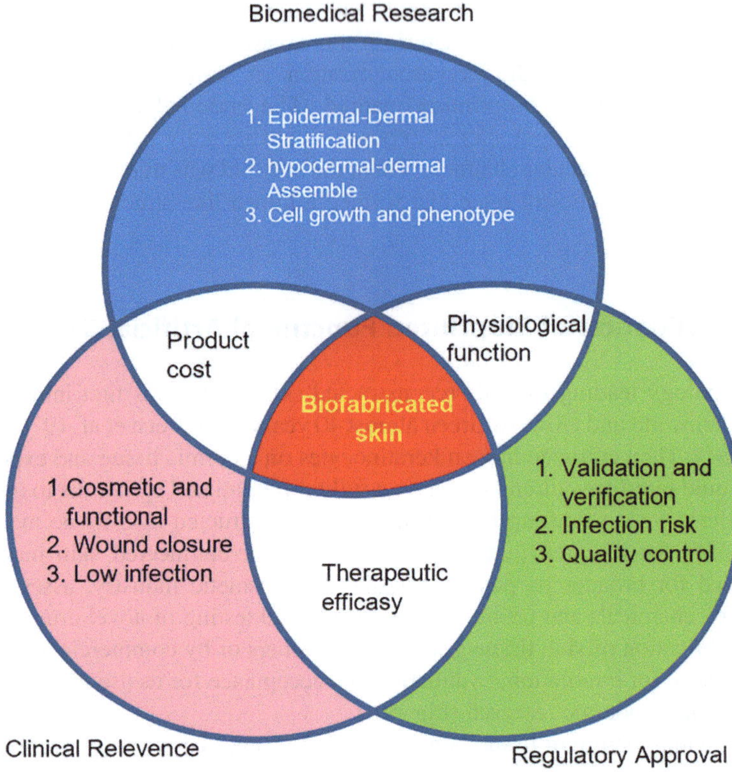

Fig. 5.6 Consideration for successful biofabricated skin product in perspective to research, regulatory agencies, and clinic

5.6 Challenges and Future Directions of Skin Tissue Fabrication

Although the primary goal of tissue engineering and regenerative medicine focuses on achieving closure and healing of wounds at a reasonable rate, further additional considerations are to be made from the perspective of patients and evaluators. By traditional manufacturing, 3D tissue analogs are often made by molding. Later, the fabricated tissues are curved out or removed by a sharp cutting tool till the desired shape is formed. However, there are still challenges associated with these methods, especially when creating or controlling the internal structure with organ-specific microarchitecture of a specific organ.

To date, the generation of laboratory-grown skin substitutes only partially addressed the requirement for stable wound closure, lacking several appendages and vasculature deemed to be accepted equally by industry, regulatory bodies, clinicians, and patients. The formation of blood vessels and neural growth in native tissues generally occurs on the interface between the hypodermis and dermis, which matured to a deep vascular plexus; capillaries spread into the dermis layer provide a gaseous exchange; however, the epidermis get oxygen and nutrients through diffusion only. A lack of organized synergistic layer makes unviable the growth of nerve and basement membrane formation, impairing sensing, protecting, and thermoregulatory functions. As mentioned earlier, the inclusion of skin appendages has the additional complication to fabricating by current strategies as the microenvironment needed for each of the appendages depends upon growing other tissue architecture (Moroni et al. 2018). One further consideration for the fabrication of skin tissue is industrial-scale mass production with consistency between different lots having optimum properties. This includes the procedures used in processing, manufacturing, and characterizing the products with repeatable and reliable results. Finally, if the cells used in therapy are autologous, consistent methods for harvesting cells are required for mass-scale production, as the phenotypic properties may change, and effectiveness may get reduced when those cells grow with extensive passage.

The current biofabrication technology and advances in tissue engineering can reduce and to some extent replace the need for skin autograft in clinical translation. Patients with life-threatening burns or major accidental cases would no longer suffer from painful morbidity if bioprinted skin grafts were available. It can also reduce the total time in ICU and hospitalization and would further reduce the need for reconstructive surgery and overall cost (Yu et al. 2019). The tissue-engineered construct also has implications in congenital defect surgery, certain surgical reconstruction, and epidermolysis bullosa treatment contributing to the importance and requirement of fabricated skin tissue. The requirement of epidermal appendages having naturally matched skin pigmentation, sweat glands, vascular plexus, and sensory nerves further opens new challenges to investigate homeostasis and abnormality (Dearman et al. 2021). The advancement of tissue engineering and regenerative medicine has a high degree of confidence that many of the appendages can be incorporated into a further skin substitute model as a full-thickness skin graft for clinical application.

5.7 Conclusion

Ultimate goal of skin tissue engineering by biofabrication is to meet the requirement or replace the current gold standard autologous skin graft. With respect to clinical application, there is a need for additional requirement of minimizing or eliminating scar formation for broad range of patients, and wound types are reliable. The regenerated skin is not only functionally or looks like native tissue but also could match patients' specific pigmentation for a wide range of pigmentation across the population. Despite some initial successful transplantation of 3D bioprinted skin tissue grafts, there is an unfulfilled need for further incorporation of different skin appendages to treat full-thickness skin wounds. However, the market's current biofabricated skin substitute options remain limited by too often a trade-off between efficacy and too high cost. Further efforts are needed to achieve an ideal skin substitute by continuous collaboration, exchange of opinion among researchers, regulatory bodies, and clinicians to ensure the final product attains a wide range of perspectives to use both in clinics and for drug and toxicity testing.

References

Abaci HE, Guo Z, Coffman A, Gillette B, Lee WH, Sia SK et al (2016) Human skin constructs with spatially controlled vasculature using primary and iPSC-derived endothelial cells. Adv Healthc Mater 5(14):1800–1807

Abaci HE, Coffman A, Doucet Y, Chen J, Jacków J, Wang E et al (2018) Tissue engineering of human hair follicles using a biomimetic developmental approach. Nat Commun 9(1):1–11

Barros NR, Kim HK et al (2021) Biofabrication of endothelial cell, dermal fibroblast, and multilayered keratinocyte layers for skin tissue engineering. Biofabrication 13(3). https://doi.org/10.1016/j.snb.2014.04.005

Bera AK, Sriya Y, Pati F (2022) Formulation of dermal tissue matrix bioink by a facile Decellularization method and process optimization for 3D bioprinting toward translation research. Macromol Biosci 22(8):1–15

Bikle DD (2014) Vitamin D metabolism, mechanism of action, and clinical applications. Chem Biol 21(3):319–329

Bikle DD, Xie Z, Tu CL (2012) Calcium regulation of keratinocyte differentiation. Expert Rev Endocrinol Metab 7(4):461–472

Cannon CL, Neal PJ, Southee JA, Kubilus J, Klausner M (1994) New epidermal model for dermal irritancy testing. Toxicol in Vitro 8(4):889–891. https://www.sciencedirect.com/science/article/pii/0887233394900957

Chang HY, Chi JT, Dudoit S, Bondre C, van de Rijn M, Botstein D et al (2002) Diversity, topographic differentiation, and positional memory in human fibroblasts. Proc Natl Acad Sci USA 99(20):12877–12882

Chen X, Yue Z, Winberg PC, Lou YR, Beirne S, Wallace GG (2021) 3D bioprinting dermal-like structures using species-specific ulvan. Biomater Sci 9(7):2424–2438

Connor NEO, Mulliken JB, Banks-Sciilegel S, Kehinde O, Green H (1981) Grafting of burns with cultured epithelium prepared from autologous epidermal cells. Lancet 1:75–78

Cross VL, Zheng Y, Won Choi N, Verbridge SS, Sutermaster BA, Bonassar LJ et al (2010) Dense type I collagen matrices that support cellular remodeling and microfabrication for studies of tumor angiogenesis and vasculogenesis in vitro. Biomaterials 31(33):8596–8607

Dearman BL, Boyce ST, Greenwood JE (2021) Advances in skin tissue bioengineering and the challenges of clinical translation. Front Surg 8:640879

Derr K, Zou J, Luo K, Song MJ, Sittampalam GS, Zhou C et al (2019) Fully three-dimensional bioprinted skin equivalent constructs with validated morphology and barrier function. Tissue Eng Part C Methods 25(6):334–343

Diani J, Gall K (2006) Finite strain 3D Thermoviscoelastic constitutive model. Society 1–10:486

Duan L, Li J, Li C, Li G (2013) Effects of NaCl on the rheological behavior of collagen solution. Korea Aust Rheol J 25(3):137–144

Eaglstein WH, Falanga V (1998) Tissue engineering and the development of Apligraf, a human skin equivalent. Cutis 62(1 Suppl):1–8

Fries KM, Blieden T, Looney RJ, Sempowski GD, Silvera MR, Willis RA et al (1994) Evidence of fibroblast heterogeneity and the role of fibroblast subpopulations in fibrosis. Clin Immunol Immunopathol 72(3):283–292

Green H, Kehinde O, Thomas J (1979) Growth of cultured human epidermal cells into multiple epithelia suitable for grafting. Proc Natl Acad Sci USA 76(11):5665–5668

Hafezi F, Shorter S, Tabriz AG, Hurt A, Elmes V, Boateng J et al (2020) Bioprinting and preliminary testing of highly reproducible novel bioink for potential skin regeneration. Pharmaceutics 12(6):1–21
Hart CE, Loewen-Rodriguez A, Lessem J (2012) Dermagraft: use in the treatment of chronic wounds. Adv Wound Care (New Rochelle) 1(3):138–141
Hospodiuk M, Dey M, Sosnoski D, Ozbolat IT (2017) The bioink: a comprehensive review on bioprintable materials. Biotechnol Adv 35(2):217–239
Intini AC, Elviri L, Cabral J, Bergonzi C, Bianchera A, Flammini L et al (2018) 3D-printed chitosan-based scaffolds: an in vitro study of human skin cell growth and an in-vivo wound healing evaluation in experimental diabetes in rats. Carbohydr Polym 199:593
Jahoda CAB (2003) Cell movement in the hair follicle dermis - more than a two-way street? J Invest Dermatol 121(6):ix–xi
Kandarova H, Hayden PJ (2021) Standardised reconstructed skin models in toxicology and pharmacology: state of the art and future development. Handb Exp Pharmacol 265:57–71
Kim G, Ahn S, Yoon H, Kim Y, Chun W (2009) A cryogenic direct-plotting system for fabrication of 3D collagen scaffolds for tissue engineering. J Mater Chem 19(46):8817–8823
Kim G, Ahn S, Kim Y, Cho Y, Chun W (2011) Coaxial structured collagen–alginate scaffolds: fabrication{,} physical properties{,} and biomedical application for skin tissue regeneration. J Mater Chem 21(17):6165–6172. https://doi.org/10.1039/C0JM03452E
Kim BS, Lee JS, Gao G, Cho DW (2017) Direct 3D cell-printing of human skin with functional transwell system. Biofabrication 9(2):025034
Koch L, Kuhn S, Sorg H, Gruene M, Schlie S, Gaebel R et al (2010) Laser printing of skin cells and human stem cells. Tissue Eng Part C Methods 16(5):847–854
Koch L, Deiwick A, Schlie S, Michael S, Gruene M, Coger V et al (2012) Skin tissue generation by laser cell printing. Biotechnol Bioeng 109(7):1855–1863
Lai G, Li Y, Li G (2008) Effect of concentration and temperature on the rheological behavior of collagen solution. Int J Biol Macromol 42(3):285–291
Lee V, Singh G, Trasatti JP, Bjornsson C, Xu X, Tran TN et al (2014) Design and fabrication of human skin by three-dimensional bioprinting. Tissue Eng Part C Methods 20(6):473–484
Lee HR, Park JA, Kim S, Jo Y, Kang D, Jung S (2021) 3D microextrusion-inkjet hybrid printing of structured human skin equivalents. Bioprinting 22:e00143. https://www.sciencedirect.com/science/article/pii/S2405886621000166
Liu N, Huang S, Yao B, Xie J, Wu X, Fu X (2016) 3D bioprinting matrices with controlled pore structure and release function guide in vitro self-organization of sweat gland. Sci Rep 6:34410
Min D, Lee W, Bae IH, Lee TR, Croce P, Yoo SS (2018) Bioprinting of biomimetic skin containing melanocytes. Exp Dermatol 27(5):453–459
Montagna W, Kligman AM, Carlisle KS (eds) (1992) Atlas of normal human skin. University of Texas Library
Moroni L, Burdick JA, Highley C, Lee SJ, Morimoto Y, Takeuchi S et al (2018) Biofabrication strategies for 3D in vitro models and regenerative medicine. Nat Rev Mater 3:21
Ng WL, Yeong WY, Naing MW (2016) Polyelectrolyte gelatin-chitosan hydrogel optimized for 3D bioprinting in skin tissue engineering. Int J Bioprint 2(1):53–62
Pruniéras M, Régnier M, Woodley D (1983) Methods for cultivation of keratinocytes with an air-liquid interface. J Invest Dermatol 81(1 Suppl):28s–33s
Rhee S, Puetzer JL, Mason BN, Reinhart-King CA, Bonassar LJ (2016) 3D bioprinting of spatially heterogeneous collagen constructs for cartilage tissue engineering. ACS Biomater Sci Eng 2(10):1800–1805
Rinn JL, Bondre C, Gladstone HB, Brown PO, Chang HY (2006) Anatomic demarcation by positional variation in fibroblast gene expression programs. PLoS Genet 2(7):e119
Roguet R, Régnier M, Cohen C, Dossou KG, Rougier A (1994) The use of in vitro reconstituted human skin in dermotoxicity testing. Toxicol in Vitro 8(4):635–639. https://www.sciencedirect.com/science/article/pii/0887233394900337

Schneider MR, Schmidt-Ullrich R, Paus R (2009) The hair follicle as a dynamic Miniorgan. Curr Biol 19(3):132–142

Shpichka A, Butnaru D, Bezrukov EA, Sukhanov RB, Atala A, Burdukovskii V et al (2019) Skin tissue regeneration for burn injury. Stem Cell Res Ther 10(1):1–16

Sorrell JM, Caplan AI (2004) Fibroblast heterogeneity: more than skin deep. J Cell Sci 117(Pt 5): 667–675

Sorrell JM, Baber MA, Caplan AI (2004) Site-matched papillary and reticular human dermal fibroblasts differ in their release of specific growth factors/cytokines and in their interaction with keratinocytes. J Cell Physiol 200(1):134–145

Sriram G, Bigliardi PL, Bigliardi-Qi M (2015) Fibroblast heterogeneity and its implications for engineering organotypic skin models in vitro. Eur J Cell Biol 94(11):483–512

Taylor G, Lehrer MS, Jensen PJ, Sun TT, Lavker RM (2000) Involvement of follicular stem cells in forming not only the follicle but also the epidermis. Cell 102(4):451–461

Todd L (2015) Accepted manuscript Development Accepted manuscript. (June): 1–33

Toyoshima KE, Asakawa K, Ishibashi N, Toki H, Ogawa M, Hasegawa T et al (2012) Fully functional hair follicle regeneration through the rearrangement of stem cells and their niches. Nat Commun 3:784

Vaezi M, Yang S (2015) Extrusion-based additive manufacturing of PEEK for biomedical applications. Virtual Phys Prototyp 10(3):123–135

Wang Y, Beekman J, Hew J, Jackson S, Issler-Fisher AC, Parungao R et al (2018) Burn injury: challenges and advances in burn wound healing, infection, pain and scarring. Adv Drug Deliv Rev 123:3–17

Xiong S, Zhang X, Lu P, Wu Y, Wang Q, Sun H et al (2017) A gelatin-sulfonated silk composite scaffold based on 3D printing technology enhances skin regeneration by stimulating epidermal growth and dermal neovascularization. Sci Rep 7(1):4288

Yoon H, Lee JS, Yim H, Kim G, Chun W (2016) Development of cell-laden 3D scaffolds for efficient engineered skin substitutes by collagen gelation. RSC Adv 6(26):21439–21447. https://doi.org/10.1039/C5RA19532B

Yoon J, Xing TL, Zhang HB, Yin RX, S.M. Yang JW and WJZ. (2018) Tyrosinase doped bioink for 3D bioprinting of living skin constructs. Sensors Actuators B Chem 199:369. https://doi.org/10.1088/1748-605X/aaa5b6

Yu JR, Navarro J, Coburn JC, Mahadik B, Molnar J, Holmes JH et al (2019) Current and future perspectives on skin tissue engineering: key features of biomedical research, translational assessment, and clinical application. Adv Healthc Mater 8(5):1–19

Zhu W, Qu X, Zhu J, Ma X, Patel S, Liu J et al (2017) Direct 3D bioprinting of prevascularized tissue constructs with complex microarchitecture. Biomaterials 124:106–115

Zidarič T, Milojević M, Gradišnik L, Kleinschek KS, Maver U, Maver T (2020) Polysaccharide-based bioink formulation for 3D bioprinting of an in vitro model of the human dermis. Nano 10(4):1–19

Solar Radiation and Phototoxicity of Cosmetics: Avenues of In Vitro Skin Models

6

Mohd. Danish Kamar, Madhu Bala, Apeksha Vikram, Sunil Kumar Patel, Gaurav Prajapati, Vinay Shakti Shukla, and Saroj Kumar Amar

Abstract

Skin is the body's largest organ, performing several critical functions, including photoprotection and thermoregulation. The skin comprises three distinct layers, each having different types of cells. The epidermis, the outermost layer of skin, contains melanin—a photoprotective pigment to protect from the hazardous effect of UV radiation. Keratinocytes and melanocytes are the primary cell type in the epidermis and provide the barrier against UV radiation and pathogens. Fibroblasts are the major connective tissue in the dermis region that is essential in regulating skin physiology and wound healing. Skin is a protective tool against the external environment, getting exposed to solar radiation, environmental contaminants, and the chemical ingredients of skin care products. Some chemical elements of cosmetics are photosensitive and cause phototoxicity under ambient sunlight. Photosensitized chemicals damage various cell organelles directly or indirectly through photosensitizer-induced oxidative damage. Several models have been developed to study the phototoxicity potential of chemicals/ingredients

M. D. Kamar · M. Bala · A. Vikram · S. K. Patel · G. Prajapati
Photobiology Laboratory, Systems Toxicology & Health Risk Assessment Group, CSIR-Indian Institute of Toxicology Research (CSIR-IITR), Lucknow, Uttar Pradesh, India

Academy of Scientific and Innovative Research, AcSIR Headquarters, CSIR-HRDC Campus, Ghaziabad, Uttar Pradesh, India

V. S. Shukla
Fragrance & Flavour Development Centre, Kanauji, Uttar Pradesh, India

S. K. Amar (✉)
Photobiology Laboratory, Systems Toxicology & Health Risk Assessment Group, CSIR-Indian Institute of Toxicology Research (CSIR-IITR), Lucknow, Uttar Pradesh, India

Environmental Lab, US Army Engineer Research and Development Center, Vicksburg, MS, USA

School of Bioscience and Bioengineering, Lovely Professional University, Phagwara, Punjab, India

A. B. Pant et al. (eds.), *Skin 3-D Models and Cosmetics Toxicity*,
https://doi.org/10.1007/978-981-99-2804-0_6

of cosmetic formulation. 2D skin models contain only one cell type and are widely used by researchers. 3D skin models mimic the physiology of human skin by having multiple cell types that are layered like actual skin. 2D skin models are cost-efficient and easy to maintain. At the same time, the 3D skin models represent the environment of the in vivo processes and also can be used to study barrier penetration of cosmetics ingredients. As per a previous study and OECD recommendation, the dermal phototoxicity potential of cosmetics ingredients in humans can be effectively assessed by in vitro approaches.

Keywords

Phototoxicity · Solar radiation · 3D skin models · Cosmetics · UV radiation

Abbreviations

ECM	Extracellular matrix
IC50	concentration at which 50% decrease in cell viability
IL-1α	Interleukin 1 alpha
LDH	Lactate dehydrogenase
MPE	Mean photo-effect
MTT	3-(4, 5-dimethylthiazol-2-yl)-2, 5-diphenyltetrazolium bromide
NHEK	Normal, human-derived epidermal keratinocytes
NRU	Neutral red uptake
OECD	Organisation for Economic Cooperation and Development
PIF	Photo-irritation factor
RHE	Reconstructed human epidermis
ROS	Reactive oxygen species
UVR	Ultraviolet radiation

6.1 Introduction

In sunlight/UV exposure, some cosmetic formulations applied topically or systemically may lead to phototoxic skin irritation. Cosmetics are generally applied onto the skin, and systemic exposure occurs through percutaneous absorption. When a photoreactive chemical gets excited by absorbing UV or visible light, it becomes phototoxic (Lee et al. 2017). The cosmetic formulation contains numerous chemicals, and the safety evaluation of these cosmetic chemicals was previously done by oral or topical exposure of various formulations on animals for toxicity testing (Vinardell and Mitjans 2017). However, the EU has banned the use of animals for testing and safety evaluation of cosmetics (Couteau and Coiffard 2010). The in vitro assays have been developed and validated by the Organization for Economic Cooperation and Development (OECD). The OECD has provided test guidelines for in vitro skin testing (OECD 2019). The main cosmetic ingredients related to light absorption and skin irritation are hair dyes, preservatives, sunscreen UV filters, etc. (Mujtaba et al. 2021). Over the years, many models have been developed for the in vitro testing of chemicals. Researchers have widely used 2D models for phototoxicity testing. However, 2D models do not represent the actual human skin. To better mimic the human skin condition, 3D skin cell culture was developed, which contains all the skin layers and the significant skin cell types. To better understand the phototoxicity of cosmetics and its testing, this chapter aims to explain 2D and 3D models for phototoxicity and the advantages and disadvantages of each model.

6.2 Solar Radiation

Solar radiation is optical radiation that comprises a range of radiation such as infrared radiation (IR), visible light, and ultraviolet radiation (UVR). Even though longer wavelengths (radiofrequency and microwave) and shorter wavelengths, both radiations are present. The life of terrestrial organisms mainly depends on the radiation energy generated by the sun. UV radiation wavelength lies in the range of 100–400 nm, and this UV range is further subdivided into the other group of radiation based on wavelengths such as UVA (320–400 nm), UVB (290–320 nm), and UVC (100–290 nm) (de Paula Corrêa 2015). Total UV radiation contains 95% of UVA and 5% UVB, and UVC does not reach the earth's surface due to stratospheric ozone. Initially, the sun was the only source of exposure, but as advancements emerged in science, some artificial sources were present for UV exposure. Due to the advent of this artificial source of radiation, opportunity for exposure to radiation has increased. Frequently repeated exposure to UV radiation can cause skin cancer because ultraviolet radiation that damages the skin is a carcinogenic factor in sunlight (de Paula Corrêa 2015). The accumulation of mutation induced by UV damage is the leading cause of skin cancer initiation. The abundance of UV radiation in the environment contributes to various skin illnesses, including skin cancer, aging, and inflammation. Initially, humans were mainly exposed to UV radiation through the occupational source of sunlight (Pfeifer 2021). Recently, UV exposure has increased due to outdoor activities and tan cosmetics as an artificial source of UV advent. Photons of UV fall under the wavelength of visible light and the wavelength of gamma radiation because UV radiation is a component of the electromagnetic spectrum (Malis et al. 2007). Based on electro-physical properties, the physical energy of UV radiation is divided into UVA, UVB, and UVC (Gallagher et al. 2014). As there is an inverse relationship between wavelength and energy, UVC radiation has the shortest wavelength (100–290 nm) with the highest energy and UVA has the longest wavelength (320–400 nm) but the least energy. At the same time, UVB lies in between UVA and UVC radiation. UV radiations affect molecules, cells, and tissues (D'Orazio et al. 2014). UV exposure varies according to geographical location and sunlight

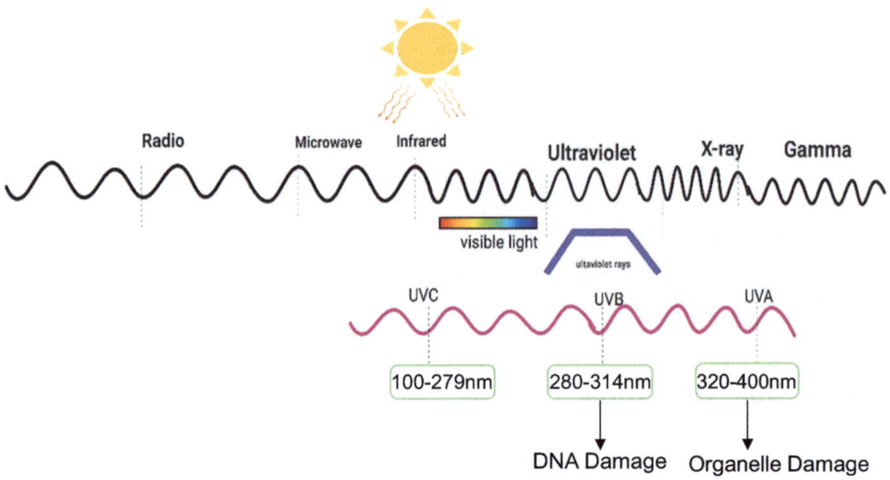

Fig. 6.1 Electromagnetic radiation spectrum showing types of UV, i.e., UVA, UVB, and UVC and the damage on cell

intensity because the sun's intensity varies at a particular spot on the planet (Fig. 6.1).

6.3 Skin Physiology

Skin is the largest organ of the body accounting for 15% of total body weight. It performs several vital functions, including thermoregulation, providing a physical barrier against pathogens, synthesizing vitamin D, and preventing the body from excessive water loss (Zouboulis and Makrantonaki 2011).

The skin is made of three layers: epidermis, dermis, and hypodermis. Each layer has a different structure and cellular composition. The outermost layer, the epidermis, mainly contains keratinocyte cells, which synthesize keratin protein with a protective role. The middle layer, the dermis, constitutes collagen, a fibrillar protein that lies on the panniculus or subcutaneous tissue consisting of lobes of fat cells called lipocytes. The thickness of each layer depends upon the body region, such as the palm and soles having the thickest epidermis and the eyelid having the thinnest epidermis layer (Baroni et al. 2012).

6.3.1 Epidermis

The epidermis is a stratified epithelium layer mainly composed of keratinocytes and dendritic cells and also consists of other cells like Langerhans cells, melanocytes, and Merkel cells. The epidermis is divided into several structural and functional layers, stratum germinativum (basal cell layer), stratum (squamous cell layer), stratum granulosum (granular cell layer), and stratum corneum (horny or cornified cell layer) (Lai-Cheong and McGrath 2017).

Keratinocytes: Around 80% of cells in the epidermis are keratinocytes. Their appearance changes from one layer to another. It starts from the basal layer of the epidermis and migrates upward in the skin through the differentiation process, keratinization, in which keratinocytes go through a synthetic and degradative phase. The basal layer contains column-shaped keratinocytes that are attached to the basement membrane. Basal cells are clonogenic and can get affected by mutagenic chemicals, affecting the rate of cell division and cell proliferation machinery. Above the basal layer resides a 5–10 cell thick squamous cell layer, spinous cells differ in shape and structure based on their location. Basal spinous cells are polyhedral shape, while the upper layer cells become flattened as they are pushed upward (Choi and Lee 2015). The granular layer is composed of cells with a high amount of keratohyalin granules in the cytoplasm that are required for synthesizing and modifying proteins responsible for keratinization. The thickness of the granular layer depends upon overlying cornified cells. Above the fine layer are the horny cells that provide mechanical protection to underlying cells and a barrier against pathogens to prevent water loss. The cells are flattened and have lost their nuclei during terminal differentiation. It takes 14 days to reach the cell from the basal layer to the topmost layer. Keratinocytes protect against invading pathogens such as microbes, viruses, and fungi from UV radiation and minimize solute, heat, and water loss (Massoud and Rezaei 2014).

Melanocytes: Melanocytes are pigment-synthesizing cells originating from the neural crest cells and mainly located in the basal layer. Melanocytes produce melanin that gives color to the epidermis, hair, and iris. Melanocytes have a round membrane-bound organelle that synthesizes melanin, known as melanosome, and transfer it to keratinocytes (Cichorek et al. 2013). In white skin, melanosomes tend to be released slower in keratinocytes than in dark skin and the rate of degradation of melanosomes compared to dark skin. UV radiation increases the process of melanogenesis and transfer of melanosomes to keratinocytes, leading to skin tanning (Abdel-Malek et al. 2010).

Langerhans cells are derived from bone marrow and migrate to the epidermis during embryonic development. Langerhans cells are members of the dendritic cell/macrophage family. They are specialized to sense the environment and interpret the microenvironmental context to generate an appropriate immune response (inflammation or tolerance). Langerhans cells account for 2–8% of the total cell population in the epidermis (Otsuka et al. 2018).

6.3.2 Dermis

The dermis is a connective tissue layer that gives the skin elasticity and strength. The dermis is rich in nerves and vascular networks and has mast cells, macrophages, and leukocytes alongside the fibroblast cells.

The dermis sits between the two layers, the epidermis and subcutaneous tissue. It interacts with the epidermis at a dermo-epithelial junction that repairs and remodels the skin during wound healing. The dermis contains two layers: reticular dermis (bottom layer) and the papillary dermis (top layer). The reticular dermis is thick and contains blood vessels, nerves, glands, and fat cells. A network of elastin and collagen fibers surrounds the reticular dermis and provides skin with its overall structure and elasticity. The papillary dermis is thinner and contains fibroblast cells, fat cells, touch receptors, macrophages, and blood vessels. The papillary dermis interlocks with the basement membrane of the epidermis. The primary function of the dermis is to provide strength and elasticity, and the blood vessels transport nutrients to the epidermis. The nerve endings allow us to feel pain, pressure, heat, cold, and itchiness (Rippa et al. 2019).

Fibroblast: Fibroblasts are the most abundant cells in connective tissues. They contribute to the secretion of the extracellular matrix that maintains the structural integrity of connective tissues. Fibroblasts produce various products, such as collagen, proteoglycans, laminins, metalloproteinases, and fibronectin. Fibroblast regulates skin physiology and wound repair (Driskell and Watt 2015).

6.3.3 Hypodermis

The innermost layer of the skin is also known as the subcutaneous layer. Hypodermis sits directly beneath the dermis layer and above the other tissue, such as muscle and bone. Fibrous septa separate fat lobules or lipocytes. Hypodermis provides insulation, support, mechanical integrity, buoyancy, and as a storehouse of energy. Hypodermis produces a hormone called leptin, which helps regulate body weight through the hypothalamus (Gilaberte et al. 2016) (Fig. 6.2).

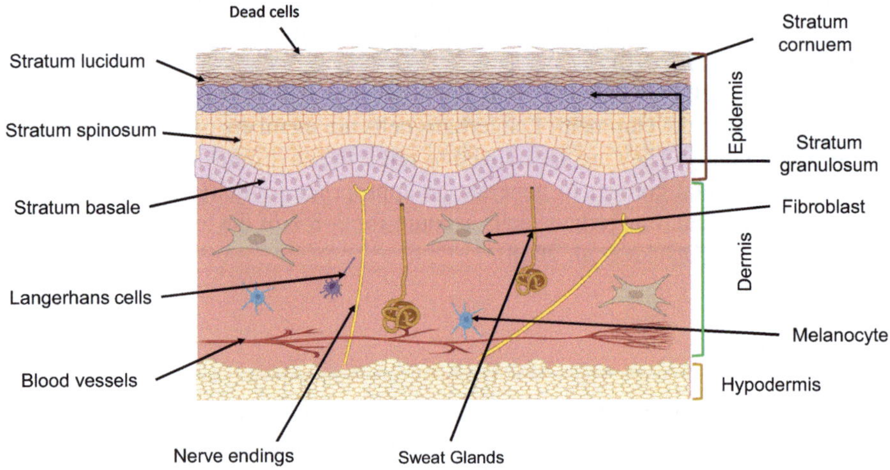

Fig. 6.2 Schematic view of human skin showing different layers of skin (Xiao et al. 2021)

6.4 Phototoxicity

Phototoxicity is defined as "*a toxic response from a substance applied to the body which is either elicited or increased (apparent at lower dose levels) after subsequent exposure to light, or that is induced by skin irradiation after systemic administration of a substance*" (Guidelines 2019). Skin is the largest organ with the body's surface area and is regularly exposed to several man-made products and environmental pollutants. Phototoxicity develops when chemicals come in contact with the skin and get activated by the sunlight, forming cytotoxic products in the skin cells. Skin irritation, erythema, and edema are phototoxicity symptoms similar to the symptoms of exaggerated sunburn. A variety of chemicals have the potential to induce phototoxicity because their structure probably contains benzene rings and heterocyclic rings that can absorb sunrays (Glatz and Hofbauer 2012).

6.4.1 Mechanism of Phototoxicity

A molecule/compound (photosensitizer) needs to absorb photons for a phototoxic reaction to occur. After absorbing the photon, the molecule promotes from its ground state to an excited state. This is the singlet/triplet state of the photosensitizer. The singlet/triplet state has higher energy than the ground state and depends on the spin state of the two electrons with the highest energy. When the two electrons have opposite spin, it is the singlet state, and when the electrons have the same spin, it is the triplet state. This singlet/triplet excited state is only stable for a short time; the triplet state is steady for 10–6 s, and the singlet state is stable only for 10–10 s. When the molecule returns from the excited state to the ground state, the energy absorbed is released by the emission of radiation, heat, or the formation of a photoproduct (chemical reaction) (Baier et al. 2006).

The mechanism of phototoxicity can be divided into direct and indirect modes of action. The direct mode of phototoxicity arises when a chemical absorbs light and gets converted into an excited state, and combines with a crucial cell constituent or transfers electron or hydrogen atoms. This transfer may convert into a toxic radical that may be toxic in subsequent reactions. At the same time, an indirect mode of phototoxicity arises when a chemical absorbs light, goes from an excited singlet state to a triplet state, and reacts with molecular oxygen to generate singlet excited oxygen. Sometimes, a chemical gets excited and transfers an electron to oxygen to create a superoxide anion (Kim et al. 2021).

The exact target of the cell's phototoxic reaction depends on the phototoxic agent's physical and chemical properties. Cosmetics applied topically are more likely to damage the keratinocytes in the epidermis layer, and drugs that are used systemically cause phototoxicity to the dermis layer (Glatz and Hofbauer 2012). The phototoxic reaction can damage cellular organelles. Lipophilic photosensitizers can quickly diffuse in the cell and destroy many cell organelles, such as mitochondria, nuclei, or lysosomes. Hydrophilic photosensitizers can damage the lipid bilayer. Direct damage to cell organelles causes the release of mediators such as histamine and eicosanoids and initiates the inflammatory response. Some cosmetic ingredients exposed under UVR lead to reactive oxygen species (ROS) generation, which causes the oxidation of cellular organelles and cell membranes. Cosmetics such as lipsticks, creams, lotions, and hair dye have different levels of toxicity due to having specific absorption spectra. P-phenylenediamine, a hair dye ingredient that is photosensitized, leads to apoptosis due to lysosome and mitochondria dysfunction (Goyal et al. 2015). Methylparaben, a preservative used in cosmetics when photosensitized, generates ROS and leads to cell death. Skin inflammation disease occurs when these photosensitized chemicals are used in excess (Dubey et al. 2017).

OECD officially approves 3 T3 neutral red uptake assay for phototoxicity testing. The test evaluates photocytotoxicity by determining the cell viability reduction after exposure to the test chemical in the absence and presence of UV/visible light irradiation. Criteria for any chemical for conducting a 3 T3 NRU test is that the chemical shows absorption in UV/visible region and dissolved in an appropriate solvent. Neutral red dye is a weak cationic dye that quickly penetrates the cell membrane and accumulates intracellularly in the lysosomes of viable cells. To check the phototoxicity potential of the test chemical, the photo-irritation factor (PIF) or mean photo-effect (MPE) is calculated. Photo-irritation factor is the ratio of IC50 (concentration at 50% decrease in cell viability) of non-radiated test chemical over irradiated test chemical. The mean photo-effect is the difference between the dark and light curves at arbitrary doses. PIF value less than 2 or MPE value less than 0.1 represents that the test chemical is non-phototoxic, PIF >2 and < 5 or MPE > 0.1 and < 0.15 predicts that the test chemical is probable phototoxic, and PIF over 5 or MPE > 0.15 represents that test chemical is phototoxic (Guidelines 2019).

6.5 Skin Models for Phototoxicity

Several models have been developed to assess the phototoxicity potential of cosmetic formulation. Earlier, in vivo models were used for phototoxicity testing, but testing was banned due to ethical concerns. In vitro models are currently being used, which can be divided into two based on the number of skin cell types.

6.5.1 2D Skin Models

Cell cultures have been used since the early 1990s by scientists. Cell culture refers to cells obtained from living tissues grown under a controlled environment. If grown in flat climates such as petri dish are called 2D cell cultures. Initially, keratinocytes present in the epidermis layer of the skin were used as the primary cell type for the cell culture. The growth of cells outside the organism's body requires specific nutrients and equipment to survive and divide. Cells are provided with complete media that contains amino acids, buffers, vitamins, and antibiotics (Breslin and

O'Driscoll 2013). Various types of 2D skin models have been developed, which use specific cell types to study the toxicity of chemicals in our environment.

6.5.1.1 NIH 3 T3 Cell Line

Primary fibroblast cell line recommended by OECD for in vitro phototoxic studies, established by scientists George Todaro and Howard Green in 1963, is derived from mouse embryonic fibroblasts. NIH 3 T3 cells are named after their "3-day transfer, inoculum 3 × 105 cells" culturing protocol. Cells are cultured in flat cell culture flask coated with poly-L lysine or collagen for adhesion (Littlefield 1982). Phototoxicity testing is based on the cytotoxicity of a chemical in the presence and absence of a non-cytotoxic dose of UVR/visible light. Cytotoxicity is expressed by dependent concentration reduction in uptake of neutral red dye. Cells are first incubated with test chemicals for an hour, and a half of the plates are exposed to a non-cytotoxic dose of UVR/visible light, and cells are incubated for 18–24 h. NRU determines cell viability. The NIH 3 T3 cell line is also used for various cosmetic ingredients' phototoxicity testing.

6.5.1.2 HaCaT Cells

HaCaT cell line is immortalized human keratinocytes cells and keratinocytes representing the significant cell type present in the epidermis, used as a screening tool for predicting the phototoxicity and irritation potential of surfactants, cosmetic ingredients, drugs, and herbal formulations. MTT and NRU assays are performed to test any phototoxic compound. HaCaT provides reproducible and reliable results (Shukla et al. 2022).

6.5.2 3D Skin Models

To overcome the limitations of 2D skin models for phototoxicity, researchers are investigating the application of a 3D reconstructed human epidermis model. The phototoxicity potential of any chemical is evaluated by comparing the viability of reconstructed human epidermis tissue (RHE tissue) with test chemicals in the presence and absence of UV/Vis radiation or sunlight. The significant advantage of 3D RHE tissue is that the test chemical is applied topically to the tissue (Tavares et al. 2020).

Hydrogel systems are the most dominant technique for creating 3D cultures. It serves as a scaffold for dermal fibroblast co-cultured with keratinocytes on the top. Generally, the hydrogel material is collagen I, but other extracellular matrix proteins can also be used. The hydrogel system allows the differentiation of keratinocytes into different layers using high Ca^{2+}, low temperature, and a unique protocol for air exposure. Cell types are generally primary keratinocytes and primary human dermal fibroblast, but HaCaT can also be used. Hydrogel models can be used to study the phototoxicity of cosmetics and other chemicals (Stanton et al. 2015).

3D skin models can be produced by 3D bioprinting; they can create different layers of the desired material in the form of hydrogel or biodegradable scaffold. Cells

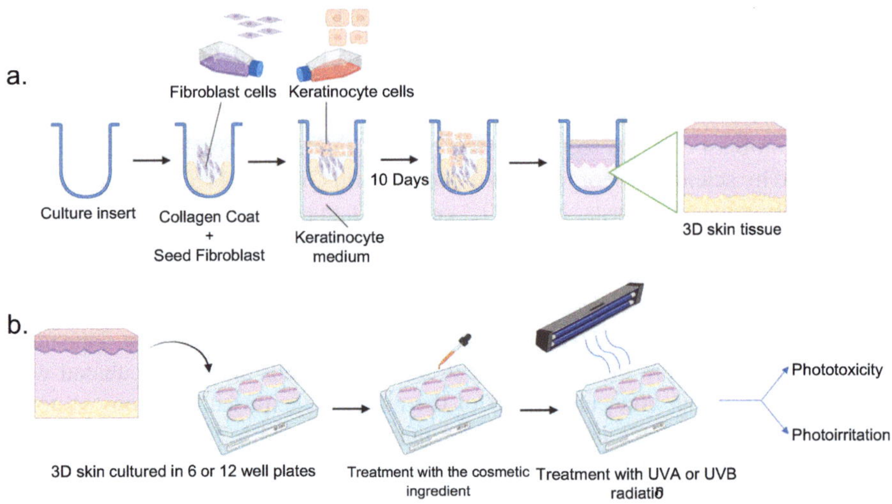

Fig. 6.3 3D skin model. (**a**) Development of 3D skin model. (**b**) 3D skin model used for phototoxicity assessment

or biomolecules can later be added to the scaffold to form biological structures (Murphy and Atala 2014). Bio-printed skin was generated, having complex systems with epidermal and dermal layers containing keratinocytes, fibroblasts, melanocytes, and collagen. 3D bioprinting is automated and has high reproducibility and throughput, but the cost of production is increased. 3D bio-print can be used to assess the cytotoxicity potential of cosmetic formulations and further for the phototoxic potential of cosmetics. Another method for 3D cell culture is a microfabricated system or cell on a chip that provides the required nutrients using microfluidics. It comprises a microfabricated cell culture of keratinocytes and fibroblasts with microfluidic channels. Three major models are currently used in phototoxicity testing (Fig. 6.3).

6.5.2.1 SkinEthic 3D Model

The SkinEthic model has a high similarity to the human epidermis. Layers of the epidermis, such as stratum corneum, stratum granulosum, and stratum spinosum, can be found in the 3D tissue. The model can be defined as "epidermis reconstituted by air lifted culture of normal human keratinocytes for 17 days in chemically inert polycarbonate filters." The number of layers of stratum corneum differs from the native skin issue. Major ceramides and their precursor, glucosylceramides, are present in the SkinEthic model. The general composition of lipids in the SkinEthic model is similar to native skin. In native skin tissue, the stratum corneum is maintained at constant thickness due to desquamation, but in the culture model, the stratum corneum becomes progressively thicker (Pellevoisin et al. 2018).

Phototoxicity testing was assessed by comparing the results with in vivo data. Several phototoxic and non-phototoxic compounds were tested with the model, and it could discriminate between phototoxic and non-phototoxic compounds. Leakage of LDH was used as a marker for the decrease in cell viability, and increased release of IL-8 and expression of IL-8 mRNA was used to quantify the phototoxicity of test compounds. It is shown that the SkinEthic model was working correctly.

6.5.2.2 EpiSkin 3D Model

EpiSkin model can be defined as a "type I bovine collagen matrix, representing the dermis, surfaced with a film of type IV human collagen, upon which is laid, after 13 days in culture, stratified differentiated epidermis derived from second passage human keratinocytes." EpiSkin has two models: the epic skin irritation model and the EpiSkin penetration model. The stratum corneum of the EpiSkin model has more layers than native skin. The organization of cells in the epidermis layer differs slightly from the native epidermis. The shape of cells in different layers is somewhat different. Lipid composition is close to that of the human epidermis. Phospholipids were comparatively less in the penetration model but almost the same in the irritation model. Precursors of ceramides and glycosphingolipids were also comparable to the human epidermis.

To examine the model for its phototoxicity testing, several known phototoxic compounds such as chlorpromazine, ofloxacin, and 6-methyl coumarin were used as controls. They were exposed to UVA at a non-cytotoxic dose for 1 h. MTT cell viability test was performed after incubation, and IL-1α release into the culture medium was quantified. Increased cell mortality and a rise in IL-1α release confirmed the model's ability to identify phototoxic compounds. Irritation and corrosivity testing was performed, and the results demonstrated that the model could distinguish between irritants/corrosive and non-irritant/non-corrosive chemicals (Lelièvre et al. 2007).

6.5.2.3 EpiDerm

EpiDerm was created by MatTek corporation and introduced in the market in 1993. The model can be described as "normal, human-derived epidermal keratinocytes (NHEK) which have been cultured to form a multilayered, highly differentiated model of the human epidermis." General morphology is similar to that of native epidermis tissue. The only difference is that the rete ridges present in the native tissue anchor the epidermis and dermis, which are absent due to being grown on polycarbonate filters. The cell shape is columnar to round and flattened in the stratum spinosum. Lipid content is similar to that of the regular human epidermis.

Phototoxicity testing of the EpiDerm skin model was done by topically applying test materials with five different concentrations and then exposed to a non-cytotoxic dose of UVA. In visible light, 1 day after the irradiation, cytotoxicity was determined by MTT assay, and the model was successfully able to differentiate between phototoxic and non-phototoxic compounds. Irritation testing was done using a formulation containing surfactants, and IL-1α mRNA levels determined the level of irritation. Results differ from human skin. The barrier function of the model is suboptimal compared to human skin. Therefore, the model can be used for screening possible irritants. Comparison studies of 22 cosmetic formulations for irritancy were done in vivo human skin and the EpiDerm model. The results indicated that the model could be used to assess the irritancy potential of cosmetic formulations. 3D skin models provide a more accurate representation of the environment of human skin (Cannon et al. 1994).

6.5.3 In Vitro 2D Vs 3D Models

2D cell culture differs from 3D cell culture with respect to cell morphology and other features. Primarily, 2D culture is grown in flat environments such as Petri plates, while the 3D cell culture is shaped into 3D spheroids using specialized conditions for culturing.

The significant difference between the two is that 2D cell culture does not represent the environment of in vivo processes. The single-cell type culture, such as fibroblast or keratinocytes, lacks the interaction, such as cell-to-cell or cell-to-extracellular matrix. At the same time, 3D cell culture gives a good representation of in vivo processes and provides a better understanding of cell-to-cell or cell-to-ECM interactions. 3D models also have gene expression capabilities, thus mimicking the in vivo process (Teimouri and Agu 2016).

The presence of stratum corneum in 3D models allows the test of cosmetics products or drugs to study skin barrier penetration. This also allows for the application of cosmetics directly onto the top layer (stratum corneum). At the same time, the 2D model lacks the barrier function, and only soluble compounds in an appropriate solvent can be used.

2D cell cultures are cheap, have ease of use, and a vast resource of scientific literature makes them more advantageous over 3D. 3D cell culture is more costly, harder to maintain, and not widely used worldwide. Another problem associated with 3D cell culture is the inability to remove cellular waste that can get accumulated in the culture. This problem can be solved by using a microfabricated cell culture that provides micro channels that enable the flow of nutrients and remove cellular waste (Kapałczyńska et al. 2018).

6.6 Conclusion

Solar radiation contains UV, visible light, and infrared radiation that interact with the skin. The cosmetic formulation consists of numerous ingredients that may be toxic to the skin. Some of the components may be photolabile. When these chemicals are exposed to sunlight, they become phototoxic. It is essential to check the safety of the chemicals in cosmetic formulations. Several models are being used to evaluate the phototoxicity potential of cosmetic ingredients. 2D models have been used for this purpose for several years. It is cheap, easy to maintain, and provides reliable results, but the major problem is that it does not represent the actual human skin. The skin has several layers, each having a specific function. To overcome the problem, 3D skin models are developed that mimic the human skin. Using a 3D skin model, we can also study the absorption and penetration of chemicals in the skin layers and how it affects the physiology of the skin. 3D models having several advantages are less popular than 2D models because of high maintenance and cost. Each skin model has its advantages and disadvantages. It depends on the study which model is best suited for it.

References

Abdel-Malek ZA, Kadekaro AL, Swope VB (2010) Stepping up melanocytes to the challenge of UV exposure. Pigment Cell Melanoma Res 23(2):171–186

Baier J et al (2006) Singlet oxygen generation by UVA light exposure of endogenous photosensitizers. Biophys J 91(4):1452–1459

Baroni A et al (2012) Structure and function of the epidermis related to barrier properties. Clin Dermatol 30(3):257–262. https://www.sciencedirect.com/science/article/pii/S0738081X11002112

Breslin S, O'Driscoll L (2013) Three-dimensional cell culture: the missing link in drug discovery. Drug Discov Today 18(5–6):240–249

Cannon CL et al (1994) New epidermal model for dermal irritancy testing. Toxicol in Vitro 8(4): 889–891. https://www.sciencedirect.com/science/article/pii/0887233394900957

Choi M, Lee C (2015) Immortalization of primary keratinocytes and its application to skin research. Biomol Ther 23(5):391–399

Cichorek M, Wachulska M, Stasiewicz A, Tymińska A (2013) Skin melanocytes: biology and development. Postepy Dermatol Alergol 30(1):30–41. https://doi.org/10.5114/pdia.2013.33376

Couteau C, Coiffard L (2010) Regulation no 1223/2009 on cosmetic products. Nouvelles Dermatologiques 29(5 PART 1):274–278

D'Orazio J, Jarrett S, Amaro-ortiz A, Scott T (2014) UV radiation and the skin. Int J Mol Sci 14(1): 12222

de Paula Corrêa M (2015) Solar ultraviolet radiation: properties, characteristics and amounts observed in Brazil and South America. An Bras Dermatol 90(3):297–310

Driskell RR, Watt FM (2015) Understanding fibroblast heterogeneity in the skin. Trends Cell Biol 25(2):92–99. https://www.sciencedirect.com/science/article/pii/S0962892414001779

Dubey D et al (2017) Photosensitized methyl paraben induces apoptosis via caspase dependent pathway under ambient UVB exposure in human skin cells. Food Chem Toxicol 108:171–185. https://doi.org/10.1016/j.fct.2017.07.056

Gallagher RP, Lee TK, Bajdik CD, Borugian M (2014) Ultraviolet radiation. Chronic Dis Can 29 Suppl 1:51–68

Gilaberte Y, Prieto-Torres L, Pastushenko I, Juarranz Á (2016) Nanoscience in dermatology anatomy and function of the skin. Elsevier, Amsterdam. https://doi.org/10.1016/B978-0-12-802926-8.00001-X

Glatz M, Hofbauer GFL (2012) Phototoxic and Photoallergic cutaneous drug reactions. Chem Immunol Allergy 97:167–179

Goyal S et al (2015) Involvement of Cathepsin B in mitochondrial apoptosis by P-Phenylenediamine under ambient UV radiation. J Hazard Mater 300:415–425. https://doi.org/10.1016/j.jhazmat.2015.07.032

Kapałczyńska M et al (2018) 2D and 3D cell cultures–a comparison of different types of cancer cell cultures. Arch Med Sci 14(4):910–919

Kim KB et al (2021) Current opinion on risk assessment of cosmetics. J Toxicol Environ Health B Crit Rev 24(4):137–161. https://doi.org/10.1080/10937404.2021.1907264

Lai-Cheong JE, McGrath JA (2017) Structure and function of skin, hair and nails. Medicine 45(6): 347–351. https://www.sciencedirect.com/science/article/pii/S1357303917300695

Lee YS et al (2017) Phototoxicity evaluation of pharmaceutical substances with a reactive oxygen species assay using ultraviolet a. Toxicol Res 33(1):43–48

Lelièvre D et al (2007) The Episkin Phototoxicity assay (EPA): development of an in vitro tiered strategy using 17 reference chemicals to predict phototoxic potency. Toxicol in Vitro 21(6): 977–995

Littlefield JW (1982) NIH 3T3 cell line. Science 218(4569):214–216. https://doi.org/10.1126/science.218.4569.214.c

Malis J et al (2007) Attenuation of UVA-induced damage to human keratinocytes by silymarin. J Dermatol Sci 46(1):21–30

Massoud A, Rezaei N (2014) Immunology of aging. Springer, Champ, pp 1–406

Mujtaba SF et al (2021) Oxidative-stress-induced cellular toxicity and Glycoxidation of biomolecules by cosmetic products under sunlight exposure. Antioxidants 10(7):1–10

Murphy SV, Atala A (2014) 3D bioprinting of tissues and organs. Nat Biotechnol 32(8):773–785. https://doi.org/10.1038/nbt.2958

OECD (2019) Test Guideline no. 432 in vitro 3T3 NRU phototoxicity test.

Otsuka M, Egawa G, Kabashima K (2018) Uncovering the mysteries of Langerhans cells, inflammatory dendritic epidermal cells, and monocyte-derived Langerhans cell-like cells in the epidermis. Front Immunol 9:1768

Pellevoisin C et al (2018) SkinEthic™ RHE for in vitro evaluation of skin irritation of medical device extracts. Toxicol in Vitro 50:418–425. https://www.sciencedirect.com/science/article/pii/S0887233318300092

Pfeifer GP (2021) Mechanisms of UV–induced mutations and skin cancer. Genome Instab Dis 1(3): 99. https://doi.org/10.1007/s42764-020-00009-8

Rippa AL, Kalabusheva EP, Vorotelyak EA (2019) Regeneration of dermis: scarring and cells involved. Cell 8(6):607

Shukla S et al (2022) Superoxide anion radical induced Phototoxicity of 2, 4, 5, 6-Tetraminopyrimidine sulfate via mitochondrial-mediated apoptosis in human skin keratinocytes at ambient UVR exposure. Food Chem Toxicol 164:112990

Stanton MM, Samitier J, Sánchez S (2015) Bioprinting of 3D hydrogels. Lab Chip 15(15): 3111–3115. https://doi.org/10.1039/C5LC90069G

Tavares RSN et al (2020) Fucoxanthin for topical administration, a phototoxic vs. Photoprotective potential in a tiered strategy assessed by in vitro methods. Antioxidants 9(4):328

Teimouri A, Agu R (2016) Provisional chapter 2D vs. 3D cell culture models for in vitro topical (dermatological) medication testing 2D vs. 3D cell culture models for topical (dermatological) medication testing. www.intechopen.com

Vinardell MP, Mitjans M (2017) Alternative methods to animal testing for the safety evaluation of cosmetic ingredients: an overview. Cosmetics 4(3)

Xiao Q et al (2021) Hydrogen sulfide in skin diseases: a novel mediator and therapeutic target. Oxidative Med Cell Longev 2021:1

Zouboulis CC, Makrantonaki E (2011) Clinical aspects and molecular diagnostics of skin aging. Clin Dermatol 29(1):3–14. https://www.sciencedirect.com/science/article/pii/S0738081X10001161

Cosmetic Ingredients: Various Efficacy-Based Testing Methods in the 3D Skin-Based Model Systems

7

Shruti Goyal, Deepti Bhat, S. Usha, Havisha M. H. Rao, Snehal Desai, and Shiv Poojan

Abstract

The 3D skin model is a highly physiological, three-dimensional cellular system of human epidermal keratinocytes (HEK) for in vitro investigations. It provides a great way to investigate various aspects of epithelial function and disease, particularly those connected to skin biology and toxicology. Over time, there has been a substantial evolution in cosmetic compositions. These are no longer just considered to be cosmetics. The misclassification of chemicals and skin corrosion seen in animal systems can also be avoided with the aid of skin models. Phototoxicity, percutaneous absorption and penetration, wound healing, and metabolism are further used. Cosmetics are a necessary component of human culture. Modern cosmetic science looks for naturally occurring cosmetics because the usage of cosmetics may pose health risks. The focus of cellular microbiology is on how bacteria and host cells interact. In the past, cutaneous defense against microorganisms that cause serious skin infections has been studied in skin research. There is a need to create fresh, experimental techniques to validate various efficacy tests based on worldwide standards. With the aid of new techniques, we could delve deeply into our microbial ecosystem and learn about the beautiful diversity of our microbiota. We also discuss numerous assays for the impact of natural substances in cosmetic formulations on acne, hair development, aging, skin rejuvenation, wound healing, and skin pigmentation.

S. Goyal · D. Bhat · S. Usha · H. M. H. Rao · S. Desai · S. Poojan (✉)
bioTox Solution Pvt. Ltd., Institute of Engineering and Technology, Lucknow, Uttar Pradesh, India

A. B. Pant et al. (eds.), *Skin 3-D Models and Cosmetics Toxicity*,
https://doi.org/10.1007/978-981-99-2804-0_7

7.1 Introduction

All tribes and civilizations have a long history of using cosmetics. In the prehistoric era, man used color to entice prey animals for hunting or to conceal the body for safety and to terrorize adversaries. Global cosmetics markets have developed in terms of quality, safety, and efficacy over time as humans have. Consumers increasingly expect cosmetic goods to provide additional benefits to improve skin health; they no longer just expect them to enhance beauty (Ribeiro et al. 2015). "A substance intended to be applied to the human body for cleansing, beautifying, boosting attractiveness, or altering the look" (Gagliardi and Dorato 2007) is how the Food and Drug Administration (FDA) defines a cosmetic. A cosmetic product is defined as "any substance or mixture intended to be placed in contact with the external parts of the human body or with the teeth and the mucous membranes of the oral cavity with a view exclusively or mainly to cleaning, perfuming, changing their appearance, protecting them, keeping them in good condition, or correcting body odors" (Pratiwi et al. 2022) (Article 2.1(a) of Regulation European Commission (EC) No 1223/2009). Consumer acceptability of cosmetics made with botanical components has increased. Furthermore, the demand for cosmetics made with herbs has expanded as consumer awareness has grown. The characteristics of several herbal compounds are favorable, including mildness, effectiveness, biodegradability, and low toxicity (Bonifácio et al. 2014). The herbal ingredients have many benefits for consumers, including delaying the aging process of the skin, preventing acne and UV ray damage, reducing hyperpigmentation, reducing wrinkles, improving hydration, improving skin elasticity and firmness, reducing scarring, reducing hair loss, treating dandruff, etc. For cosmetic goods, efficacy and safety are also crucial factors (Bonifácio et al. 2014). Cosmetic goods are not meant to treat dermatological conditions, and many regulatory organizations regulate cosmetics and pharmaceuticals differently. Some substances used in cosmetics, however, can help the skin heal. The term "cosmeceutical" is frequently used to refer to a class of cosmetics that also promotes skin health. In the early 1980s, medications that had a pharmacological therapeutic effect, but no biological therapeutic benefit were referred to as "cosmeceuticals." Atopic dermatitis, contact dermatitis, eczema, and other skin problems have all been studied in relation to the health benefits of cosmetic components. (Brandt et al. 2011).

The need to assess cosmetic goods for their impact on skin health is growing along with technological advancements, consumer expectations, and competitiveness. Systems using animal models offer a transferrable validation for efficacy-based claims. However, a number of businesses forego using animal models for validating the safety and efficacy of personal care products, either because of internal policy or legislative prohibitions on their usage (Ferreira et al. 2019). As a result, systems based on 3D models have emerged as useful tools for such validations. Due to their low cost, high-throughput adaptability, and ability to produce data much quicker than traditional in vivo techniques, 3D tissue testing systems have advantages over alternative techniques for efficacy validation. Many personal care products are intended for topical use only and do not include oral consumption. Therefore, issues like the components' metabolic conversion or their absorption through tight junctions like internal epithelia may not be considered by testing procedures for evaluating personal care products (Ferreira et al. 2019; Hua 2019). To validate topical components rather than those needing systemic absorption and metabolism, cell-based model systems are more valuable.

The ability to create tissues that closely resemble their in vivo counterparts is required to create human, three-dimensional (3D) models that will further understand the biology of skin keratinocytes. When cells are spatially arranged to exhibit the architectural elements present in vivo, the relationship between growth and differentiation in tissues is ideal; it is lost in two-dimensional culture systems. All through the twentieth century, testing on animals was a common practice (Ferdowsian and Gluck 2015). However, politicians in Europe rigorously regulated and ultimately outlawed the use of animals for testing cosmetics due to rising public concern about what happens to laboratory animals as well as ethical and scientific issues (Szymański et al. 2020). The laws governing animal testing have tightened up since the 1990s. 2013 saw the full implementation of the European Union's ban on animal testing of finished cosmetics and their ingredients, as well as the sale of such items.

In the past two decades, human skin equivalents have been successfully created in vitro, mostly by utilizing keratinocytes cultivated on a dermal replacement. Keratinocytes and fibroblasts are used to construct 3D skin models. The dermis is mirrored by fibroblasts, which are embedded in an extracellular matrix. The epidermis is then created by keratinocytes (Edmondson et al. 2014). Compared to two-dimensional cell culture, three-dimensional skin analogs more closely approximate the structure of human skin tissue. Every cell in a monolayer is in intimate touch with chemicals to examine the cells' reactions to them. This means that the amounts that cause irritations in monolayer cell assays may differ significantly from in vivo circumstances (Brohem et al. 2011).

Furthermore, numerous researches have contrasted cells developed in 2D versus 3D environments. Such cells have important phenotypic, cellular signaling, cell motility, and medication response variations (Mazzoleni et al. 2009). Cosmetics are not evenly absorbed in all sections of a 3D tissue-like skin model, and not all cells can directly interact with them. Therefore, 3D models of cellular reactions more closely resemble in vivo conditions. Evaluating the anti-inflammatory and antiaging effects of cosmetic ingredients, skin aging is a complicated biological process that is regulated by several intrinsic and extrinsic variables that alter the architecture of the skin, especially in areas exposed to the sun (Brohem et al. 2011; Ganceviciene et al. 2012).

Developing new chemicals with anti-inflammatory characteristics is an essential focus of development since inflammatory processes can significantly speed up skin aging (Estrada et al. 2012). An inflammation must first be produced in the skin model employing irritants to test novel chemicals. Ex vivo skin or skin equivalent-based Techniques for a test of efficacy Critical Parameters for Cell-Based and In Vitro Assays (Suhail et al. 2019). The solubility of the test material is a crucial factor that should be carefully considered when assessing the effectiveness of any cosmetic product. Since they are compatible with different cell culture mediums and buffer systems, aqueous soluble compounds do not present too much of a challenge. However, since organic solvents themselves might have certain unintended consequences in vitro systems, compounds soluble in them present some difficulties (Cevallos et al. 2017). For instance, it has been demonstrated that high quantities of DMSO, which is frequently used as a vehicle for dissolving test chemicals, have an impact on cell-based systems (Cevallos et al. 2017; Timm et al. 2013). Another important factor to consider before integrating components in efficacy-based validations is the cytotoxicity of the substances. In general, the component concentration for all cell-based assays should be non-toxic. This is crucial because cytotoxicity can have various impacts on cells that can skew the interpretation of the assay results. The assay interference caused by the test substances is another crucial factor to consider. Some test substances may interact with the assay in an unspecific way, which might result in false positive results (Ng et al. 2019; Tate and Ward 2004). For the following reasons, test chemicals interfere with assays: Test substance turbidity and color can obstruct assay signals (e.g., absorbance and fluorescence). The test chemicals may frequently display characteristics of redox cycling and metal ion chelation. The test material may demonstrate non-specific protein reactivity or create protein aggregates, which may be mistaken for an inhibition unique to a certain mechanism (Rajan et al. 2021). Therefore, selecting the right test controls is crucial for deciphering the assay results. We discuss skin similar model-based approaches to assess cosmetic chemicals for the following activities in the following sections. Consumer acceptability of cosmetic goods made with natural substances is rising, and numerous herbal compounds are being looked at in this regard. With this in mind, we have included examples of herbal substances in each section that have the specific biological activity needed for the efficacy of cosmetic products (Zappelli et al. 2016; Romes et al. 2021). The scope of this chapter does not allow for a full discussion of the pathophysiology and dermatological issues. This chapter primarily focuses on the numerous assay types that can be used to test the efficacy of cosmetic components in formulations (Table 7.1).

Table 7.1 Various methods and efficacy tests for cosmetic ingredients

International validation method	Standardized 3D in vitro tissue models
ECVAM/OECD TG 431 skin corrosion test ECVAM/OECD TG 439 skin irritation test COLGATE/IIVS eye irritation validation ECVAM/COLIPA eye irritation validation COLIPA genotoxicity pre-validation ECVAM phototoxicity pre-validation German skin penetration validation COLIPA/BMBF skin metabolism study Allergenicity studies Microbicide research Airway toxicity validation	Normal (non-transformed) human cell-derived tissues Highly reproducible, lot-to-lot and year-to-year Structurally, biochemically comparable to in vivo Mitotically and metabolically active No cross-species extrapolation errors

7.2 Industrial Cosmetic Ingredient Efficacy Solutions

- Antiaging
- Skin whitening
- UV protection
- Skin pigmentation
- Cosmetic evaluations
  - Skin irritation
  - Skin corrosion
  - Skin sensitization
- The skin microbiome test and efficacy

7.2.1 Antiaging

There are two causes of skin aging: intrinsic and extrinsic. Chronological aging that is brought on by genetic and hormonal factors is known as intrinsic aging. On the other hand, challenges to the skin from outside sources including UV radiation, air pollution, and nutritional variables result in extrinsic aging (Zhang and Duan 2018; Bocheva et al. 2022; Pincemail and Meziane 2022). The phenotypic traits of these two aging processes differ slightly. The scientific community is aware of the importance of the interaction between external influences and the ensuing internal biological reaction by organisms. To explain this phenomenon, the term "skin aging exposome" has been coined. This phrase refers to all of the exposures that a person experiences from conception to death (Krutmann et al. 2017). It considers how internal and environmental factors interact to cause biological and clinical aging indications. According to the literature, environmental and dietary factors influence the expression of inflammatory cytokines, reactive oxygen species (ROS), and matrix metalloproteases (MMPs), which result in the onset of early indications of skin aging (Liguori et al. 2018; Hajam et al. 2022). Additionally, dietary supplements and particulate matter stimulate the aryl-hydrocarbon receptor (AhR) in keratinocytes, which promotes early aging and compromises the integrity of the skin. The antioxidant and anti-inflammatory activities, capacity to affect the expression levels of genes involved in aging, effect on collagenase, and elastase and hyaluronidase activities are the basis for the cell-based assays used to assess the skin aging and skin rejuvenation potential of cosmetic ingredients (Pittayapruek et al. 2016). These platforms are further described in the sections that follow. Aloe vera, edelweiss, Withania somnifera, Triphala, Ginkgo biloba, Curcuma longa, and Centella asiatica are a few examples of high herbs in flavonoids and have anti-inflammatory and antioxidant characteristics (Papakonstantinou et al. 2012).

7.2.2 Skin Whitening

A study on skin whitening revealed that repeated UV exposure could cause pigment development to last for a very long time. Additionally, numerous investigations found that pigmentation patches persisted at increased levels through day 13. Due to the equipment's inability to distinguish between different colors, clinical subjective evaluations and combined objective instrument measurements were still required to evaluate a material's capacity for whitening and spot removal. The effectiveness of whitening or spot removal products can be evaluated using this pigmented spot model. Additionally, combining subjective and objective approaches may be a wise reference when evaluating the effectiveness of whitening solutions (Saleem et al. 2020).

7.2.3 UV Protection

Test subjects are exposed to UV protection either in explant settings or using the RHE model. In many ways, the antiquated concept of UV burn where UVB rays were thought to harm is outdated. The hazards associated with UVA and UVB are now well known (Couteau and Coiffard 2016; D'Orazio et al. 2013; Mohania et al. 2017). The light or sun protection factor refers to the ability to prevent sunburn, often called medically "erythema," and other direct and obvious UV-related skin damage. The SPF level is calculated in accordance with ISO 24444:2019. The in vivo test, now the gold standard for assessing sunscreen products, can conclusively measure the UV protection factor. What factors must be considered and how is this kind of examination conducted? An ISO 24444 method guarantees that specific prerequisites and parameters are upheld so that findings may be compared regardless of the laboratory or location where they are measured (Juzeniene and Moan 2012). The amount to be applied, the kind of radiation, the UV spectrum, the skin types of the test subjects, and even the ambient temperature in the testing area are among these aspects. In this laboratory test without using a test subject as a biological test base, there is also a standard here for defining the level of protection against skin damage caused by UVA irradiation. ISO 24443:2021 regulates how UVA protection factors are measured. According to the recommendation of the European Commission, UVA protection must represent at least one-third of the sun protection factor (Majeed et al. 2020; Union 2009).

7.2.4 Skin Pigmentation

Melanin, a natural pigment, is produced in the body and distributed throughout the skin and hair follicles, which results in the pigmentation of human skin. Tyrosinase, TYRP1, and dopachrome tautomerase play a major role in the enzymatic process of melanin formation, also known as melanogenesis (DCT). Tyrosine can be biotransformed into melanin inside melanocytes, specifically in lysosomal vesicles called melanosomes (Moreiras et al. 2021; Cichorek et al. 2013; Kang et al. 2020; Lajis et al. 2012). A microtubule network transports mature melanosomes with melanin to the melanocyte dendritic extremities, where they are transmitted to keratinocytes and gives skin pigmentation and photoprotection. Melanin is crucial for the evenness of skin tone, hair pigmentation, and photo-defense against UV-induced DNA damage. Some internal causes (inflammatory or hormonal reactions) or environmental factors can change the production of melanin (sun exposure) (D'Mello et al. 2016). Age spots, melasma, acne lesions, and other hyper-pigmented marks can result from this altered melanin production and need to be treated with skincare products. An active ingredient or cosmetic compound's pro-pigmenting or depigmenting (whitening/lightening/anti-spot) actions can be assessed using a panel of complementary assays (Yamaguchi and Hearing 2009).

7.3 Skin and Eye Irritation Assessment of Cosmetic Products

An essential component of evaluating the safety of cosmetic ingredient formulations is assessing the risk for skin and eye irritation. According to OECD 404 (Kim et al. 2018), dermal irritation is the development of reversible skin injury after applying a test material for up to 4 h. According to the OECD (405) (Vinardell and Mitjans 2017), eye irritation is the development of changes in the eye after the application of a test material to the anterior surface of the eye. It is totally reversible within 21 days of administration. Using test methods based on rebuilt human tissue, skin and ocular irritation are evaluated. Commercially accessible 3D models with human epidermis reconstructions (RhE) are used for evaluating skin irritation (OECD Test Method 439) (Cosmetics n.d.; FDA 2021; ECETOC 1990; OECD 1996, 2000) and 3D models with human cornea-like epithelium reconstructions (RhCE) are used for testing eye irritation (OECD Test Method 492) (Takahashi et al. 2009, 2011). Other in vitro models address substances that do not cause classification for eye irritation or severe eye injury; it should be mentioned (Kojima et al. 2013; Barthe et al. 2021). However, we shall just pay attention to the RhCE model. However, using reconstructed human tissues, we create 3D models that replicate the biochemical and physiological characteristics of the higher layers of the human skin and eye. Living human keratinocytes have been cultivated to create a multilayered, highly differentiated epidermis for the RHE skin model (SCCS (Scientific Committee on Consumer Safety) 2021). The model contains a functional skin barrier with an in vivo-like lipid profile and well-structured basal cells. RhCE is a corneal model made of human live cells grown to develop a differentiated, multilayered corneal epithelium. Similar to the typical human in vivo corneal epithelium, the model comprises highly structured basal cells that gradually flatten out as the apical surface of the tissue is approached (Pistollato et al. 2021; Kim et al. 2015).

The cells in both models are metabolically and mitotically active, and they also release a number of cytokines known to play a key role in irritation and inflammation (Kaluzhny et al. 2015). At the air–liquid interface, reconstructed human tissues are developed on specialized platforms (Kolle et al. 2011; Lock-Andersen et al. 1997;

Clippinger et al. 2021). The test item is directly put to the tissue surface, simulating "real life" exposure well. The reduction of MTT (3-(4,5)-dimethyl-2-thiazolyl-2,5-dimethyl-2H-tetrazolium bromide) by cells into a blue formazan salt, which is quantitatively evaluated following extraction from tissues, is the endpoint utilized in both the RhCE test procedures. Interleukin-1 (IL-1) production measurement is a second endpoint that can be utilized to boost sensitivity (Clippinger et al. 2021; Jensen and Teng 2020; Breslin and O'Driscoll 2013; Bos and Meinardi 2000). The test item is categorized as non-irritant if the viability is greater than 50% (RhE) or 60% (RhCE) (no-label or UN GHS No Category). The test item is categorized as irritant if the viability in the instance of the model is less than or equal to 50% (UN GHS Category 2). This cannot be predicted if the viability is less than or equal to 60%, and more testing may be necessary (Jensen and Teng 2020). No single in vitro test or testing array has been authorized as a stand-alone replacement for the in vivo test. New test systems are being developed using stem cells. These could lead to new possibilities for researching in vitro ocular toxicity (Breslin and O'Driscoll 2013). Cosmetic evaluations of substances enters the body through the skin, it is said to have undergone dermal absorption. Human skin has a thickness range of 200 um to 400 um, which helps with skin absorption. The epidermal membrane is manufactured by chemical and heat separation (60 °C for 1–2 min). Through its ability to prevent molecules from entering and leaving the skin, the stratum corneum acts as a barrier (Bos and Meinardi 2000). The skin's bottom layers are shielded by this barrier. A radiolabeled test material is put to the surface of a skin sample dividing the two chambers of a diffusion cell as part of an in vitro skin integrity test to ensure an intact stratum corneum is kept throughout skin preparation. Solid applications range from 1 to 5 mg/cm^2. To ensure that there is continuous passive diffusion of substances, the chemical remains on the skin for 24 h at 32°. At various times during the experiment, the receptor fluid is taken and examined for the test substances and/or metabolites. One major drawback is that although it has been demonstrated that some compounds are metabolized by the skin after percutaneous absorption, the metabolites of test substances can still be measured (Benigni et al. 2018).

7.4 Safety Testing Skin Corrosion

The OECD 431: In vitro Skin, Corrosion 3D-Model Test guideline evaluates a substance's likelihood of corroding skin. Cell viability is measured after the test agent is applied to the 3D model's reconstructed epidermis and functional stratum corneum, which simulate human skin. The test is based on the idea that corrosive materials can enter the cell, and because they are cytotoxic to the deeper layers, lower cell viability below a predetermined threshold. This makes it possible to identify and categorize corrosive substances, which are crucial for ensuring the secure handling, packaging, and delivery of chemicals, insecticides, and cosmetics that may harm skin tissue that causes the production of irreversible damage to the skin (Zhang et al. 2021; Gorzelanny et al. 2020; Riss et al. 2004).

7.5 Skin Sensitization

In order to identify skin sensitizer hazards, a number of defined methods (DAs) have been presented. Their constituent parts, data integration processes (DIPs), and performances have been compiled in Ref. (Worth et al. 1998). Importantly, based on the empirical data from this article, the proposed DAs' accuracy for forecasting human skin sensitization dangers, which ranged from 75.6% to 85.0%, was better than the LLNA's (74.2%). In addition to the OECD-adopted assays currently in use, a number of additional and novel assays are being validated and adapted as official TGs (Kolle et al. 2011), some of which have predictive performances comparable to those of the proposed DAs even when viewed as stand-alone assays (Lock-Andersen et al. 1997; Worth et al. 1998).

Testing for skin sensitization is therefore a shifting aim. The Scientific Committee on Consumer Safety (SCCS) publishes "Notes of Guidance for the Testing of Cosmetic Ingredients and Their Safety Evaluation" (Fentem et al. 1998; EU 2000) to provide guidance to testing and safety evaluation for the cosmetic industry, ensuring that testing can be carried out in compliance with EU cosmetic laws.

Despite the advancements noted above, further work is still required to overcome some drawbacks of the NAM-based approaches now in use. For instance, it has been acknowledged that some chemicals relevant to the cosmetic industry may be challenging to evaluate using standard OECD-certified assays (Kleinstreuer et al. 2018). As far as is known, these restrictions are listed in each TG and may apply to testing hydrophobic compounds, pre–pro-haptens, and complex substances, such as natural extracts, where the constituent of concern is frequently present in tiny amounts within a complex mixture.

The Genomic Allergen Rapid Detection (GARD) assay (Kleinstreuer et al. 2018; Andres et al. 2020), based on the measurements of a biomarker signature of genes associated with immunologically relevant pathways to the sensitization process, has shown promise to address some of these limitations. New state-of-the-art scientific methods are currently in the OECD Test Guideline Program (TGP) and are under evaluation for official TG adaption (Kumar et al. 2014). A protocol is also available for testing solid materials, such as medical devices, using both polar and nonpolar extraction vehicles in accordance with ISO-10993:12 (Ashikaga et al. 2010). For instance, the GARD assay is compatible with various solvents that can be used to increase the bioavailability of a test item.

These discoveries may also prove useful for cosmetic-related test items, such as UVCBs or natural extracts, which have a restricted solubility in DMSO or water, two common assay solvents. To overcome some of the solubility restrictions, many 3D models based on the reconstructed human epidermis (RHE) have also been created (Liu et al. 2021). While some assays are less transparent, most offer a well-defined readout of well-established biomarkers (such as IL-18). When investigating a small number of "difficult-to-test" substances in comparison to human reference data, the majority of the RHE-based assays demonstrated similar or marginally improved performances (depending on the specific RHE-assay) to the best performing OECD-validated assay, the h-CLAT assay (Strickland et al. 2022), demonstrating that such assays may constitute a viable source of information within a weight-of-evidence framework.

The most glaring drawback of the current OECD-validated assays is probably that they have only been validated for skin sensitization hazard identification and not for assessment of sensitizing potency, which is an essential part of risk assessment of cosmetic ingredients when used in consumer products. This is in addition to the assays' limited applicability domain. Skin sensitivity is a threshold phenomenon, and a quantitative risk assessment (QRA) of each constituent seeks to identify the highest dosage of the chemical that does not cause sensitivity (also known as the NESIL value) (Gilmour et al. 2019; Api et al. 2020).

The general method for QRA has been described for fragrances (Gilmour et al. 2022), and its applicability to general cosmetic ingredients is currently under discussion. It involves continuously predicting skin-sensitizing potency as a point of departure (POD), which is then adjusted by applying uncertainty factors. The development of NAM-based strategies for continuous assessment of skin-sensitizing potency for use as the point of departure in the QRA is ongoing. Examples include the recently proposed GARDskin dose–response model and the DA-based artificial neural network model for predicting LLNA EC3 (Gradin et al. 2021b; Barthe et al. 2021).

Last but not least, as new NAM-based methods for evaluating cosmetic ingredients are developed to replace conventional animal models, the ultimate arbiter of these tests' ability to protect human health must be evaluated based on how well they correlate with trustworthy data on the skin-sensitizing activity of chemicals in humans, and not how well they recapitulate the flaws of the "gold" standard animal tests, irrespective of their historical consideration as valid and adapted OECD methods. Preclinical testing of cosmetic ingredients using the NAM strategies outlined above is a crucial and important first step in ensuring the safety profile of cosmetics for chemicals with as-yet-unknown sensitization potential. Still, as stated in (Amerongen et al. 2021), post-market surveillance, also known as cosmetovigilance, will continue to play a significant role in ensuring the use of cosmetic ingredients and their concentration in formulated products remain safe.

7.6 3D Skin Model on the Skin Microbiome Cosmetic Ingredients Test and Efficacy

Cellular microbiology focuses on how host cells and bacteria interact. In terms of research on the skin, this has focused on studying the cutaneous defense against microorganisms that cause serious skin infections. When working with skin-derived microorganisms, it is essential to remember that the bacteria's internal functions will change when they are removed from their natural environment and placed in conventional culture media (Gradin et al. 2021a; Chen et al. 2018). Van der Krieken et al. created a human stratum corneum in vitro model for bacterial growth to study skin-derived microorganisms and prevent physiological stress associated with environmental changes. This rather straightforward model comprises a 2% layer of agar in a 24-well plate and a 2% suspension of calluses from the heels of healthy volunteers. Different bacteria can be injected to the top surface of dead corneocytes after the top callus layer has dried. As opposed to using ordinary agar plates, this skin stratum corneum model has the benefit of allowing the study of the growth rates of various skin microorganisms under various more precise settings (Jung et al. 2022; Suzuki et al. 2021; Egert and Simmering 2016; Tomic-Canic et al. 2014). Additionally, the authors discussed several benefits of this model over its three-dimensional (3D) skin equivalents. For instance, bacteria can be cultured on a dry surface at 32 °C, which more closely resembles the in vivo environment than the standard 37 °C incubation temperature for 3D skin models. The stratum corneum model does not need any expensive equipment to set up and is rather quick and simple (Mikolajczyk and Roesner 2019; Grogan et al. 2019). The lack of functional keratinocytes in this model is a drawback since it prevents research on how the microbiota and keratinocytes interact. However, it was shown that the stratum corneum model was helpful for analyzing the interactions between components of the usual abundant microbiota like Staphylococcus epidermidis and Propionibacterium acnes and likely skin pathogens like Staphylococcus aureus and Pseudomonas aeruginosa (Castelino et al. 2017). More clearly, the scientists showed that this methodology could be used to culture an entire in vivo skin microbiota obtained via skin swabbing. Remember that over the course of the culture's 7 days, the injected microbiota's diversity and composition remained unchanged (Grogan et al. 2019; Castelino et al. 2017; Rademacher et al. 2018).

To acquire more individualized disease-oriented insight into the influence of the microbiota on human skin biology, it may be helpful to employ cells with suppressed or improved expression of different genes of interest in 3D skin models and other cell types. In this regard, it has already been successfully established to use 3D skin models to assess the effectiveness of RNAi-based topical treatments (Van Gele et al. 2011; Grzymajlo 2022; Emmert et al. 2020; Rodrigues 2017). It makes sense that most published research on host-microbiota interactions using 3D models has concentrated on the impact of specific microbiota species (Muszer et al. 2015; Di Domizio et al. 2016). Therefore, it is necessary to create and advance experimental designs for the colonization of 3D skin models with thorough methodologies to validate efficacy tests to more effectively validate cosmetic ingredients to simulate the in vivo skin condition more nearly. This could make the 3D skin models a go-to alternative to animal models in subsequent cutaneous research.

References

Amerongen CCA, Ofenloch RF, Cazzaniga S, Elsner P, Gonçalo M, Naldi L, Svensson Å, Bruze M, Schuttelaar MLA (2021) Skin exposure to scented products used in daily life and fragrance contact allergy in the European general population–The EDEN Fragrance Study. Contact Dermatitis 84:385–394. https://doi.org/10.1111/cod.13807

Andres E, Barry M, Hundt A et al (2020) A new prediction model for distinguishing skin sensitisers based on IL-18 release from reconstructed epidermis: enhancing the assessment of a key event in the skin sensitisation adverse outcome pathway. J Dermat Cosmetol 4(5):123–137. https://doi.org/10.15406/jdc.2020.04.00164

Api AM, Basketter D, Bridges J, Cadby P, Ellis G, Gilmour N, Greim H, Griem P, Kern P, Khaiat A, O'Brien J, Rustemeyer T, Ryan C, Safford B, Smith B, Vey M, White IR (2020) Updating exposure assessment for skin sensitization quantitative risk assessment for fragrance materials. Regul Toxicol Pharmacol 118:104805. https://doi.org/10.1016/j.yrtph.2020.104805. Epub 2020 Oct 17

Ashikaga T, Sakaguchi H, Sono S, Kosaka N, Ishikawa M, Nukada Y, Miyazawa M, Ito Y, Nishiyama N, Itagaki H (2010) A comparative evaluation of in vitro skin sensitisation tests: the human cell-line activation test (h-CLAT) versus the local lymph node assay (LLNA). Altern Lab Anim 38(4):275–284. https://doi.org/10.1177/026119291003800403

Barthe M, Bavoux C, Finot F, Mouche I, Cuceu-Petrenci C, Forreryd A, Chérouvrier Hansson A, Johansson H, Lemkine GF, Thénot JP, Osman-Ponchet H (2021) Safety testing of cosmetic products: overview of established methods and New Approach Methodologies (NAMs). Cosmetics 8:50. https://doi.org/10.3390/cosmetics8020050

Benigni M, Pescina S, Grimaudo MA, Padula C, Santi P, Nicoli S (2018) Development of microemulsions of suitable viscosity for cyclosporine skin delivery. Int J Pharm 545(1–2): 197–205. https://doi.org/10.1016/j.ijpharm.2018.04.049. Epub 2018 Apr 23

Bocheva G, Slominski RM, Janjetovic Z, Kim TK, Böhm M, Steinbrink K, Reiter RJ, Kleszczyński K, Slominski AT (2022) Protective role of melatonin and its metabolites in skin aging. Int J Mol Sci 23(3):1238. https://doi.org/10.3390/ijms23031238. PMID: 35163162; PMCID: PMC8835651

Bonifácio BV, Silva PB, Ramos MA, Negri KM, Bauab TM, Chorilli M (2014) Nanotechnology-based drug delivery systems and herbal medicines: a review. Int J Nanomedicine 9:1–15. https://doi.org/10.2147/IJN.S52634. Epub 2013 Dec 9. PMID: 24363556; PMCID: PMC3862741

Bos JD, Meinardi MM (2000) The 500 Dalton rule for the skin penetration of chemical compounds and drugs. Exp Dermatol 9(3):165–169. https://doi.org/10.1034/j.1600-0625.2000.009003165.x

Brandt FS, Cazzaniga A, Hann M (2011) Cosmeceuticals: current trends and market analysis. Semin Cutan Med Surg 30(3):141–143. https://doi.org/10.1016/j.sder.2011.05.006

Breslin S, O'Driscoll L (2013) Three-dimensional cell culture: the missing link in drug discovery. Drug Discov Today 18:240–249. https://doi.org/10.1016/j.drudis.2012.10.003

Brohem CA, Cardeal LB, Tiago M, Soengas MS, Barros SB, Maria-Engler SS (2011) Artificial skin in perspective: concepts and applications. Pigment Cell Melanoma Res 24(1):35–50. https://doi.org/10.1111/j.1755-148X.2010.00786.x. Epub 2010 Nov 9. PMID: 21029393; PMCID: PMC3021617

Castelino M, Eyre S, Moat J, Fox G, Martin P, Ho P, Upton M, Barton A (2017) Optimisation of methods for bacterial skin microbiome investigation: primer selection and comparison of the 454 versus MiSeq platform. BMC Microbiol 17(1):23. https://doi.org/10.1186/s12866-017-0927-4. PMID: 28109256; PMCID: PMC5251215

Cevallos AM, Herrera J, López-Villaseñor I, Hernández R (2017) Differential effects of two widely used solvents, DMSO and ethanol, on the growth and recovery of *Trypanosoma cruzi* epimastigotes in culture. Korean J Parasitol 55(1):81–84. https://doi.org/10.3347/kjp.2017.55.1.81. Epub 2017 Feb 28. PMID: 28285511; PMCID: PMC5365264

Chen YE, Fischbach MA, Belkaid Y (2018) Skin microbiota-host interactions. Nature 553(7689): 427–436. https://doi.org/10.1038/nature25177. Erratum in: Nature. 2018 Mar 21; 555 (7697): 543. PMID: 29364286; PMCID: PMC6075667

Cichorek M, Wachulska M, Stasiewicz A, Tymińska A (2013) Skin melanocytes: biology and development. Postepy Dermatol Alergol 30(1):30–41. https://doi.org/10.5114/pdia.2013.33376. Epub 2013 Feb 20. PMID: 24278043; PMCID: PMC3834696

Clippinger AJ, Raabe HA, Allen DG, Choksi NY, van der Zalm AJ, Kleinstreuer NC, Barroso J, Lowit AB (2021) Human-relevant approaches to assess eye corrosion/irritation potential of agrochemical formulations. Cutan Ocul Toxicol 40(2):145–167. https://doi.org/10.1080/15569527.2021.1910291. Epub 2021 Apr 20

Cosmetics (n.d.). https://cdsco.gov.in/opencms/opencms/en/Cosmetics/cosmetics/

Couteau C, Coiffard L (2016) Overview of skin whitening agents: drugs and cosmetic products. Cosmetics 3:27. https://doi.org/10.3390/cosmetics3030027

D'Mello SAN, Finlay GJ, Baguley BC, Askarian-Amiri ME (2016) Signaling pathways in melanogenesis. Int J Mol Sci 17:1144

Di Domizio J, Pagnoni A, Huber M, Hohl D, Gilliet M (2016) Le microbiote cutané: le poids lourd sort de l'ombre [The skin microbiota: a colossus steps into the spotlight]. Rev Med Suisse 12(512):660–664

D'Orazio J, Jarrett S, Amaro-Ortiz A, Scott T (2013) UV radiation and the skin. Int J Mol Sci 14(6): 12222–12248. https://doi.org/10.3390/ijms140612222. PMID: 23749111; PMCID: PMC3709783

ECETOC (1990) Monograph no. 15, "Skin Irritation", European Chemical Industry, Ecology and Toxicology Centre, Brussels

Edmondson R, Broglie JJ, Adcock AF, Yang L (2014) Three-dimensional cell culture systems and their applications in drug discovery and cell-based biosensors. Assay Drug Dev Technol 12(4): 207–218. https://doi.org/10.1089/adt.2014.573. PMID: 24831787; PMCID: PMC4026212

Egert M, Simmering R (2016) The microbiota of the human skin. Adv Exp Med Biol 902:61–81. https://doi.org/10.1007/978-3-319-31248-4_5

Emmert H, Rademacher F, Gläser R, Harder J (2020) Skin microbiota analysis in human 3D skin models-"free your mice". Exp Dermatol 29(11):1133–1139. https://doi.org/10.1111/exd.14164. Epub 2020 Aug 27

Estrada R, Chavez-Lopez G, Estrada-Chavez G, Paredes-Solis S (2012) Specialized dermatological care for marginalized populations and education at the primary care level: is community dermatology a feasible proposal? Int J Dermatol 51:1345–1350

EU (2000). Official Journal of the European Communities L136/91 of 8 June 2000, Method B.40 Skin Corrosion

FDA (2021) Importing cosmetics. https://www.fda.gov/industry/importing-fda-regulated-products/importingcosmetics#cosmetic

Fentem JH, Archer GEB, Balls M, Botham PA, Curren RD, Earl LK, Edsail DJ, Holzhutter HG, Liebsch M (1998) The ECVAM international validation study on in vitro tests for skin corrosivity. 2. Results and evaluation by the management team. Toxicol in Vitro 12:483–524

Ferdowsian HR, Gluck JP (2015) The ethical challenges of animal research. Camb Q Healthc Ethics 24(4):391–406. https://doi.org/10.1017/S0963180115000067

Ferreira GS, Veening-Griffioen DH, Boon WPC, Moors EHM, Gispen-de Wied CC, Schellekens H, van Meer PJK (2019) A standardised framework to identify optimal animal models for efficacy assessment in drug development. PLoS One 14(6):e0218014. https://doi.org/10.1371/journal.pone.0218014. Erratum in: PLoS One. 2019; 14(7): e0220325. PMID: 31194784; PMCID: PMC6563989

Gagliardi L, Dorato S (2007) 1.1—General Concepts. In: Current legislation on cosmetics in different countries, analysis of cosmetic products. Elsevier, Amsterdam, pp 3–28

Ganceviciene R, Liakou AI, Theodoridis A, Makrantonaki E, Zouboulis CC (2012) Skin anti-aging strategies. Dermatoendocrinol 4(3):308–319. https://doi.org/10.4161/derm.22804. PMID: 23467476; PMCID: PMC3583892

Gilmour N, Kimber I, Williams J, Maxwell G (2019) Skin sensitization: Uncertainties, challenges, and opportunities for improved risk assessment. Contact Dermatitis 80(3):195–200. https://doi.org/10.1111/cod.13167. Epub 2018 Dec 7. PMID: 30525211; PMCID: PMC6587935

Gilmour N, Reynolds J, Przybylak K, Aleksic M, Aptula N, Baltazar MT, Cubberley R, Rajagopal R, Reynolds G, Spriggs S, Thorpe C, Windebank S, Maxwell G (2022) Next generation risk assessment for skin allergy: decision making using new approach methodologies. Regul Toxicol Pharmacol 131:105159. https://doi.org/10.1016/j.yrtph.2022.105159. Epub 2022 Mar 17

Gorzelanny C, Mess C, Schneider SW, Huck V, Brandner JM (2020) Skin barriers in dermal drug delivery: which barriers have to be overcome and how can we measure them? Pharmaceutics 12(7):684. https://doi.org/10.3390/pharmaceutics12070684. PMID: 32698388; PMCID: PMC7407329

Gradin R, Forreryd A, Mattson U, Jerre A, Johansson H (2021a) Quantitative assessment of sensitizing potency using a dose-response adaptation of GARDskin. Sci Rep 11(1):18904. https://doi.org/10.1038/s41598-021-98247-7. PMID: 34556744; PMCID: PMC8460622

Gradin R, Forreryd A, Mattson U et al (2021b) Quantitative assessment of sensitizing potency using a dose–response adaptation of GARDskin. Sci Rep 11:18904. https://doi.org/10.1038/s41598-021-98247-7

Grogan MD, Bartow-McKenney C, Flowers L, Knight SAB, Uberoi A, Grice EA (2019) Research techniques made simple: profiling the skin microbiota. J Invest Dermatol 139(4):747–752.e1. https://doi.org/10.1016/j.jid.2019.01.024. PMID: 30904077; PMCID: PMC7439564

Grzymajlo K (2022) The game for three: Salmonella–host–microbiota interaction models. Front Microbiol 13:854112

Hajam YA, Rani R, Ganie SY, Sheikh TA, Javaid D, Qadri SS, Pramodh S, Alsulimani A, Alkhanani MF, Harakeh S, Hussain A, Haque S, Reshi MS (2022) Oxidative stress in human pathology and aging: molecular mechanisms and perspectives. Cells 11(3):552. https://doi.org/10.3390/cells11030552. PMID: 35159361; PMCID: PMC8833991

Hua S (2019) Physiological and pharmaceutical considerations for rectal drug formulations. Front Pharmacol 10:1196. https://doi.org/10.3389/fphar.2019.01196. PMID: 31680970; PMCID: PMC6805701

Jensen C, Teng Y (2020) Is it time to start transitioning from 2D to 3D cell culture? Front Mol Biosci 7:33

Jung DR, Yoo HY, Kim MJ, Singh V, Park SH, Jeong M, Park BJ, Shin JH (2022) Comparative analysis of scalp and gut microbiome in androgenetic alopecia: a Korean cross-sectional study. Front Microbiol 13:1076242. https://doi.org/10.3389/fmicb.2022.1076242. PMID: 36578576; PMCID: PMC9791053

Juzeniene A, Moan J (2012) Beneficial effects of UV radiation other than via vitamin D production. Dermatoendocrinol 4(2):109–117. https://doi.org/10.4161/derm.20013. PMID: 22928066; PMCID: PMC3427189

Kaluzhny Y, Kandárová H, d'Argembeau-Thornton L, Kearney P, Klausner M (2015) Eye Irritation Test (EIT) for hazard identification of eye irritating chemicals using reconstructed Human Cornea-like Epithelial (RhCE) tissue model. J Vis Exp 102:e52979. https://doi.org/10.3791/52979. PMID: 26325674; PMCID: PMC4692559

Kang MC, Lee JW, Lee TH, Subedi L, Wahedi HM, Do SG, Shin E, Moon EY, Kim SY (2020) UP256 inhibits hyperpigmentation by tyrosinase expression/dendrite formation via rho-dependent signaling and by primary cilium formation in melanocytes. Int J Mol Sci 21:5341

Kim K, Park H, Lim KM (2015) Phototoxicity: its mechanism and animal alternative test methods. Toxicol Res 31(2):97–104. https://doi.org/10.5487/TR.2015.31.2.097. Erratum in: Toxicol Res. 2015 Sep; 31(3): 321. PMID: 26191378; PMCID: PMC4505355

Kim CS, Noh SG, Park Y, Kang D, Chun P, Chung HY, Jung HJ, Moon HR (2018) A potent tyrosinase inhibitor, (E)-3-(2,4- Dihydroxyphenyl)-1-(thiophen-2-yl)prop-2-en-1-one, with anti-melanogenesis properties in α-MSH and IBMX-Induced B16F10 melanoma cells. Molecules 23:2725

Kleinstreuer NC, Hoffmann S, Alépée N, Allen D, Ashikaga T, Casey W, Clouet E, Cluzel M, Desprez B, Gellatly N, Göbel C, Kern PS, Klaric M, Kühnl J, Martinozzi-Teissier S, Mewes K, Miyazawa M, Strickland J, van Vliet E, Zang Q, Petersohn D (2018) Non-animal methods to predict skin sensitization (II): an assessment of defined approaches. Crit Rev Toxicol 48(5): 359–374. https://doi.org/10.1080/10408444.2018.1429386. Epub 2018 Feb 23. PMID: 29474122; PMCID: PMC7393691

Kojima H, Hayashi K, Sakaguchi H, Omori T, Otoizumi T, Sozu T, Kuwahara H, Hayashi T, Sakaguchi M, Toyoda A, Goto H, Watanabe S, Ahiko K, Nakamura T, Morimoto T (2013)

Second-phase validation study of short time exposure test for assessment of eye irritation potency of chemicals. Toxicol in Vitro 27(6):1855–1869. https://doi.org/10.1016/j.tiv.2013.05.013. Epub 2013 Jun 7

Kolle SN, Kandárová H, Wareing B, van Ravenzwaay B, Landsiedel R (2011) In-house validation of the EpiOcular(TM) eye irritation test and its combination with the bovine corneal opacity and permeability test for the assessment of ocular irritation. Altern Lab Anim 39(4):365–387. https://doi.org/10.1177/026119291103900410

Krutmann J, Bouloc A, Sore G, Bernard BA, Passeron T (2017) The skin aging exposome. J Dermatol Sci 85(3):152–161. https://doi.org/10.1016/j.jdermsci.2016.09.015. Epub 2016 Sep 28

Kumar S, Poojan S, Verma V, Verma MK, Lohani M (2014) Rapid isolation of integrin rich multipotent stem cell pool and reconstruction of mouse epidermis equivalent. Am J Stem Cells 3(1):27–36. PMID: 24660112; PMCID: PMC3960755

Lajis AFB, Hamid M, Ariff AB (2012) Depigmenting effect of Kojic acid esters in hyperpigmented B16F1 melanoma cells. J Biomed Biotechnol 2012:952452

Liguori I, Russo G, Curcio F, Bulli G, Aran L, Della-Morte D, Gargiulo G, Testa G, Cacciatore F, Bonaduce D, Abete P (2018) Oxidative stress, aging, and diseases. Clin Interv Aging 13:757–772. https://doi.org/10.2147/CIA.S158513. PMID: 29731617; PMCID: PMC5927356

Liu C, Chu D, Kalantar-Zadeh K, George J, Young HA, Liu G (2021) Cytokines: from clinical significance to quantification. Adv Sci 8:2004433. https://doi.org/10.1002/advs.202004433

Lock-Andersen J, Therkildsen P, de Fine OF, Gniadecka M, Dahlstrøm K, Poulsen T, Wulf HC (1997) Epidermal thickness, skin pigmentation and constitutive photosensitivity. Photodermatol Photoimmunol Photomed 13(4):153–158. https://doi.org/10.1111/j.1600-0781.1997.tb00220.x

Majeed M, Majeed S, Jain R, Mundkur L, Rajalakshmi HR, Lad P, Neupane P (2020) A randomized study to determine the sun protection factor of natural pterostilbene from Pterocarpus marsupium. Cosmetics 7:16. https://doi.org/10.3390/cosmetics7010016

Mazzoleni G, Di Lorenzo D, Steimberg N (2009) Modelling tissues in 3D: the next future of pharmaco-toxicology and food research? Genes Nutr 4(1):13–22. https://doi.org/10.1007/s12263-008-0107-0. Epub 2008 Dec 18. PMID: 19104883; PMCID: PMC2654048

Mikolajczyk R, Roesner LM (2019) Grundlegende Aspekte zum Hautmikrobiom [General aspects regarding the skin microbiome]. Hautarzt 70(6):400–406. https://doi.org/10.1007/s00105-019-4412-x

Mohania D, Chandel S, Kumar P, Verma V, Digvijay K, Tripathi D, Choudhury K, Mitten SK, Shah D (2017) Ultraviolet radiations: skin defense-damage mechanism. Adv Exp Med Biol 996: 71–87. https://doi.org/10.1007/978-3-319-56017-5_7

Moreiras H, Seabra MC, Barral DC (2021) Melanin transfer in the epidermis: the pursuit of skin pigmentation control mechanisms. Int J Mol Sci 22:4466

Muszer M, Noszczyńska M, Kasperkiewicz K, Skurnik M (2015) Human microbiome: when a friend becomes an enemy. Arch Immunol Ther Exp 63(4):287–298. https://doi.org/10.1007/s00005-015-0332-3. Epub 2015 Feb 15. PMID: 25682593; PMCID: PMC4499106

Ng NL, Tuet WY, Chen Y, Fok S, Gao D, Tagle Rodriguez MS, Klein M, Grosberg A, Weber RJ, Champion JA (2019) Cellular and acellular assays for measuring oxidative stress induced by ambient and laboratory-generated aerosols. Res Rep Health Eff Inst (197):1–57

OECD (1996). OECD test guidelines programme: final report of the OECD workshop on harmonization of validation and acceptance criteria for alternative toxicological test methods. Held in Solna, Sweden, 22–24 January 1996 (http://www1.oecd.org/ehs/test/background.htm)

OECD (2000) Guidance document on the recognition, assessment and use of clinical signs as humane endpoints for experimental animals used in safety evaluation. OECD environmental health and safety publications. Series on testing and assessment no. 19 (http://www1.oecd.org/ehs/test/monos.htm)

Papakonstantinou E, Roth M, Karakiulakis G (2012) Hyaluronic acid: a key molecule in skin aging. Dermatoendocrinol 4(3):253–258. https://doi.org/10.4161/derm.21923. PMID: 23467280; PMCID: PMC3583886

Pincemail J, Meziane S (2022) On the potential role of the antioxidant couple vitamin E/selenium taken by the oral route in skin and hair health. Antioxidants (Basel) 11(11):2270. https://doi.org/10.3390/antiox11112270. PMID: 36421456; PMCID: PMC9686906

Pistollato F, Madia F, Corvi R, Munn S, Grignard E, Paini A, Worth A, Bal-Price A, Prieto P, Casati S et al (2021) Current EU regulatory requirements for the assessment of chemicals and cosmetic products: challenges and opportunities for introducing new approach methodologies. Arch Toxicol 1:3

Pittayapruek P, Meephansan J, Prapapan O, Komine M, Ohtsuki M (2016) Role of matrix metalloproteinases in photoaging and photocarcinogenesis. Int J Mol Sci 17(6):868. https://doi.org/10.3390/ijms17060868. PMID: 27271600; PMCID: PMC4926402

Pratiwi R, Auliya As NN, Yusar RF, Shofwan AAA (2022) Analysis of prohibited and restricted ingredients in cosmetics. Cosmetics 9:87. https://doi.org/10.3390/cosmetics9040087

Rademacher F, Simanski M, Gläser R, Harder J (2018) Skin microbiota and human 3D skin models. Exp Dermatol 27:489–494. https://doi.org/10.1111/exd.13517

Rajan R et al (2021) Review of the current state of protein aggregation inhibition from a materials chemistry perspective: special focus on polymeric materials. Mater Adv **2**:1139–1176. https://doi.org/10.1039/D0MA00760A

Ribeiro AS, Estanqueiro M, Oliveira MB, Sousa Lobo JM (2015) Main benefits and applicability of plant extracts in skin care products. Cosmetics 2:48–65. https://doi.org/10.3390/cosmetics2020048

Riss TL, Moravec RA, Niles AL et al (2004) Cell viability assays. In: Markossian S, Grossman A, Brimacombe K et al (eds) Assay guidance manual. Eli Lilly & Company and the National Center for Advancing Translational Sciences, Bethesda, MD. https://www.ncbi.nlm.nih.gov/books/NBK144065/

Rodrigues HA (2017) The cutaneous ecosystem: the roles of the skin microbiome in health and its association with inflammatory skin conditions in humans and animals. Vet Dermatol 28(1):60–e15. https://doi.org/10.1111/vde.12408

Romes NB, Abdul Wahab R, Abdul Hamid M (2021) The role of bioactive phytoconstituents-loaded nanoemulsions for skin improvement: a review. Biotechnol Biotechnol Equip 35:711–730. https://doi.org/10.1080/13102818.2021.1915869

Saleem S, Muhammad G, Hussain MA, Altaf M, Bukhari SNA (2020) Withania somnifera L.: Insights into the phytochemical profile, therapeutic potential, clinical trials, and future prospective. Iran J Basic Med Sci 23(12):1501–1526. https://doi.org/10.22038/IJBMS.2020.44254.10378. PMID: 33489024; PMCID: PMC7811807

SCCS (Scientific Committee on Consumer Safety) (2021) SCCS notes of guidance for the testing of cosmetic ingredients and their safety evaluation 11th revision, 30–31 March 2021, SCCS/1628/21. https://ec.europa.eu/health/sites/default/files/scientific_committees/consumer_safety/docs/sccs_o_250.pdf

Strickland J, Truax J, Corvaro M, Settivari R, Henriquez J, McFadden J, Gulledge T, Johnson V, Gehen S, Germolec D, Allen DG, Kleinstreuer N (2022) Application of defined approaches for skin sensitization to agrochemical products. Front Toxicol 4:852856. https://doi.org/10.3389/ftox.2022.852856. PMID: 35586187; PMCID: PMC9108145

Suhail S, Sardashti N, Jaiswal D, Rudraiah S, Misra M, Kumbar SG (2019) Engineered skin tissue equivalents for product evaluation and therapeutic applications. Biotechnol J 14(7):e1900022. https://doi.org/10.1002/biot.201900022. Epub 2019 May 17. PMID: 30977574; PMCID: PMC6615970

Suzuki K, Inoue M, Cho O, Mizutani R, Shimizu Y, Nagahama T, Sugita T (2021) Scalp microbiome and sebum composition in Japanese male individuals with and without androgenetic alopecia. Microorganisms 9(10):2132. https://doi.org/10.3390/microorganisms9102132. PMID: 34683453; PMCID: PMC8536999

Szymański Ł, Jęderka K, Cios A, Ciepelak M, Lewicka A, Stankiewicz W, Lewicki S (2020) A simple method for the production of human skin equivalent in 3d, multi-cell culture. Int J Mol

Sci 21(13):4644. https://doi.org/10.3390/ijms21134644. PMID: 32629914; PMCID: PMC7369873

Takahashi Y, Hayashi T, Watanabe S, Hayashi K, Koike M, Aisawa N, Ebata S, Sakaguchi H, Nakamura T, Kuwahara H, Nishiyama N (2009) Inter-laboratory study of short time exposure (STE) test for predicting eye irritation potential of chemicals and correspondence to globally harmonized system (GHS) classification. J Toxicol Sci 34(6):611–626. https://doi.org/10.2131/jts.34.611

Takahashi Y, Hayashi K, Abo T, Koike M, Sakaguchi H, Nishiyama N (2011) The Short Time Exposure (STE) test for predicting eye irritation potential: intra-laboratory reproducibility and correspondence to globally harmonized system (GHS) and EU eye irritation classification for 109 chemicals. Toxicol in Vitro 25(7):1425–1434. https://doi.org/10.1016/j.tiv.2011.04.012. Epub 2011 Apr 12

Tate J, Ward G (2004) Interferences in immunoassay. Clin Biochem Rev 25(2):105–120. PMID: 18458713; PMCID: PMC1904417

Timm M, Saaby L, Moesby L, Hansen EW (2013) Considerations regarding use of solvents in in vitro cell based assays. Cytotechnology 65(5):887–894. https://doi.org/10.1007/s10616-012-9530-6. Epub 2013 Jan 18. PMID: 23328992; PMCID: PMC3967611

Tomic-Canic M, Perez-Perez GI, Blumenberg M (2014) Cutaneous microbiome studies in the times of affordable sequencing. J Dermatol Sci 75(2):82–87. https://doi.org/10.1016/j.jdermsci.2014.05.001. Epub 2014 May 23

Union (2009) Regulation (EC) no 1223/2009 of the European parliament and of the council of 30 November 2009 on cosmetic products (recast) (Text with EEA relevance). Off J Eur Union 342:59. https://ec.europa.eu/health/sites/health/files/endocrine_disruptors/docs/cosmetic_1223_2009_regulation_en.pdf

Van Gele M, Geusens B, Speeckaert R, Dynoodt P, Vanhoecke B, Van Den Bossche K, Lambert J (2011) Development of a 3D pigmented skin model to evaluate RNAi-induced depigmentation. Exp Dermatol 20(9):773–775. https://doi.org/10.1111/j.1600-0625.2011.01319.x. Epub 2011 Jun 24

Vinardell MP, Mitjans M (2017) Alternative methods to animal testing for the safety evaluation of cosmetic ingredients: an overview. Cosmetics 4:30. https://doi.org/10.3390/cosmetics4030030

Worth AP, Fentem JH, Balls M, Botham PA, Curren RD, Earl LK, Esdail DJ, Liebsch M (1998) An evaluation of the proposed OECD testing strategy for skin corrosion. Altern Lab Anim 26:709–720

Yamaguchi Y, Hearing VJ (2009) Physiological factors that regulate skin pigmentation. Biofactors 35:193–199

Zappelli C, Barbulova A, Apone F, Colucci G (2016) Effective active ingredients obtained through biotechnology. Cosmetics 3:39. https://doi.org/10.3390/cosmetics3040039

Zhang S, Duan E (2018) Fighting against skin aging: the way from bench to bedside. Cell Transplant 27(5):729–738. https://doi.org/10.1177/0963689717725755. Epub 2018 Apr 25. PMID: 29692196; PMCID: PMC6047276

Zhang C, Duan J, Huang Y, Chen M (2021) Enhanced skin delivery of therapeutic peptides using spicule-based topical delivery systems. Pharmaceutics 13(12):2119. https://doi.org/10.3390/pharmaceutics13122119. PMID: 34959402; PMCID: PMC8709454

8 Animal Component-Free Medium for Long-Term Maintenance of Human Skin Explants and Its Application in Toxicity Studies of Cosmetics

N. S. Remya, Renjith P. Nair, and Anugya Bhatt

Keywords

Cosmetic toxicity · Explant culture · Serum-free medium · Reconstruct skin model

8.1 Introduction

Developing skin models for dermatological and cosmetic toxicity studies is a relatively challenging area of research, considering the limitations of long-term preservation/maintenance of the phenotypically stable test system (Adler et al. 2011). Global legislation regulating animal usage for cosmetic studies has forcefully impacted the quest for an ideal skin model system for cosmetic toxicity and other dermatological studies. Various in vitro skin models like monolayer cell culture models, co-culture models, organotypic culture models, and 3D reconstructed skin models are available for the said purpose (Randall et al. 2018). However, skin explants are better because their unique characteristics closely reflect the physiological outcome (Eberlin et al. 2020). A comparison of the skin explant model and the reconstructed skin model is given in Table 8.1.

Skin explants are derived from excised skin tissue. After removing subcutaneous fat and other contaminants, explant cultures are initiated in a suitable culture

N. S. Remya
Division of Toxicology, Department of Applied Biology, Biomedical Technology Wing, Sree Chitra Tirunal Institute for Medical Sciences and Technology, Thiruvananthapuram, Kerala, India

R. P. Nair · A. Bhatt (✉)
Division of Thrombosis Research, Department of Applied Biology, Biomedical Technology Wing, Sree Chitra Tirunal Institute for Medical Sciences and Technology, Thiruvananthapuram, Kerala, India
e-mail: anugyabhatt@sctimst.ac.in

A. B. Pant et al. (eds.), *Skin 3-D Models and Cosmetics Toxicity*,
https://doi.org/10.1007/978-981-99-2804-0_8

Table 8.1 A comparison of skin explant model and c

Parameters	Skin explant model	Reconstructed skin model
Cells represented	Most of the skin cells	One or two
Maturation time	No control	Can be controlled
Skin information obtained	Extensive	No information obtained
Applications possible	Many	Many
Appendages	Present	Lacking
Innervation and circulation	Lacking	Lacking
Production	Easy to produce	Difficult and long
Cost of maintenance	Cost-effective	High cost
Maintenance duration	Short	Short

medium. This helps in the maintenance of the structure of native skin, including the distinct skin cell types and skin-specific extracellular matrix (Neil et al. 2020). The unique characteristic of this ready-to-use system is that natural stratification of the skin layers is maintained, including skin appendages. In the skin explant model, the entire native skin cell population, including keratinocytes, melanocytes, Langerhans cells, and dermal fibroblast cells, is present in a niche of relevant skin-specific skin extracellular matrix comprising of collagen, elastin, glycosaminoglycans (GAGs), etc. Even though these processed tissues lack blood circulation and innervation, they are still used as an ex vivo model for studying the impact of toxic exposure on the skin. Skin explants serve as organotypic models, providing a 3-dimensional culture environment. This helps in the effective cell–cell interactions by making the model close to the physiological conditions. When cultured at the air–liquid interface (ALI), skin explants can be developed as test systems for understanding the effect of topically applied substances.

8.2 Human Skin Explants—Applications

The tremendous usage of cosmetics in the modern world has prompted the need to understand the toxicity issues associated with human exposure in a detailed manner (Mishra and Rahi 2022). The various chemicals widely employed in cosmetic formulation, including active ingredients, preservatives, fragrances, heavy metals, pose a severe threat to consumers and the environment where they are disposed of. Hence, it is imperative to estimate the toxic adverse effects of components of a cosmetic product. Most of the toxicity studies of chemicals performed are in vivo animal studies. However, endorsing the principles of the 3Rs in regulatory toxicity testing has incited the scientific world to seek alternative toxicity methods (Almeida et al. 2017). Skin explants have been an ideal choice for cosmetic toxicity testing from then onwards. Modifying the conventional testing strategies to include additional biological endpoints is possible with such in vitro systems. Human skin explants are particularly interesting as they closely match the subject of interest. Some of the additional endpoints introduced are modeling various skin diseases,

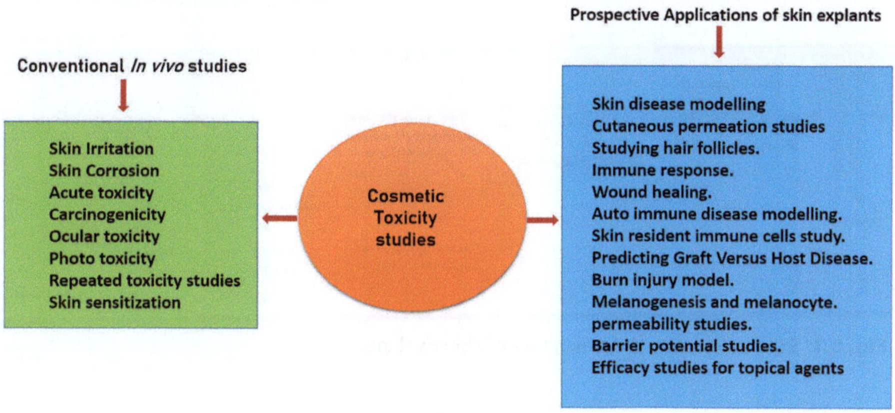

Fig. 8.1 Prospective applications of skin explants in cosmetic toxicity testing

studying cutaneous permeation, studying hair follicles, skin infections, dermal and epidermal specific studies, skin resident immune cells and immune responses, melanogenesis and melanocyte permeability studies, etc. (Sutterby et al. 2022). Figure 8.1 summarizes the conventional cosmetic toxicity testing and the prospective applications of skin explants for cosmetic toxicity studies.

8.3 Generation of Skin Explants

Human skin explants are generally obtained from elective plastic surgery/bariatric surgery (weight loss surgery) that involves excess skin removal procedures. Clinical wastes from surgical procedures like panniculectomy, brachioplasty, and abdominoplasty provide sources for skin explants. Skin from the abdominal region is of particular interest. The region's representative skin has more basal characteristics preserved well because of the low exposure to external toxic perturbances such as solar radiation and pollution. The tissue collection site is cleaned with 10% povidone-iodine solution, and skin with the underlying subcutaneous tissue is harvested. After removing adipose tissue, skin tissue with a thickness of about 0.5 ± 0.1 mm is sliced from the excised skin tissue using an electric dermatome. They are collected in ice-cold physiological saline. Hair appendages are removed, and the skin is soaked in 0.1% benzalkonium bromide for 15 min to sterilize. They are washed extensively with ice-cold saline with antibiotics/antimycotics. All the procedures are carried out under aseptic conditions. From the slices, circular sections of 10–12 mm diameter are cut out and cultured at an air–liquid interface (ALI) in a cell culture insert till further analyses (Fig. 8.2).

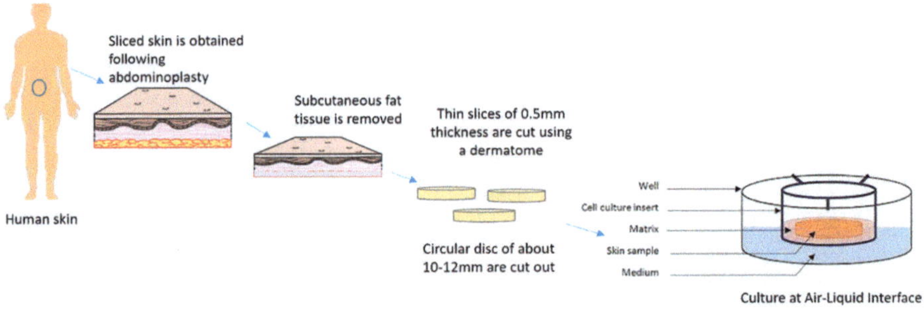

Fig. 8.2 Figure depicting the generation of skin explants

8.4 Preservation of Skin Tissue for Explant Generation

Often skin sources are only sometimes available for the generation of explant cultures. For this reason, it is imperative to preserve the clinically sourced skin tissue/explants generated from it as and when available for further use. This could be achieved by skin banking. Skin preservation strategies rely on whether the harvested skin is required to maintain its viability (Kearney 2005). Proper antiseptic measures should be taken for the collection of skin as skin tissue is a rich source of microbial contamination, and sterilization techniques cannot be used to maintain viable cultures. However, they could be collected and stored in solutions with antibiotics and antibiotics to reduce the bioburden.

8.5 Low-Temperature Preservation

Preservation of skin for explant cultures remains a challenge. Even though the tissue is placed in a nutrient-rich medium, ischemic tissue necrosis can occur at the center of the tissue due to insufficient diffusion of oxygen and nutrients from the periphery to the center and inadequate removal of toxic metabolites. Hence, even though the tissue is placed in a nutrient-rich medium, ischemic tissue necrosis can occur at the center of the tissue due to insufficient diffusion of oxygen and nutrients from the periphery to the center and inadequate removal of toxic metabolites. The skin's moisture content is about 70%, so the quality of skin preserved under low-temperature conditions dramatically depends on the phase transition of water (Mojumdar et al. 2017). Refrigeration at 4 °C is a simple and convenient technique for short-term preservation. However, reports show that cell viability decreases directly proportional to the storage time (Fahmy et al. 1993). Ideally, the preservation time of fresh skin slices should not exceed 72 h when stored in ice-cold physiological saline. It should be used for explant culture generation within a week if stored in tissue culture media with serum.

With the introduction of cryoprotective agents such as DMSO, polyethylene glycol, propylene glycol, and glycerin, it is possible to store at ultra-low

Table 8.2 Comparison of different temperature on skin explant storage

Temperature	4 °C	−20 °C	−80 °C	−196 °C
Equipment	Refrigerator	Freezer	Ultra-low-temperature freezer	Liquid nitrogen
Need for cryoprotective agent	Nil	Yes	Yes	Yes
Cell viability following recovery	30–80%	50–60%	50–60%	60-&0%
Duration of storage	3–7 days	30–60 days	60–1 year	1–2 year
Cost of maintenance	Moderate	Costly	Costly	Moderate

temperatures. The cryoprotectants prevent ice crystal formation (Elliott et al. 2017). During the freezing process, two types of damage can occur to the cells. The formation of intracellular ice crystals can damage the cell membrane and subcellular organelles. This type of damage often occurs concerning quick freezing.

On the other hand, the cells will be subjected to solution damage when the water present in the extracellular solution where cells are suspended gets frozen. This type of damage is frequent in the case of slow-freezing methods. Hence, it is possible to overcome the above issues with adequate cryoprotective agents (Karlsson and Toner 1996). So a technique called "vitrification" could be employed in the cryopreservation of skin tissue for long-term storage of skin tissues. Vitrification is the instant solidification of a solution by increasing the solution's viscosity during the cooling process (Costa et al. 2020). This is achieved by adding anti-freeze or cryoprotective agents that modulate the phase transition process of water. However, revival procedures of this cryopreserved frozen tissue need to be optimized as the incorporated cryoprotective agents can be detrimental at a higher temperature. Also, the chance of getting cryogenic injury to the tissue component cells is higher under cryopreservation. A comparison of different temperature conditions on the preservation of skin is given in Table 8.2.

8.6 Dynamic Culture Conditions

The advantage of an explant culture system is that the skin-specific extracellular matrix (ECM) is not dissociated/disrupted. So the cell–cell interaction and cell–matrix interaction are well preserved so that the functional phenotype of the skin cells is maintained to a more considerable extent (Randall et al. 2011). The ECM provides biomechanical and biochemical cues to modulate the resident cells' morphogenesis, differentiation, and homeostasis. The ECM proteins and the growth factor receptors on the cells are also well preserved to respond to extraneous supplementation of growth factor, thereby allowing the possibility of manually modulating the cellular response. Three-dimensional (3D) culture methods involving a rotary cell culture system bioreactor benefit the long-term maintenance of explant cultures (Astashkina and Grainger 2014). The rotary cell culture system bioreactor is a rotating culture vessel with a centrally placed co-axial oxygenator.

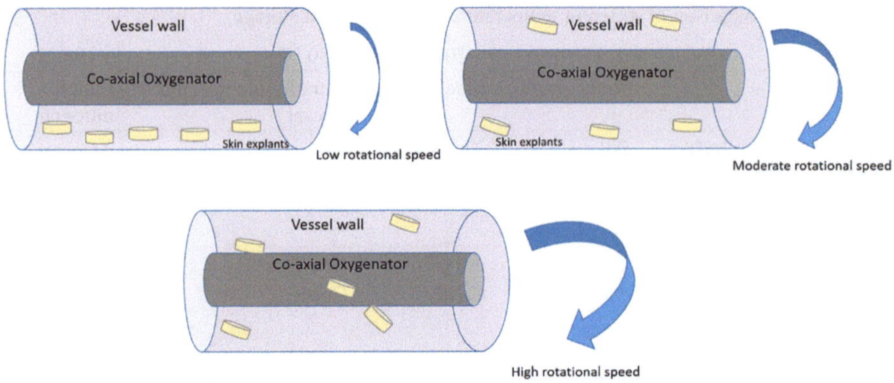

Fig. 8.3 Representation of rotary cell culture bioreactor system for skin explant cultures

The concurrent rotation of vessel walls and the centrally placed oxygenator creates laminar flow and a minimum shear force so that sufficient diffusion of nutrients and oxygen to the tissue is ensured (Fig. 8.3).

8.7 Culture Techniques for Human Keratinocytes

The human skin implants can be directly used for toxicity studies or as a source for propagating human epithelial cells in culture. However, when used as a source of keratinocytes, the culture conditions need to be optimized to get a good yield of keratinocytes from the explants. Keratinocytes have distinct nutrient requirements, so they are quickly overgrown by other cell populations, such as dermal fibroblasts (Sorg et al. 2017). When attempting to propagate keratinocytes from skin explants, careful separation of the epidermis from the dermis should be tried to avoid contamination with dermal fibroblast. Keratinocytes from the basal layer of the epidermis can proliferate and form colonies. Two methods are currently employed to propagate the keratinocyte cell population from explant culture as follows: (a) culturing in the presence of a feeder layer and serum-containing medium and (b) culturing in the absence of a feeder layer and serum-free medium.

8.7.1 In Serum-Containing Medium with a Feeder Layer

Rheinwald and Green (1975) proposed the method of co-culturing keratinocytes on top of irradiated, non-proliferating fibroblast. Murine fibroblast NH3T3 exposed to gamma rays (6000 rad) or subjected to mitomycin C treatment is shown to give a better result (Rheinwald and Green 1975). The primary explant culture of keratinocytes is plated at a suitable cell density onto the feeder layer. Furthermore, the culture medium for sustaining both cells is provided with 10% fetal bovine serum

Table 8.3 Composition of keratinocyte propagation medium

Constituents		Concentration
Base	Dulbecco's minimum essential medium	90%
	Fetal bovine serum	10%
Supplements	Adenine	24 μg/mL
	Epidermal growth factor	10 ng/mL
	Hydrocortisone	0.5 μg/mL
	Insulin	5 μg/mL
	Cholera toxin	6 ng/mL
	Transferrin	10 μg/mL
	3,3′5'-triiodothyronine	1.3 ng/mL
	Glutamine	0.29 mg/mL

and special supplements. A typical composition of keratinocyte propagating medium is given in Table 8.3.

Epidermal growth factor (EGF) is added to increase keratinocytes' proliferation and growth rate. Cholera toxin stimulates adenylate cyclase's enzymatic activity, thereby increasing intracellular levels of cyclic AMP (Green 1978). In about 10 days, a multilayered sheet of keratinocytes is obtained following this culture method. The secondary culture of keratinocyte is obtained by detaching the confluent layer using proteolytic agents such as dispase and plating into a new flask with feeder culture. As keratinocytes have a limited lifespan, they should be used within 4–5 passages from primary explant culture.

8.7.2 In Serum-Free Media Without a Feeder Layer

Although a considerably better growth rate of keratinocytes is obtained when cultured in serum-containing media and with a feeder layer, it is advisable to use serum-free media to lower the risk of interfering components and to avoid ethical issues. A chemically defined medium can culture the cells under controlled conditions. Determining the ionic concentration, nutrient compositions, and specialized supplements makes it possible to customize according to the growing requirement of keratinocytes. Table 8.4 lists supplements that could be used for keratinocyte propagation.

8.8 Culture at Air–Liquid Interface (ALI)

When the conventional cell culture technique is performed for keratinocytes, the cells are seen attached to the surface of the culture plate submerged in the liquid media. Furthermore, for this reason, the differentiation of the cells is inhibited, so stratification of epithelial layers cannot be achieved. So if the monolayer culture is done in a "lifted manner" so that the basolateral side of the monolayer is in contact

Table 8.4 Serum-free media for keratinocytes

	Supplements
Chemically undefined	Bovine brain extract
	Fraction IV albumin from human serum
	Bovine thymus extract
	Human placental extract
	Bovine hypothalamus extract
Chemically Defined	Progesterone
	Triiodothyronine
	Hydrocortisone
	Insulin
	Adenine
	Calcium ions
	Monoethanolamine
	Phosphoethanolamine
	Transferrin
	Amino acids (isoleucine, histidine, methionine, phenylalanine, tryptophan, and tyrosine)

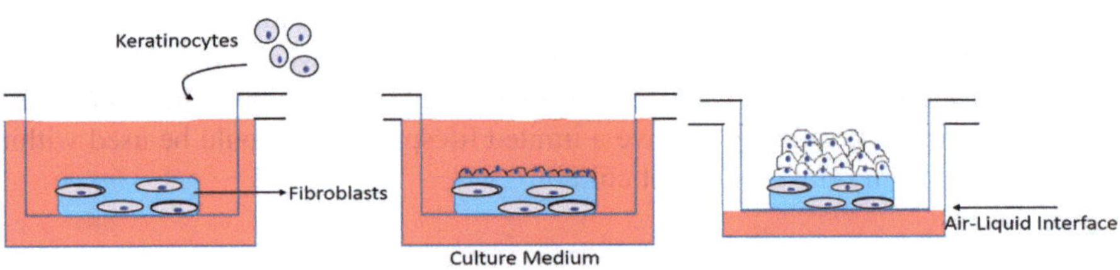

Fig. 8.4 Air–liquid interface culture for keratinocytes

with the culture media and the superficial cells are exposed to air which mimics the microenvironment of skin (Green 1978). Various substrates can be used for this purpose, such as membranes made of collagen, fibrin, and laminin floating on a liquid surface medium (Pruniéras et al. 1983). Organotypic skin equivalents could be derived by co-culturing keratinocytes on top of a dermal fibroblast-seeded permeable scaffold and then submerged in suitable media under air–liquid interface conditions (Parenteau et al. 1992). A typical representation of ALI culture is given in Fig. 8.4.

8.9 Animal Component-Free Medium

Animal component-free (ACF) media are defined as a medium that does not contain any primary raw materials derived directly from animal tissue or body fluid. However, it could include secondary or territory raw materials derived from human tissue or proteins produced by recombinant technology (Whitford et al. 2018). The difference between ACF from serum-free (SF) media is that serum-free media is devoid only of serum/plasma/or hemolymph but may contain other primary raw materials of animal origin such as tissue extract, platelet lysate, hormones, and growth factor cocktail. Nevertheless, in the case of animal component-free media, these raw materials of animal origin are not included.

In conventional cell culture methods, fetal bovine serum (FBS) is an essential constituent of cell culture media. However, there are disadvantages to using serum exclusively for culture purposes. The primary disadvantage is its batch-to-batch variability. This variability can affect the performance of culture outcomes. Also, there is a risk of potential viral and other adventitious contaminants, including mycoplasma and endotoxin, present in the serum (Froud 1999). Other animal-derived primary raw materials, such as tissue extract, growth factors, and hormones, can pose the same risk. Media containing animal tissue-derived supplements possess immunogenic potential due to the presence of xenogeneic proteins. Hence, it is not always advisable to use the cells conditioned in such a medium for immunological studies or cell-based therapies. The serum is considered "bio-reactive" and can interfere with many biological cascade pathways (Barnes and Sato 1980). For this reason, there is a potential risk of obtaining unreliable results from in vitro studies using fetal bovine serum. By utilizing an ACF medium, there is an option for a chemically defined medium with traceable ingredients for cell culture. The downstream process can be made simplified using a chemically defined animal component-free (ACF) medium. The essential components replaced in an ACF medium are growth factors and human blood derivatives such as serum albumin, platelet lysate, and protein and lipid supplements derived from non-human/non-animal sources such as plants, bacteria, and yeast. The advantages of using ACF medium are compiled in Fig. 8.5.

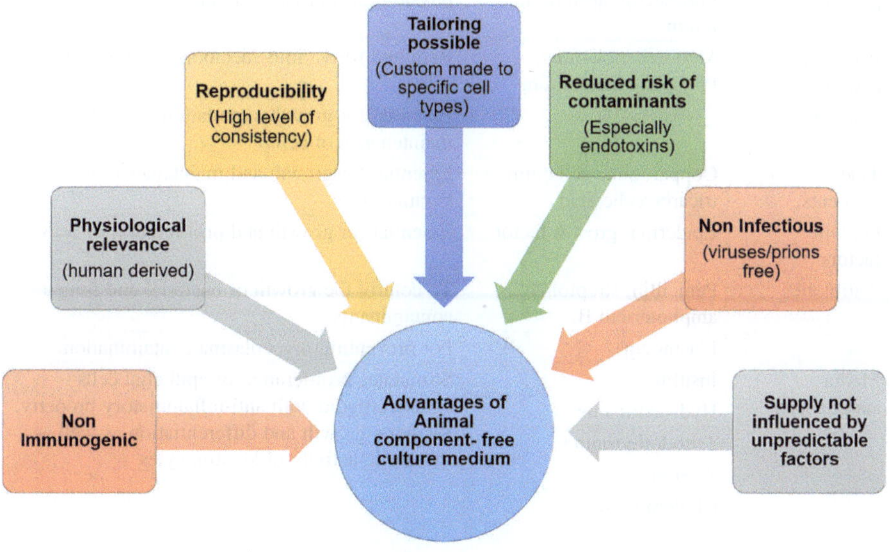

Fig. 8.5 Advantages of animal component-free culture medium

8.10 Composition of Animal Component-Free Medium for Skin Explants

With the successful translation and adaptation of KeratinoSens™ as an alternative to animal skin sensitization assay, researchers are in quest of similar animal product-free cell culture systems for the maintenance or preservation of skin explants (Riebeling et al. 2018). The essential components replaced in an ACF medium are growth factors and human blood derivatives such as serum albumin, platelet lysate, and protein and lipid supplements derived from non-human/non-animal sources such as plants, bacteria, and yeast. A typical composition of animal component-free medium for skin explants is given in Table 8.5.

Table 8.5 General composition of animal component-free medium for skin explants

Components	Specific ingredients	Purpose
Buffering systems	HCO_3, HEPES	For regulating pH
Inorganic salts	Calcium, potassium, sodium	Helps in retaining the osmotic balance and helps in regulating the membrane potential of cells
Amino acids	Essential amino acids Nonessential amino acids	Building blocks for proteins. Essential for growth and proliferation of cells
	L-glutamine	L-glutamine provides nitrogen for NAD, NADPH, and nucleotides Also serves as secondary energy source for metabolism
Carbohydrates	Glucose, galactose	Main source of energy
Proteins and peptides	Albumin, transferrin, fibronectin, aprotinin, fetuin	Required for growth, proliferation and, other physiological process of cells
Fatty acids and lipids	Monoethanolamine, Phosphoethanolamine	Required for various metabolic activities of the cells
Vitamins	–	Essential for growth stimulation as well as maintenance of cells
Trace elements	Copper, zinc, selenium, tricarboxylic acid	Essential for growth and maintenance of keratinocytes
Growth factors	Epidermal growth factor	Essential for growth and proliferation of cells
Antibiotics	Penicillin, streptomycin, amphotericin B,	To control the growth of bacterial and fungal contaminants
	Plasmocin	For preventing mycoplasma contamination
Media supplements	Insulin	Stimulates proliferation of epithelial cells
	Hydrocortisone	Glucocorticoid with anti-inflammatory property.
	Triiodothyronine	Supports growth and differentiation as well as metabolic activity of keratinocytes
	Adenine	
	Cholera toxin	

8.11 Future Directions

The prospects of using skin explants and skin explant culture as an alternative to animal testing strategies for cosmetics and other skin disease modeling studies are giving the scientific community promising results. Furthermore, incorporating animal component-free media and better storage modalities would benefit the next-level scope of expansion of cosmetic toxicity studies and enhance the clinical translation potential of skin explants.

References

Adler S, Basketter D, Creton S, Pelkonen O, van Benthem J, Zuang V, Andersen KE, Angers-Loustau A, Aptula A, Bal-Price A et al (2011) Alternative (non-animal) methods for cosmetics testing: current status and future prospects—2010. Arch Toxicol 85(5):367–485. https://doi.org/10.1007/s00204-011-0693-2

Almeida A, Sarmento B, Rodrigues F (2017) Insights on in vitro models for safety and toxicity assessment of cosmetic ingredients. Int J Pharm 519(1):178–185. https://doi.org/10.1016/j.ijpharm.2017.01.024

Astashkina A, Grainger DW (2014) Critical analysis of 3-D organoid in vitro cell culture models for high-throughput drug candidate toxicity assessments. Adv Drug Deliv Rev 69-70:1–18. https://doi.org/10.1016/j.addr.2014.02.008

Barnes D, Sato G (1980) Serum-free cell culture: a unifying approach. Cell 22(3):649–655. https://doi.org/10.1016/0092-8674(80)90540-1

Costa CAS, Borges AA, Nascimento MB, Aquino LVC, Silva AR, Oliveira MF, Pereira AF (2020) Effects of vitrification techniques on the somatic tissue preservation of agouti (Dasyprocta leporina Linnaeus, 1758). Biopreserv Biobank 18(3):165–170. https://doi.org/10.1089/bio.2019.0109. Accessed 20 Dec 2022

Eberlin S, Silva MS, Facchini G, Silva GH, Pinheiro AL, Eberlin S, Pinheiro AD (2020) The ex vivo skin model as an alternative tool for the efficacy and safety evaluation of topical products. Altern Lab Anim 48(1):10–22. https://doi.org/10.1177/0261192920914193. Accessed 20 Dec 2022

Elliott GD, Wang S, Fuller BJ (2017) Cryoprotectants: a review of the actions and applications of cryoprotective solutes that modulate cell recovery from ultra-low temperatures. Cryobiology 76:74–91. https://doi.org/10.1016/j.cryobiol.2017.04.004

Fahmy FS, Navsaria HA, Frame JD, Jones CR, Leigh IM (1993) Skin graft storage and keratinocyte viability. Br J Plast Surg 46(4):292–295. https://doi.org/10.1016/0007-1226(93)90005-V

Froud SJ (1999) The development, benefits and disadvantages of serum-free media. Dev Biol Stand 99:157–166. PubMed

Green H (1978) Cyclic AMP in relation to proliferation of the epidermal cell: a new view. Cell 15(3):801–811. https://doi.org/10.1016/0092-8674(78)90265-9

Karlsson JOM, Toner M (1996) Long-term storage of tissues by cryopreservation: critical issues. Biomaterials 17(3):243–256. https://doi.org/10.1016/0142-9612(96)85562-1

Kearney JN (2005) Guidelines on processing and clinical use of skin allografts. Clin Dermatol 23(4):357–364. https://doi.org/10.1016/j.clindermatol.2004.07.018

Mishra G, Rahi S (2022) Need of toxicity studies for cosmetic products and their approaches. Biol Sci 2(1):105–109. https://doi.org/10.55006/biolsciences.2022.0201

Mojumdar EH, Pham QD, Topgaard D, Sparr E (2017) Skin hydration: interplay between molecular dynamics, structure and water uptake in the stratum corneum. Sci Rep 7(1):15712. https://doi.org/10.1038/s41598-017-15921-5

Neil JE, Brown MB, Williams AC (2020) Human skin explant model for the investigation of topical therapeutics. Sci Rep 10(1):21192. https://doi.org/10.1038/s41598-020-78292-4

Parenteau NL, Bilbo P, Nolte CJM, Mason VS, Rosenberg M (1992) The organotypic culture of human skin keratinocytes and fibroblasts to achieve form and function. Cytotechnology 9(1): 163–171. https://doi.org/10.1007/bf02521744

Pruniéras M, Régnier M, Woodley D (1983) Methods for cultivation of keratinocytes with an air-liquid interface. J Investig Dermatol 81(1, Supplement):S28–S33. https://doi.org/10.1111/1523-1747.ep12540324

Randall KJ, Turton J, Foster JR (2011) Explant culture of gastrointestinal tissue: a review of methods and applications. Cell Biol Toxicol 27(4):267–284. https://doi.org/10.1007/s10565-011-9187-5

Randall MJ, Jüngel A, Rimann M, Wuertz-Kozak K (2018) Advances in the biofabrication of 3D skin in vitro: healthy and pathological models. Front Bioeng Biotechnol 6:154. https://doi.org/10.3389/fbioe.2018.00154

Rheinwald JG, Green H (1975) Serial cultivation of strains of human epidermal keratinocytes: the formation of keratinizing colonies from single cells. Cell 6:331–343. https://doi.org/10.1016/S0092-8674(75)80001-8

Riebeling C, Luch A, Tralau T (2018) Skin toxicology and 3Rs—current challenges for public health protection. Exp Dermatol 27(5):526–536. https://doi.org/10.1111/exd.13536. Accessed 20 Dec 2022

Sorg H, Tilkorn DJ, Hager S, Hauser J, Mirastschijski U (2017) Skin wound healing: an update on the current knowledge and concepts. Eur Surg Res 58(1–2):81–94. https://doi.org/10.1159/000454919

Sutterby E, Thurgood P, Baratchi S, Khoshmanesh K, Pirogova E (2022) Evaluation of in vitro human skin models for studying effects of external stressors and stimuli and developing treatment modalities. View 3(2):20210012. https://doi.org/10.1002/VIW.20210012. Accessed 20 Dec 2022

Whitford WG, Lundgren M, Fairbank A (2018) Chapter 8–cell culture media in bioprocessing. In: Jagschies G, Lindskog E, Łącki K, Galliher P (eds) Biopharmaceutical processing. Elsevier, Amsterdam, pp 147–162

Regulatory Requirements for Safety/Toxicity Assessment of Cosmetics/Nanocosmetic Products: Challenges and Opportunities

9

Sana Sarkar, Anuj Pandey, and A. B. Pant

Abstract

Cosmetics are regulated globally to maintain their safety and effectiveness. However, different regulatory frameworks adopted by each country adversely affect the competition and economic growth of the cosmetic industry. Further, animal testing for safety and efficacy purposes in cosmetics has sparked controversial debates in the last few decades. Alternative research methodologies have become increasingly popular, particularly after the release of the three R's principles (replacement, reduction, and refinement). Although many alternatives to animal testing have been introduced in the cosmetics industry, studying the safety of cosmetic products and their ingredients is still challenging. In the present chapter, we have attempted to explore the information available on the regulatory frameworks for cosmetics/nanocosmetics in various countries. We have provided a brief overview of the ban on using animals in cosmetic testing and relevant alternative approaches employed in regulatory safety testing. This chapter also covers numerous challenges encountered in substituting animals with alternatives and offers suggestions to overcome the current barriers.

Keywords

Cosmetic safety · Nanocosmetics safety · Cosmetics regulatory requirements · Animal free testing · Alternative to laboratory animals

S. Sarkar · A. Pandey
CSIR-Indian Institute of Toxicology Research (CSIR-IITR), Lucknow, Uttar Pradesh, India

A. B. Pant (✉)
CSIR-Indian Institute of Toxicology Research (CSIR-IITR), Lucknow, Uttar Pradesh, India

Academy of Scientific and Innovative Research (AcSIR), Ghaziabad, Uttar Pradesh, India
e-mail: abpant@iitr.res.in

A. B. Pant et al. (eds.), *Skin 3-D Models and Cosmetics Toxicity*,
https://doi.org/10.1007/978-981-99-2804-0_9

9.1 Introduction

Cosmetics are among the most extensively used products worldwide; hence, the scope of regulation of these products is quite broad. The high demand for cosmetics is directly related to their ability to help fulfill men's fundamental desire to look attractive and youthful. Cosmetics and related products have been used for ages to serve various purposes, from increasing appearance to improving confidence (Effiong et al. 2019). Cosmetics are defined as any product or article that is intended to be rubbed, poured, sprinkled, or sprayed on, introduced to, or otherwise applied to the human body or a portion of it, excluding soap, with the goal of cleansing, beautifying, facilitating attractiveness, or altering appearance (US-FDA 2021a). Moisturizers, hair care products, makeup, shaving creams, nail paints, perfumes, toothpaste, mouthwashes, face and body cleansers, and deodorants are a few cosmetic products.

Observable makeup was not regarded as "respectable" in ancient times. Shampoo, lotions, creams, and even makeup were commonly homemade. None were regulated, and some were even deemed risky. For instance, Laird's Bloom of Youth, made around 1860, was used for the skin and complexion but was later revealed to contain harmful amounts of lead. Unfortunately, it is clear from the fact mentioned above that humanity has apparently sacrificed their health and safety throughout history in their pursuit of beauty with various harmful handmade cosmetics.

Additionally, cosmetics were not covered under the original Food and Drug Act, popularly known as "The Pure Food and Drug Act," passed in 1906, although it did include color additives for foods and drugs. Attempts to involve cosmetics in the 1906 Act were unsuccessful because they made up a small portion of the economy, were only utilized by a small population, and were perceived as unnecessary. In the 1920s, changes in commerce started to emerge due to external factors like the use of cosmetics in the film industry and the influx of women into the workforce. These factors promoted the direct sale of cosmetics in retail establishments and beauty parlors. Until the 1930s, there was still concern about regulating foods, medicines, and cosmetics (Katz et al. 2022). Fortunately, the days of ignorance when people used risky, harmful, and even poisonous mixtures to enhance their appearances are over. However, the desire to look beautiful and youthful persists, leading to the implementation of verifiable science and technology under a strict regulatory framework to achieve this goal (Effiong et al. 2019).

Cosmetic regulatory frameworks differ significantly across countries and need to be harmonized, thus posing challenges to the production of cosmetics for sale in the global market. The significant markets adhere to similar regulatory standards, but current discrepancies are substantial enough to affect the cosmetic sector by limiting innovation, lowering the market's growth, and hindering international trade. Several international organizations have been working cooperatively to harmonize the regulatory frameworks for cosmetics in various countries. Examples include the International Cooperation on Cosmetics Regulation (ICCR), the Organization for Economic Cooperation and Development (OECD), and the International Organization for Standardization (ISO) (Ferreira et al. 2022). Furthermore, cosmetics are not considered a necessary good; they are luxury products. Thus, it is unethical to test cosmetics/cosmetic ingredients on animals to evaluate their safety profiles. Thus, in recent years, there has been increased concern about using animals in cosmetic testing among animal welfare agencies, researchers, and even the general public. Many nations have established laws banning the use of animals in cosmetic product testing to protect animals from unnecessary suffering and harm. These nations are now focusing on developing alternatives to animal testing to assess cosmetics' safety.

This chapter overviews cosmetic and nanocosmetic product regulations worldwide and recent developments in alternative testing approaches. In particular, it discusses the cosmetics regulatory frameworks in different countries, the ban on animal testing, relevant alternatives currently employed in regulatory safety testing, the challenges encountered in substituting animals, and how they can be overcome.

9.2 Nanotechnology in Cosmetics

A nanotechnology is an innovative tool used extensively in the production of cosmetics. The ability of nanotechnology to improve the qualities of cosmetic products has made it a promising addition to the cosmetics sector. For over 30 years, the cosmetic industry has extensively employed nano-based compounds (Pastrana et al. 2018; Carrouel et al. 2020; Revia et al. 2019). Nanotechnology has the potential to alter and enhance properties like absorption, texture, protection for active substances, and the overall effectiveness of cosmetics (Revia et al. 2019). Nanotechnology uses nanoparticles or nanomaterials that are produced artificially or naturally and range in size from 1 to 100 nm (Khezri et al. 2018). Cosmetics made from nanomaterials have distinct advantages over cosmetics made on a micro-scale. The cosmetic industry uses nanoparticles to produce results that persist for a long time and have greater durability. The large surface area of nanomaterials enables the ingredients to be transported through the skin more efficiently (Ahmad et al. 2018). Some of the critical goals in employing nanotechnology in cosmetics include effective penetration of components into the skin for improved product delivery, new color components (such as in lipsticks and nail polishes), transparency (such as in sunscreens), and long-lasting benefits (such as in makeup). The ultimate goal of the cosmetics industry when employing nanomaterials is to achieve long-term stability and deliver the proper amount of ingredients to the desired body areas. The anti-aging lotion Capture™, based on liposomes, was introduced by Christian Dior in 1986. Over the years, nanomaterials have been used in many cosmetic products, and several internationally well-known cosmetic brands have adopted them (Raj et al. 2012). L'Oréal S.A., which invests a significant amount of money in nanotechnology, utilizes up to four nano-ingredients (titanium dioxide (TiO_2), zinc oxide (ZnO), silica (SiO_2), and carbon black) in some of its formulas and ranks sixth in the United States in terms of the number of patents linked to nanotechnology (Rigano and Lionetti 2016). Shiseido employs nano-TiO_2 and nano-ZnO in wet-based formulas (such as emulsions) but avoids using them in aerosols due to the potential risk of inhalation hazards (Shiseido n.d.). Generally, well-known cosmetic companies worldwide gradually incorporate nanomaterials into their products (Fytianos et al. 2020).

Nevertheless, over the past 10 years, there have been growing concerns regarding the potential effects of cosmetic items incorporating nanomaterials on human health and the environment. The rapid diffusion of cosmetic products containing nanomaterials onto the market has raised alarms about their possible impact on human health and the environment. Concerns about the safety of nanomaterials and their application in consumer products, including cosmetics, have been raised by the World Health Organization (WHO), nongovernmental government agencies, political institutions, and agencies (Pastrana et al. 2018). The Food and Drug Administration (FDA) has established its own guidelines for the use of nanotechnology in industrial products, and the European Commission (EC) has updated the recommendations on the safety evaluation of nanomaterials in cosmetic products (Bernauer et al. 2019). Since animal testing is explicitly forbidden by the EC Cosmetic Regulation No. 1223/2009, future toxicological findings for risk assessments in Europe must not involve animal testing. Instead, safety evaluation must be done employing alternative approaches. According to the 2020 announcement of the European Union Observatory for Nanomaterials (EUON), all manufacturers that produce, utilize, or import nanomaterials will need to be registered under the REACH (Registration, Evaluation, Authorisation, and Restriction of Chemicals) program. This encourages nanomaterial-based companies to provide consumers with proper product safety information. The use of nanomaterials in consumer products that are not registered under REACH is regarded as illegal. The European Commission (EC) is provided with a priority list of nanomaterials by the Scientific Committee on Consumer Safety (SCCS) for risk assessment of nanomaterials employed in cosmetics (Fytianos et al. 2020).

9.3 Current Regulatory Framework of Cosmetics

The cosmetics sector is a global, dynamic, and expanding industry. Over the last few decades, massive industrial innovation has led to a wide range of new cosmetic products and increased sales. The worldwide cosmetic industry had a value of USD 341.1 billion in 2020 alone, and it is predicted to rise to USD 560.50 billion by 2030, with a compound annual growth rate of 5.1% from 2021 to 2030 (Ferreira et al. 2022). In order to assure the safety and quality of cosmetic products and prevent negative consequences for consumer health, the cosmetic industry must be regulated owing to its highly inventive, dynamic, and complex nature. The ability of the global initiative to sell the same cosmetic product across all markets is substantially hampered by the fact that regulatory frameworks vary significantly between markets and nations and need to be harmonized. The significant markets broadly adhere to similar regulatory standards, but the current discrepancies are substantial enough to affect the cosmetics sector by limiting innovation and lowering the market's potential for growth. These variations may also impact international trade and hamper the ability of regulatory bodies to ensure that every product complies with the local laws used by individual nations (Ferreira et al. 2022). Therefore, it is crucial to identify solutions that can converge regulatory frameworks for cosmetics, foster innovation, boost market growth, and remove trade obstacles. Several international groups have been working together to attain this goal. One example is the International Cooperation on Cosmetics Regulation (ICCR), formed in 2007, which is a voluntary group of cosmetic regulatory authorities from the United States of America (USA), Brazil, Chinese Taipei, Canada, the European Union (EU), Japan, and the Republic of Korea. This group meets annually to discuss various subjects related to cosmetic safety and regulation (for instance, substitutes for animal testing, nanotechnology, and microbiological restrictions) (US-FDA n.d.). Other organizations that play essential roles in developing global standards for cosmetics and the mutual acceptance of testing method guidelines include the Organization for Economic Cooperation and Development (OECD) and the International Organization for Standardization (ISO). However, there is still more that can be done to deepen the current collaboration efforts between the various nations and promote the ongoing dialog. The laws and regulations governing the manufacture and marketing of cosmetic products in the European Union (EU), the USA, Canada, Japan, China, Brazil, Australia, Korea, and India represent some of the leading global markets, which are discussed below.

9.3.1 European Union (EU)

The definition of a cosmetic in the EU includes several additional categories, such as pharmaceuticals, biocides, and medical devices, and is centered on the area of application and possible uses. According to the EU, a cosmetic is defined as "Any substance or mixture intended to be placed in contact with the external parts of the human body (epidermis, hair system, nails, lips, and external genital organs) or with the teeth and the mucous membranes of the oral cavity with a view exclusively or mainly to cleaning them, perfuming them, changing their appearance, protecting them, keeping them in good condition, or correcting body odors." Before a product is put on the market in the EU, the responsible person (RP), typically the producer or the importer, must guarantee its safety. For this reason, the RP must ensure that a cosmetic product safety report (CPSR) is established and that the cosmetic product passes a safety evaluation based on the relevant data (European Union 2009a). According to Regulation (E.C.) 1223/2009, the safety assessor (SA), who is appointed by the RP and qualified in pharmacy, toxicology, medicine, or a related field, or who has completed a course that is accepted as equivalent by a Member State, conducts the safety assessment. The regulation for the SA enlists just one criterion; it does not include any further prerequisites or a definition. As a result, even though the same laws are enforced throughout the EU, the CPSR may differ because it was prepared by experts with various educational backgrounds, professional experiences, and depth of knowledge. The CPSR is divided into two parts and may be obtained in the product information file (PIF) of the cosmetic—Part A:

cosmetic product safety information, which contains all the information required for the safety assessment of the product; Part B: cosmetic product safety assessment, which is the opinion of the cosmetic safety assessor about the product's safety.

In contrast, the PIF includes the following details: the cosmetic product description, the cosmetic product safety report, a detail of the manufacturing process and a statement of compliance with GMP, proof of the effects claimed for the cosmetic product, and information on any animal test carried out by the producer, his agents, or suppliers in relation to the development or safety analysis of the cosmetic product or its ingredients to meet legislative requirements (European Union 2009a). Whenever product modifications or new information becomes available, the CPSR and PIF must be constantly updated and revised. The RP must also provide some information via the cosmetic products notification portal (CPNP), such as the product category and identity, probable exposure conditions, and the framework formulation. Except for cosmetic goods containing nanomaterials, which are subject to an additional procedure, the notification method is the same for all cosmetic items (Ferreira et al. 2022).

Specific criteria for marketing cosmetic goods containing nanoparticles were set by Regulation (C.E.) No 1223/2009. According to Article 16 of Regulation (CE) No. 1223/2009, manufacturers must notify the EC in advance of their intention to use nanoscale ingredients by sending product-related information to the Cosmetic Products Notification Portal (CPNP) 6 months before releasing the product for sale. An estimate of the amount of nanomaterial in the cosmetic product intended to be marketed annually, its toxicological profile, and safety data of the nanomaterial used in a product, depending on the category of the cosmetic product and its exposure conditions, should all be included in the notification. It must also include information about the nanomaterial identification and its specifications, such as particle size and physical and chemical properties. The Scientific Committee on Consumer Safety (SCCS) published recommendations in 2012 outlining the standards to undertake physicochemical characterization, identify the toxicological profile, and determine a nanomaterial's highly probable exposure conditions (SCCS 2012). In order to ensure that consumers can use cosmetic products containing nanomaterials more safely, the regulation also mandates that manufacturers clearly identify nanomaterials on the label by placing the word "nano" after the INCI name of the ingredient (European Union 2009b). However, there is still debate regarding the effectiveness of such a legal obligation.

9.3.2 United States of America (USA)

For instance, a product may fall under two classifications in the USA simultaneously. For example, an antidandruff shampoo may be both a cosmetic and a medication because it has two purposes: to clean the hair (cosmetic) and to treat dandruff (drug). In these situations, the item in question must adhere to both rules (US-FDA 2022). The Federal Drug and Cosmetic Act (FD&C Act) defines the two major product categories: cosmetics and pharmaceuticals, the latter of which includes a subcategory of over-the-counter (OTC) medications that can be marketed without a prescription (US-FDA 2018). Cosmetics are defined under the Federal Food, Drug, & Cosmetic Act (FD&C Act) as "articles intended to be rubbed, poured, sprinkled, or sprayed on, introduced into, or otherwise applied to the human body for cleansing, beautifying, promoting attractiveness, or altering the appearance." Products, including skin moisturizers, lipsticks, nail paint, eye and facial makeup formulations, shampoos, permanent waves, hair dyes, toothpaste, and deodorants, are all covered by this definition with any substance manufactured to be used as a cosmetic product component. Both ingredients and finished cosmetics imported into the USA must adhere to the same safety and labeling standards as products manufactured locally in the USA. Except for color additives (other than the coloring substances used in coal-tar hair colors), which must be approved for the specific intended application, the FDA does not need pre-market approval of cosmetics. Therefore, the product's manufacturers or distributors have to guarantee its safety. Cosmetic companies must advertise safe, correctly labeled cosmetics, use no prohibited substances, and abide by limitations on restricted ingredients. Additionally, it is considered best to strictly adhere to safety recommendations and criteria issued by the industry.

The FDA in the USA regulates the use of nanomaterials in cosmetics. The FDA evaluated scientific and regulatory considerations for the safety and efficacy of goods incorporating nanomaterials in the FDA Nanotechnology Task Force Report of 2007 (Fytianos et al. 2020). Guidelines outlining safety concerns for cosmetic items containing nanomaterials were suggested by the Task Force. Based on it, producers ought to take safety measures to ensure the safety of nanomaterial-based cosmetic goods. Nanotechnology and nanomaterials are still not subject to a specific regulatory definition. In 2014, the FDA released guidelines for the industry called "Final Guidance for Industry—Safety of Nanomaterials in Cosmetic Products" (FDA 2014) which evaluates safety concerns and offers guidance to the cosmetic industries (Katz et al. 2015).

9.3.3 Canada

In Canada, a cosmetic product is defined by the Food and Drugs Act as "any substance or mixture of substances, manufactured, sold, or represented for use in cleansing, improving, or altering the complexion, skin, hair, or teeth, and includes deodorants and perfumes." This includes items purchased in bulk by institutions (such as hand soap in schools) and utilized by professional esthetic services, as well as handmade cosmetics offered at craft fairs or products made by home-based companies. In Canada, the manufacturer has to guarantee the cosmetic product's safety. Health Canada must be notified of the sale of any cosmetics in Canada. Within 10 days of the product's initial sale, manufacturers must submit a Cosmetic Notification Form (CNF) for each product. This online notice form provides details such as the manufacturer's address and phone number, the function and type of the cosmetic, and the concentration of ingredients. The producer is mainly responsible for ensuring that the product complies with all regulatory obligations; thus, the notification does not signify approval for sale or any other kind of guarantee about the product's safety (Canada.ca 2017). The use of nanoparticles in various Canadian cosmetic markets is expanding. According to Health Canada, a nanomaterial is "any substance or product manufactured, and any component material, ingredient, device, or structure if: (1) it is comprised within the nanometric dimensions in at least one external dimension or has an internal dimension or surface structure within the nanoscale, or (2) it is smaller or larger than the nanoscale in all dimensions, but exhibits one or more nanoforms properties or phenomena." A list of hazardous cosmetic compounds, more precisely, a list of cosmetic ingredients that are restricted or forbidden, was created by Health Canada in 2007 (Kumud and Sanju 2018).

9.3.4 Japan

In Japan, cosmetics are described as "articles with mild action on the human body, which are intended to be applied to the human body through rubbing, sprinkling, or other methods, aiming to clean, beautify, increase the attractiveness, alter the appearance, or to keep the skin or hair in good condition" (Japan Ministry of Health 2014). There are six categories of cosmetics: perfume and eau de cologne, makeup, skincare items, hair care products, special-purpose cosmetics, and cosmetic soaps. In order to register a cosmetic product in Japan, the authorities must first obtain cosmetic manufacturing and marketing licenses. Each license has specific demands. However, to preserve their products' integrity, marketing license holders must adhere to the Good Vigilance Practice (GVP) and the Good Quality Practice (GQP) standards. After receiving the necessary licenses, the manufacturers must submit a cosmetic marketing notification to the same prefecture that issued the permit. The product can subsequently be placed on the market once all the previously mentioned standards have been met (Crevedo 2022).

9.3.5 China

According to the definition of cosmetic products in China, these are "daily chemical products intended to be applied on the external part of the human body (such as skin, hair, nails, lips, etc.) by spreading, spraying, or other similar ways for cleansing, protecting, beautifying, or grooming purposes." (Su et al. 2020). The State Administration for Market Regulation (SAMR) and the National Medical Products Administration (NMPA), an independent Drug Administration Bureau governed by SAMR, are the two primary competent bodies in China that oversee cosmetic rules. The NMPA has nine subsidiary departments, one of which is the Cosmetic Safety Supervision Department. Medical Products Administrations (MPAs), located at the provincial level and under NMPA, are in charge of filing domestic, non-special use cosmetics and issuing production licenses to cosmetics firms. The 1989 Regulations concerning the hygiene supervision of cosmetics served as the basis for China's current regulatory framework. In China, multiple laws must be followed and considered, but the Technical Safety Standard for Cosmetics 2015, which replaced the Hygiene Standard for Cosmetics 2007, is the most significant. Special cosmetics must be registered and approved by the NMPA before manufacture, but regular cosmetics can be put on the market immediately following notice under China's new legislation. From January 1, 2022, prior to registration or notification, the registrant or notifier must either conduct a self-assessment safety review or delegate this duty to a qualified agency. They must also disclose the product safety assessment results at the registration or notification time (Su et al. 2020).

9.3.6 Brazil

Personal care items, including cosmetics and perfumes, are described in Brazil as "Preparations consisting of natural or synthetic substances, for external use on various parts of the human body, including the skin, capillary system, nails, lips, external genital organs, teeth, and mucous membranes of the oral cavity, with the sole or main purpose of cleaning them, perfuming them, altering their appearance, correcting body odours, and or protecting or maintaining them in good condition" (Pomela 2015). The registration processes vary depending on the type of product in Brazil. Pre-market approval processes are required for some of the products classified as grade II cosmetics and mentioned in Annex VIII of Resolution RDC 07/2015. Following the publication date in the Brazilian Official Gazette, these procedures are effective for 5 years and may be renewed for further equal-length periods. Pre-market approval is optional for cosmetic items not listed in Annex VIII of Resolution RDC 07/2015; the Brazilian Health Regulatory Agency (Anvisa) must be notified. The Cosmetic Automation System (SGAS System) is used for the online notification process, which is effective for 5 years from the day the online protocol is finalized and can be extended for further equal-length periods.

Nanomaterials and nanotechnology-specific regulations do not exist in Brazil. A discussion on nanotechnology and security surveillance was encouraged in 2012 by ANVISA (National Agency for Sanitary Vigilance). The Internal Committee of Nanotechnology (CIN) was founded in 2013 to validate the state of our understanding of nanomaterials. They prepared a document outlining the initiatives and regulatory frameworks related to nanotechnologies in other nations and alternative principles and frameworks (Melo et al. 2015).

9.3.7 Australia

Cosmetics, according to the Australian Government, are "substance that is designed to be used on any external part of the human body – or inside the mouth – to change its odors, change its appearance, cleanse it, keep it in good condition, perfume it or protect it." The import, production, marketing, and delivery of cosmetics are tightly controlled and complex in Australia. Before introducing a cosmetic product into the

Australian market, it must undergo various product assessments to determine the necessary approvals and registrations. Trademark and/or patent clearances must also be obtained to reduce the risk of infringing on the intellectual property rights of others. Labeling, packaging, and advertising must also be checked for compliance with the Australian Consumer Law and applicable advertising codes. The Therapeutic Goods Administration (TGA), the Department of Health under the Australian Industrial Chemicals Introduction Scheme (AICIS), and the Australian Competition and Consumer Commission (ACCC) are the three government regulators in charge of monitoring cosmetics regulation in Australia. The National Industry Chemicals Notification and Assessment Scheme (NICNAS) monitors the safety of the ingredients in cosmetics and personal care items in Australia, while the Therapeutic Goods Administration (TGA) regulates sunscreens, which are therefore regarded as drugs. However, neither of these associations makes a distinction between bulk materials and nanoparticles (Raj et al. 2012). In Australia, the NICNAS defines a nanomaterial as "an industrial material intentionally produced, manufactured, or designed to have specific properties or a specific composition and one or more dimensions, typically between 1 and 100 nm." As mandated by the TGA, the Industrial Chemicals Notification and Assessment Act 1989 regulates all chemical ingredients (including natural items) as industrial chemicals (Kumud and Sanju 2018).

9.3.8 Korea

South Korea, which accounts for around 2.8% of the worldwide market, is among the top ten cosmetics markets in the world. "K-Beauty" is rising, and Korea is regarded as the world's center of innovation in the cosmetic industry. South Korean products dazzle with their efficiency, packaging, and sensory appeal, thus inspiring Western brands. In South Korea, cosmetics are rubbed, sprayed, or otherwise applied to the skin or hair to maintain, improve, or enhance the appearance of the skin or hair. Currently valued at around $10 billion, the South Korean cosmetics market is predicted to grow at a CAGR of 4.95% from 2017 to 2030 (Peters and Choi 2020). The South Korean Government issued the comprehensive cosmetics laws known as the Cosmetics Act 3 (Act No. 17250) in 2000 to help improve public health and expand the cosmetics business. The law, which was most recently revised in April 2020, includes provisions for manufacturing, importing, and marketing cosmetics and cosmetic ingredients and specific guidelines for product labeling and promotion (Peters and Choi 2020).

9.3.9 India

The Drugs and Cosmetics Act of 1940, guidelines from 1945, and labeling declarations issued by the Bureau of Indian Standards (BIS) govern cosmetic items in India. The BIS established the cosmetic standards listed in Schedule "s" of the Drugs and Cosmetics Rules of 1945. In addition, the BIS provided the specifications for skin creams and lipsticks in Indian Standards (IS) 6608:2004 and 9875:1990 (Nanda 2018). Each raw material must pass a heavy metal test in accordance with Indian Standard 6608:2004. If raw materials are screened early, the manufacturer may not need to test the final cosmetic product for heavy metals (CliniExperts 2016). Rule 134 of the Drugs and Cosmetics Rules contains restrictions on the use of cosmetics, including colors, pigments, and dyes besides those listed by the BIS and Schedule Q. Arsenic and lead compounds are no longer allowed to be used as coloring agents in cosmetic products, according to D&C Rule 145. The import of cosmetics containing arsenic or lead is prohibited by Rule 135.

Similarly, the manufacture and import of cosmetics with mercury-containing ingredients are prohibited by Rules 145 D and 135 A (Centre for Science and Environment n.d.). The "Nanotechnology Sectional Committee" group has been established by the Bureau of Indian Standards (BIS) and comprises 33 members from various research institutions and companies. This committee attempts to standardize laws governing nanotechnology (Kumud and Sanju 2018).

9.4 Ban of Cosmetic Testing on Animals

Animals have long been employed in biomedical research as significant experimental subjects due to their physiological resemblance to humans. Animal testing is typically necessary to determine the efficacy and safety of drugs. Animals occasionally experience injury, discomfort, suffering, and even death. Animals are extensively employed in preclinical research for many significant diseases since pharmaceuticals are a necessary commodity. Animal testing has been used for many years to evaluate cosmetic products. Cosmetics are not considered an essential commodity; instead, they are luxury goods. Using in vivo tests for cosmetic items has long raised ethical concerns due to their potential to cause skin irritation, stinging, contact urticaria, allergic sensitization, photoallergy, and phototoxicity. The issue of animal experimentation in cosmetics has received a lot of attention over the years, and consumers are becoming more aware of the issue and imposing higher demands on the sector to ensure the welfare of animals. Fortunately, the cosmetics sector is prioritizing finding alternatives to animal testing, and the number of nations with enforceable bans on animal testing is expanding. Many countries have implemented laws that forbid using animals to test cosmetics in order to prevent unnecessary animal suffering and harm. The present scenario of implementation and bans of animal experimentation in a few nations is discussed herein:

Europe—The European Union was the first to ban animal testing for cosmetics. As of March 2013, the European Union entirely prohibited the sale and import of cosmetics that have undergone animal testing or the use of ingredients that have undergone such testing. The European Union is a significant market for cosmetics businesses worldwide, and this policy has compelled various nations, including China and South America, to seek alternatives for animal testing methods employed in the cosmetic industry (Skincare n.d.; Sreedhar et al. 2020).

USA—Currently, legislation banning the use of animals in cosmetic research has been approved in eight U.S. states: California, Hawaii, Illinois, Maine, Maryland, Nevada, New Jersey, and Virginia (The Human Society of the United States n.d.). Even though the FD&C Act does not entirely ban the use of animals in safety cosmetic studies, and the FDA supports the use of alternative methods for the improvement, reduction, and replacement of animal testing, it is the manufacturers' responsibility to carry out whatever in vivo tests are deemed necessary to maintain the safety of their products in the rest of the country (US-FDA 2021b).

Canada—Animal testing for cosmetics is not prohibited in Canada. Bill S-214 (the Cruelty-Free Cosmetic Act) was introduced in Canada in 2015 to stop using animals for cosmetic research and the sale of cosmetic items produced using these methods. However, this bill has yet to become law, so it is still acceptable to utilize such procedures (Toronto Humane Society 2021).

Japan—Japan is in the process of banning animal experiments for testing cosmetics. Till today, there has been no specific legislative obligation in Japan for all cosmetic goods to be tested on animals, and there are also no laws that forbid such tests. There is no application process for approval, and each cosmetics manufacturer is urged to ensure the quality of their products in accordance with the self-responsibility principle. The manufacturers have been given the authority to conduct their own safety assessments of raw materials and final goods per their requirements.

China—Global trade between regions like the European Union and nations with "cruelty-free" testing standards has long been significantly hampered by China's mandatory animal testing requirement for cosmetics registration, But China has begun to harmonize its laws as many other nations eventually prohibit animal experiments. The need for general cosmetics, whether imported or produced in China, to undergo animal testing was officially abolished in China on May 1, 2021. Nevertheless, a few prerequisites and exceptions may exist. One requirement is to provide the GMP certification from the nation's or region's cosmetic regulatory authority. Since many nations still need to give this form of GMP certification, it is challenging to meet this criterion (RedOrangePeach 2022).

Brazil—Animal testing on cosmetics has already been prohibited in some Brazilian states, including Mato Grosso do Sul, Amazonas, Paraná, Minas Gerais, Pará, Pernambuco, Santa Catarina, Rio de Janeiro, Sao Paulo, and the Federal District. Anvisa guidelines, however, continue to acknowledge the use of animal testing to evaluate the risks associated with cosmetics and their constituents (Humane Society International 2021).

Australia—Animal testing for cosmetic safety is not permitted in Australia after the ban commenced on July 1, 2020. Cosmetics and products tested on animals outside of Australia are also prohibited from being sold in Australia (AG-Department of Health and Aged Care 2019).

Korea—The production of cosmetics involving animal testing was planned to be prohibited by the Korea National Assembly in 2018. South Korea's Ministry of Agriculture, Food, and Rural Affairs has prepared plans to ban the use of animals in cosmetic testing. The Government's Five Year Plan for Animal Welfare forbids testing of finished cosmetic products on animals, while a ban on testing ingredients is still up for discussion (Cruelty Free International 2016).

India—Animal testing on cosmetics was prohibited in India in 2014 (The Times of India 2014a). The Ministry of Health and Family Welfare has incorporated the new regulation into the already-existing Drugs and Cosmetic Rules, 1945. According to the new law, testing cosmetic products on animals is forbidden. The import of cosmetic products tested on animals is also prohibited in India (The Times of India 2014b).

9.5 Alternative Methods for Animal Testing

For a long time, there has been debate about animal suffering, distress, and death during scientific research. It is argued that as animals are living organisms, they have a right to be free from pain and suffering, and using them in research is considered unethical and ought to be discontinued. Numerous acts and legislative measures have been passed to reduce animal suffering during testing and ensure the ethical use of animals. For instance, the Royal Society for the Prevention of Cruelty to Animals founded the animal rights organization in 1824. Another law was passed in the UK in 1876 to combat animal cruelty. It was introduced in India, France, and the USA in 1960, 1963, and 1966, respectively (Doke and Dhawale 2015). To protect animals from abuse and cruelty, a number of laws and guidelines are currently observed on a global scale. Guidelines for animal housing, breeding, feeding, transportation, and, most importantly, their use in scientific experiments are provided by organizations like ICH (International Conference on Harmonization of Technical Requirements for Registration of Pharmaceuticals for Human Use), CPCSEA (Committee for the purpose of Control and Supervision on Experiments on Animals), NIH (National Institutes of Health), and OECD (Organization for Economic Cooperation and Development). In addition to the primary ethical issue, other drawbacks of animal experimentation include the need for skilled or trained personnel and time-consuming procedures. Another disadvantage is the exorbitant cost associated with housing and breeding animals and the lengthy protocols of animal research. Several alternatives have been proposed to address some of the issues with animal testing and to prevent unethical practices. The search for the development of new methods and techniques aimed at the reduction, refinement, and replacement of the use of animals in research has become a global trend since the introduction of the 3R's principle (replacement, reduction, and refinement) in 1959 by Russell et al. in their book "The principles of humane experimental technique." The development of alternatives to animals has dramatically increased during the past 30 years. The following section provides a brief summary of several contemporary alternative techniques:

Computational Approaches—The numerous possible biological and toxic effects of a chemical, and potential pharmaceutical candidate are predicted using computational models, which avoid the need for dissecting animals. Examples of such approaches include quantitative structure–activity relationships (QSARs), grouping and read across, computer-aided drug design (CADD), and physiologically based kinetic (PBK) models. QSARs are computer-based methods for estimating a substance's likelihood of being toxic based on its similarities to other drugs and our understanding of human biology. They have the potential to replace animal experiments with cosmetic testing. QSAR techniques are being used more frequently by companies and governments to avoid using animals in chemical testing. Only the most promising compounds discovered through primary screening are tested in vivo. For instance, in vivo testing is required to determine the receptor-binding site of a drug. A potential receptor-binding site of a drug molecule can be predicted using CADD software. In order to prevent undesirable testing compounds with no biological activity, CADD attempts to identify probable binding sites. These software tools can also specifically customize a novel drug for a given binding site. Finally, animal testing is carried out to get conclusive data. The computer database does a remarkable job of predicting possible properties of drug candidates, like carcinogenicity and mutagenicity. The most modern QSAR software provides more accurate results indicating a molecule's ability to cause cancer.

The procedures' speed and relatively low cost make computer models superior to traditional animal models (Doke and Dhawale 2015). In silico predictions are also a great strategy when combined with partially concluded data, such as the in vitro mutagenicity test. While the in vitro micronucleus assay is necessary to examine genotoxicity, a QSAR prediction can assist in better understanding a substance's ability to cause DNA damage before the tests are carried out. According to the ICH M7 guideline, such methods are widely approved for the regulatory assessment of pharmaceutical impurities (Fioravanzo et al. 2012). Organs simulated on a chip and vast chemical databases are now available to researchers to determine whether a cosmetic is likely to adversely affect humans. The COSMOS project has developed sophisticated computer models that can predict where a chemical will end up in the body after coming into contact with human tissue. A database of more than 5000 ingredients used in cosmetics and their effects is also being generated as part of the COSMO project (European Commission 2015).

In vitro test systems—An important substitute for animal testing is in vitro cell and tissue cultures, which involve the growth of cells outside the body in a laboratory setting. After being removed from the animal, the cells and tissues from the liver, kidneys, brain, skin, and other organs can be maintained outside the body for a few days to several months or even a few years in an appropriate growth medium. Animal and human cells are isolated and grown as a monolayer on the surface of culture plates or flasks during in vitro culture. It is also possible to use cellular components like membrane fragments and enzymes. There are many different uses for various cultures, including cell, callus, tissue, and organ culture. The advantages of in vitro methods are their simplicity, efficiency, time-saving, and low cost. To assess the toxicity and efficacy of potential therapeutic compounds and chemicals, several in vitro approaches are routinely used (Clift and Doak 2021). These in vitro tests determine the effectiveness and toxicity of almost all cosmetics, pharmaceuticals, and chemicals. Researchers at Wyss Institute, Harvard University, have developed "organs-on-chips technology" that mimics the microenvironment and physiological processes of human organs like the lung, liver, brain, and skin. Compared to animal research, they are more accurate in simulating human physiology and can substitute animals that endure painful, lengthy tests to determine whether cosmetics are toxic or likely to irritate the skin, eyes, or other body tissues. MatTek's cornea-like 3D tissue structures made from human cells can be employed instead of rubbing or dripping cosmetic products into the eyes of rabbits (Lee et al. 2017).

In vitro dermal absorption tests, capable of predicting probable dermal absorption in humans, are the gold standard method for studying skin pharmacokinetics. Through in vitro skin absorption tests, several formulation types, including hair dyes, shampoos, foundations, moisturizers, cleansers, soaps, sunscreen, suspensions, foams, patches, and aqueous formulations, can be tested. Detailed instructions for conducting in vitro skin absorption tests can be found in the OECD Guidelines 2004, 2011, and 2019 (OECD 2004a, b, 2011). A first set of "basic criteria" for the in vitro evaluation of skin absorption of cosmetic ingredients was adopted by the SCCNFP (Scientific Committee on Cosmetics and Nonfood Products) in 1999 and amended in 2003 (SCCNFP/0750/03). In 2010, the SCCS (Scientific Committee on Consumer Safety) revised this opinion and released it as (SCCS/1358/10) (Barthe et al. 2021).

For performing appropriately in vitro skin absorption tests for cosmetic chemicals, the OECD 428 guideline and the SCCS "Basic Criteria" (SCCS/1358/10) are viewed as necessary. When conducting an in vitro dermal absorption test, a skin sample is placed in a Franz-type diffusion cell between two chambers (a donor chamber and a receptor chamber), with the stratum corneum facing the donor compartment and the dermis touching the receptor compartment. Most of the time, patients undergoing plastic surgery provide human skin samples. The most convenient skin to work with is abdominal skin, owing to its large surface area. When skin viability and metabolic activity are not being explored, carefully managed frozen human skin is adequate for investigating the passive penetration of cosmetic compounds (Barbero and Frasch 2016). However, fresh skin samples are required for studies requiring the presence of live epidermal tissue, such as analyses of drug transporters (Clerbaux et al. 2019) *and skin metabolism* (Alriquet et al. 2015).

Living human keratinocytes have been cultured to produce a multi-layered, highly differentiated epidermis for the RHE skin model. The model contains a functional skin barrier with an in vivo-like lipid profile and well-structured basal cells. Eye irritation testing is conducted using a commercially available 3D model based on a reconstructed human cornea-like epithelium (RhCE) (OECD Test Method 492) (OECD 2019). Living human cells were used to develop the multi-layered, differentiated corneal epithelium that makes up the RhCE corneal model. The endpoint used in both RhE and RhCE test procedures is the reduction of MTT (3-(4,5)-dimethyl-2-thiazolyl-2,5-dimethyl-2H-tetrazolium bromide) by cells into a blue formazan salt, which is quantitatively evaluated after extraction from tissues. Interleukin-1 (IL-1) production measurement is a second endpoint that can be utilized to enhance sensitivity. If the viability of the test item is greater than 50% (RhE) or 60% (RhCE) (no label or UN GHS no category), it is classified as a non-irritant. If the viability of the test item is less than or equal to 50% in the RhE model, it is classified as an irritant (Barthe et al. 2021). No prediction can be made if the viability of the RhCE model is less than or equal to 60%; additional testing may be necessary. The hen's egg test on chorioallantoic membrane (HET-CAM) has been utilized for eye irritation and toxicity testing. These tests have described irritation levels ranging from barely irritating to severely irritating compounds (Prinsen et al. 2017). In vitro systems like the KeratinoSens™ assay, which uses immortalized human keratinocytes (HaCaT) lineage transfected with a selected plasmid, have been validated to evaluate the sensitization potential of chemical compounds (Natsch et al. 2015). EpiDerm™, EpiSkin™, and SkinEthic™ are OECD-validated models that reasonably resemble human skin (16–18).

The exposure of the skin to solar irradiation and photoreactive xenobiotic compounds, including cosmetics, may cause unusual skin problems. Phototoxicity is an acute light-induced reaction that occurs when photoreactive agents in cosmetics are activated by sunlight and converted into toxic products in skin cells. The primary focus is on non-animal test methods, such as in vitro and chemico (cell-free test tube methods), which determine the phototoxicity of cosmetics to minimize animal suffering and agony. Standard fibroblast cells derived from Swiss mouse embryonic tissue cells (3 T3) are used in the in vitro 3 T3 NRU ultraviolet experiment (95% correlation with in vivo assay) to assess the 50% mean inhibitory concentration (IC50), with and without exposure to solar radiation (Nabarretti et al. 2022). *In chemico,* methods have been employed to detect the formation of reactive oxygen species or DNA strand break activity in cosmetics with a potential for phototoxicity. Other in vitro test systems include the erythrocyte photohemolysis test and the phototoxicity test employing a human 3-dimensional (3D) epidermis model (Kim et al. 2015).

There are a few drawbacks associated with these isolated systems as well. Typically, they cannot provide all of an organism's physiological responses. When removed from the organism, the components frequently degenerate and lose the capacity to carry out their unique functions. Another disadvantage is that the impact of the exposure route, which has a significant effect on the test results, cannot be evaluated with these approaches.

Alternative Organisms—The use of higher model vertebrates for experimentation, such as guinea pigs, rats, dogs, and monkeys, have been greatly restricted by ethical concerns. The use of alternative organisms such as plants, single-celled organisms, invertebrates, and other non-animal organisms has been suggested that can be used in cosmetic testing in place of experimental animals. All of these can react to certain noxious stimuli, and some may experience pain. Nevertheless, many analysts suggest that they do not sense pain or suffering in the same manner as animals do, especially when there is no brain or neural tissue present.

Microorganisms—The use of bacteria and fungi to evaluate various genotoxic effects has received increasing attention in recent years. These organisms have the significant benefit of being much simpler and quicker to culture than most animal or human cells. Their genetic makeup is more straightforward than that of animals and humans. Furthermore, a broad understanding of their physiology and functions facilitates their use, particularly in toxicological research that leads to the development of new techniques. Genetic material alterations are relatively simple to identify and characterize. The use of fungi in mutagenicity testing has been demonstrated to be very beneficial, and they appear to be more sensitive than bacteria.

Brewing yeast, or Saccharomyces cerevisiae, is the most well-known and significant model organism due to its quick growth, ease in replica plating and mutant isolation, dispersed cells, precisely defined genetic system, and highly adaptable DNA transformation system. S. cerevisiae contributes to our understanding of the fundamental cellular biology in neurodegenerative diseases like Alzheimer's, Parkinson's, and Huntington's diseases by analyzing endogenous or heterologous proteins whose aggregation is the root cause of these ailments (Pereira et al. 2012). Slime molds, algae, and protozoa have also been shown to be beneficial. Protozoa typically have specialized functions that resemble humans, although they are generally relatively primitive. The cilia in the human bronchial tube, for example, and those of protozoa both react to smoke or phenols. Smoke toxicity tests have utilized a variety of protozoans. Protozoans are currently being considered for use in screening tests for carcinogenesis, mutagenesis, and reproductive toxicity (Doke and Dhawale 2015). These microorganisms can effectively be used for toxicity testing cosmetic products and their ingredients.

Invertebrates—Animals employed in laboratories are frequently replaced with invertebrate species. Invertebrates have been used to study various diseases, including Parkinson's disease, endocrine and cognitive disorders, muscle dystrophy, wound healing, cell aging, programmed cell death, retrovirus biology, diabetes, and toxicological testing (Castillo and de la Guardia 2017). There are several restrictions on the use of invertebrates in treating human diseases since they lack the adaptive immune system and have undeveloped organ systems. However, many invertebrates can be studied in a single experiment in a short time with fewer ethical issues because of their numerous advantages, like their short life cycle, small size, and superficial anatomy. In comparison with animals, their maintenance costs are also lower. For instance, a shelter that can only house a few mice could house thousands of flies. One of the most extensively researched invertebrate species is the fruit fly, *Drosophila melanogaster* (Allocca et al. 2018). Its genome has been extensively studied, making it possible to investigate the molecular processes that underlie human diseases. Its entire genome, which contains more than 14,000 genes on four chromosomes, has been sequenced and annotated. The majority of *Drosophila melanogaster's* genome is carried by just three genes. It is believed that the fly has functional homologs of over 75% of the genes linked to human diseases. Numerous organs in the fly, including the heart, lungs, intestines, kidneys, and reproductive system, perform similar activities to those of mammals. Various molecular and genetic methods are currently available for determining the mutagenicity, teratogenicity, and reproductive toxicity of *Drosophila melanogaster.* Many drugs that affect the central nervous system produce similar responses in flies as in humans. Fruit flies were an exceptional and sensitive model for studying human genetics and diseases because of their many similarities in development and behavioral activities. The *Drosophila melanogaster* (Meigen) somatic mutation and recombination test, also known as the "wing spot test," was used to assess the genotoxicity of 10 essential oil constituents used as flavoring agents or cosmetic ingredients as part of a screening project aimed at determining their mutagenic activity (Mademtzoglou et al. 2011).

Caenorhabditis elegans (C. elegans), another invertebrate model, has been widely used in toxicity testing. It is a tiny nematode that may be maintained using in vitro methods at a low cost. It has been frequently demonstrated that toxicity ranking screens in *C. elegans* are just as predictive of LD50 rankings in rats and mice. Furthermore, numerous cases of conservation of toxicant modes of action between *C. elegans* and mammals have been identified. These strong correlations support the use of *C. elegans* assays in early safety testing of cosmetics and as a part of tiered or integrated toxicity evaluation techniques. Still, they do not suggest that nematode data may substitute for mammalian data in assessing health hazards and toxicological assessments of cosmetics. Cosmetic safety testing studies employing *C. elegans* would provide findings from an entire organism with intact and metabolically functioning digestive, reproductive, endocrine, sensory, and neuromuscular systems, in contrast to toxicity testing utilizing in vitro cell cultures. The Complex Object Parametric Analyzer and Sorter™ (COPAS) automates the examination of several endpoints on hundreds of *C. elegans* per minute using microfluidics and laser-based technologies. Studies evaluating six or seven water-soluble compounds have revealed that the COPAS ranking for these endpoints in *C. elegans* coincides with the mouse LD50 ranking for the same compounds. The COPAS has also been used to evaluate larval growth and reproductive production (Hunt and The 2017). COPAS-based quantification of hundreds of compounds from the U.S. EPA's ToxCast™ phase I and phase II libraries on *C. elegans* larval growth correctly predicts developmental toxicity in rabbits or rats with a balanced accuracy of 45–53%, which is somehow less than the concordance for developmental toxicity between rats and rabbits, which was 58% (Boyd et al. 2016).

Lower Vertebrates—A small freshwater fish known as zebrafish (Danio rerio) has been used as a cost-efficient alternative to filling the void between fully synthetic techniques and mammalian model systems. Research on embryonic zebrafish provides an excellent middle ground for testing cosmetic products and their ingredients by enabling scientists to access the benefits of working with mammals more responsibly and ethically while still helping high-throughput cosmetics testing. During early development, the optical clarity of zebrafish enables easy screening, direct observation of gene expression, developmental stages, and phenotypic traits, and efficient evaluation of cosmetics toxicity test endpoints. Its laboratory maintenance on cell culture plates and petri dishes is favored by its small size, short life cycle, and high fecundity (Lachowicz et al. 2021). Organ-focused data could be obtained using zebrafish embryos. This information can only be provided by organoids or other non-animal models such as in vitro, tissue-based, or ex vivo. This offers a comparative benefit compared to in vitro cellular experiments. The transparency of zebrafish embryos enables noninvasive tests and permits easy monitoring, which could be particularly advantageous for developmental toxicity tests for cosmetics and their ingredients. The in vivo visualization of tissue/cells in acute toxicity studies for cosmetics is a significant advantage of zebrafish. These in vivo readouts are far more accurate and informative because the cells or tissues used in in vitro tests are just a part of a living organism.

Human Volunteers—Prior to large-scale human trials, a technique known as "microdosing" might offer crucial information about the safety of cosmetics. Volunteers are exposed to a very low, one-time dose of the test compound, and the effects are monitored using advanced technologies. Microdosing can substitute specific animal tests for cosmetic safety testing and help identify cosmetics that would not work on humans, preventing their testing on animals. High levels of safety must always be maintained. Human interests should always receive priority over scientific and societal interests. As a result, the investigator should stop conducting the study as soon as it is recognized that the risks exceed the expected benefits. An ethical committee should be consulted for compatibility tests for cosmetic items that could harm volunteers, provided that the committee complies with all applicable rules and regulations in the country where the study is being conducted. Human volunteers should be fully informed of the study's objectives, procedures, and potential discomfort (Nobile 2016).

Prior to participating in the study, free and informed written consent is required from all volunteers. Modern brain imaging and recording methods using human volunteers, such as functional magnetic resonance imaging (fMRI), could replace antiquated cosmetic testing studies using animals with brain damage. Researchers may now safely study the human brain at the level of a single neuron (using intracranial electroencephalography), and they can even use transcranial magnetic stimulation to temporarily or irreversibly induce brain diseases.

Human–Patient Simulators—Computerized human–patient simulators, which can breathe, bleed, convulse, talk, and even die, could be used to study the biological response to the application of cosmetics and their ingredients. These simulators are much more effective at teaching students about physiology and pharmacology than simple exercises involving the dissection of animals. The most advanced simulators simulate diseases and injuries and provide appropriate biological effects for pharmacological therapies. Human simulators, virtual reality platforms, and computer simulators have largely replaced animal laboratories in medical colleges across the USA, Canada, and India. Systems like TraumaMan, which simulate breathing, and bleeding human torso with realistic skin and tissue layers, ribs, and internal organs, are frequently used to teach emergency surgical techniques for more advanced medical training. These systems are more effective at imparting lifesaving skills than programs that require students to cut into live pigs, goats, or dogs (Liventsev et al. 2021).

9.6 Roadblocks to the Implementation of Animal Alternatives

Scientific Constraints—Current scientific methods for testing theories present a significant obstacle to substituting animals for cosmetic safety testing. The standard procedure mainly entails testing a chemical in models with increasing complexity while developing trust in the hypothesis as it overcomes each obstacle. The most frequent justification for using animals is the apparent requirement to test a cosmetic in a "complex, entire being" before being sufficiently confident that testing on people can be done safely (Taylor 2019). This is based on the notion that testing cosmetics on a sophisticated and complex creature will be able to identify all potential, unforeseen ways in which a cosmetic could be damaging (or ineffective), thus bypassing the damage to human volunteers. In vitro-based approaches are not considered sufficient because they are perceived as inadequate due to their apparent lack of complexity. The potential quest to record every possible interaction of cosmetics within complex animals may raise the issue of cosmetics being tested on the wrong species. Researchers who support alternative techniques find this extremely irrational, and there is a significant gulf between the two parties regarding the importance of complexity versus relevancy. The complexity vs. relevance dispute may be resolved using the adverse outcome pathway (AOP) framework. The AOP is a systematical procedure that uses the available details about a toxicological response and explains the mechanistic interconnections between an initial molecular event, several intermediate critical events, and the adverse outcome. The AOP framework offers practical recommendations to encourage the development of alternative cosmetic testing procedures (Halappanavar et al. 2020). Another solution to increase complexity and relevance is to use "lab-on-a-chip" techniques and more advanced in vitro techniques like "3D tissue constructs" and "mini-brains" (Caruso 2017).

Traditional Barriers—Despite significant advancements in reducing and improving animal testing, the scientific and regulatory communities frequently still view animal testing as the "gold standard" to which all alternative tests must comply. Additionally, researchers that use animals in their studies will attest to the challenges associated with publishing research that employs a strategy that is distinct from the standard approach and obtaining funding for developing/using innovative alternatives. Researchers often complain behind closed doors about journal editors even requesting that their proposal or research be tested in an animal model before it is published (Cronin 2017). It is challenging for new ideas to get acceptance by the current scientific community. Research groups dissatisfied with this situation are frequently unwilling to speak up because it might adversely impact their research funding or university tenure.

Absence of Strict Laws—Under EU rules, in vitro alternatives to animal testing may be used in place of in vivo testing. Validated in vitro tests can take the place of animal studies as long as the test results are of equivalent quality and value for assessing safety. This leads to the conclusion that there are no mandatory legal requirements for alternatives to animal testing in EU cosmetics law. The language is cautious; instead of using the words "preferred" or "obliged," it is "permitted to employ" alternative approaches. Furthermore, Directive 2001/83/EC of the European Parliament and the Council of the EU often communicate mixed information. On one side, it supports the 3R's principles, but on the other, it explicitly mentions animal testing, even specifying in one instance the type of animal that should be utilized (e.g., rodent or non-rodent). This raises concerns for researchers regarding the potential replacement of animal testing entirely by in vitro tests and other recent technologies. It is currently more appropriate to replace the traditional animal test with integrated testing strategies (ITS), which incorporate both in vivo and in vitro tests (Vermeire et al. 2013).

Lack of Funding—Despite extensive efforts and notable advancements in this area, raising funds to develop alternatives to animal testing is still quite challenging. For instance, the EC and the cosmetic sector each committed €25 million to develop substitutes for using animals to assess long-term toxicity following bans on animal testing for cosmetics in 2009 (Taylor 2019). In addition, the EC has invested €180 million in replacement approaches under the most recent significant scientific funding stream, Framework Project 7 (2007–2013) (EC 2021). The total budget for Framework Project 7 was €45.3 billion, and the commission only allocated 0.4% of its research budget to alternatives to animal testing. Thus, investment in alternative development is extremely low when compared to overall science funding. National funding levels are considerably lower than central funding, possibly indicating a general lack of interest in improving scientific processes due to ethical concerns. The rate of progress in developing alternatives to animal testing of cosmetics is expected to be slow until funding levels substantially increase and are proportional to the magnitude of the problem.

Bureaucratic Barriers—The adoption of alternatives is often delayed due to bureaucratic obstacles, especially when it comes to regulatory acceptance. An effort to synchronize testing requirements globally often results in bureaucratic delays. For instance, the reconstituted skin model's initial validation was confirmed by ECVAM for detecting corrosive compounds in 1998 (ECVAM Scientific Advisory Committee (ESAC) 1998), but the OECD did not approve it until 2004 (OECD, TG 431). The OECD did not adopt the skin irritation model (OECD TG 439) till 2010, despite the fact that its initial version had been validated in 2007 (ESAC 2007). Using unusual protocols, political pressure caused the EU to accept skin procedures before the OECD for corrosion in 2000 (European Parliament and Commission of the European Communities 2000) and irritation in 2009 (Commission of the European Communities 2009). The EU never seemed to adopt this method, even though other systems have experienced comparable delays. For instance, the direct peptide reactivity assay (DPRA) for skin sensitization was not made public as OECD TG 44C until 2015, despite being authorized in 2012 (European Commission, Joint Research Centre, Institute for Health and Consumer Protection (IHC) 2012). It took more than 2 years following its OECD publication for it to be published in the EU Test Methods Regulation (Commission of the European Communities 2017). The timing of the cycle for revising test guidelines is one factor causing delays at the OECD and the EU. The process at the OECD is annual; a whole year is wasted if the deadline for submitting techniques is missed. Given enough political will, this process can be expedited by increasing the frequency of meetings. Additionally, the majority of EU members in Europe are also OECD members, thus negating the need for a second round of negotiations to revise the Test Methods Regulation.

Lack of International Harmony—The most "cautious" regulatory body establishes the permissible degree of "risk" because cosmetic items are typically produced for a global market. Manufacturers frequently continue to use animal-based models, notwithstanding the availability of alternatives or the encouragement of alternatives by particular regulators to comply with the applicable regulations in most countries and reduce the risk of a delay or rejection. This obstacle will be overcome only by harmonizing regulatory requirements internationally (Vonk et al. 2015). Harmonization is desirable since it ensures that all participating nations will accept the results of a single (animal) test undertaken in a laboratory in one country for the regulatory submission of cosmetics, thus saving time and resources. This is termed "mutual acceptance of data (MAD)." Over the past 20 years, significant efforts have been made to promote global harmony in the chemical and pharmaceutical sectors. Alternatives to animal testing must also go through the same harmonizing procedures as traditional approaches. Despite several initiatives to strengthen harmonization, different national interpretations may result in additional requirements for regulatory acceptance between nations. Other regulators may sometimes share the European desire to promote the 3R's strategies more widely. Although alternatives are now acceptable in Europe, due to a lack of international harmonization of categorization and labeling standards, rabbit skin irritation tests are still performed in Europe for non-EU regulators (Taylor 2019).

Validation of Alternative Approaches—Before alternative approaches can replace animal experiments in the market authorization process, they must undergo stringent validation procedures (Kooijman 2013). Validation of alternative approaches is relatively easy but time-consuming and demands significant financial investments from manufacturers and/or the scientific community. Its scientific output is also not highly appreciated by the scientific community and only yields small scientific credits. Additionally, recovering the expenses of the validation studies is impossible due to the lack of a market for alternative methods, which is partially attributable to the unwillingness of regulatory bodies to accept data obtained through alternative testing procedures (Kooijman 2013). The additional animal testing required to obtain adequate data for the particular context in which the alternative approach will be employed occasionally makes the validation of alternatives an effort that is not worth undertaking. Manufacturers, scientists, and regulators frequently continue to use animal-based models despite the availability of alternatives to reduce the possibility of rejection or delay (Vonk et al. 2015).

9.7 Overcoming Roadblocks to Implement Alternatives

- **It is generally recognized that only one alternative method involving** a single in vitro test or in silico prediction method **could completely replace in vivo animal tests.** Hence, Integrated Approaches to Testing and Assessment (IATA) must be used to evaluate cosmetic products' safety profiles based on AOP data. AOPs are the main component of the toxicological knowledge framework that provides a current understanding of the relationship between a molecular initiating event and an adverse outcome. Several AOPs are now being developed for various complicated toxicity endpoints in the OECD AOP initiative. The

AOPs are expected to aid in developing numerous precise in vitro test procedures and innovative integrated approaches for efficiently assessing the safety of several cosmetics and their ingredients.

- Currently, it is the researcher's responsibility rather than the regulators to demonstrate that there is no alternative available to replace animal testing in their proposed projects. The regulator evaluating a project that offers to utilize animals is not often an expert in the field. When an alternative method that can prevent animal experimentation or partially replace it is available, regulatory agencies should simply take responsibility for enforcing the law. Currently, some animal protection organizations consider it the responsibility of regulators and hold them accountable. Regulators who are genuinely committed to questioning the need for animal testing, such as those with vast knowledge of alternatives or animal protection, must only be involved in conducting ethical evaluations of research projects involving animals. Furthermore, a strict attitude adopted by regulators under a tough directive from their governments would be beneficial (Taylor 2019).
- The backbone of developing or promoting alternatives to using animals in cosmetic research and product safety testing is funding. National and international regulatory agencies must allocate substantial funds to provide essential seed money to researchers or companies interested in developing novel alternative procedures. Numerous financial incentives must also be given to companies and research laboratories to encourage the use of alternatives to animal testing in cosmetics.
- International efforts to support the work of animal welfare organizations and a general shift in public attitude resulted in a number of animal testing bans being imposed globally, along with the development of new alternatives. However, many nations continue to employ animal-based models to adhere to the prevalent laws in most countries and minimize the possibility of the global market rejecting or delaying the release of cosmetic items. The main impediments to completely eliminating animal-based tests in cosmetic testing were a need for mutual acceptance and international harmonization. An internationally harmonized testing of cosmetics and their ingredients could help companies and their products be more competitive worldwide and eliminate unnecessarily repeated testing, thus saving time and resources. Several organizations with global recognition have been promoting universal values and harmonizing animal welfare in research and safety testing. However, effective global harmonization still needs to be improved, and there is a need to develop international standards and guidelines to promote alternative approaches in the worldwide market. Several global organizations like People for the Ethical Treatment of Animals (PETA), the Food and Agricultural Organization of the United Nations (FAO), the World Organization for Animal Health (OIE), WSPA (World Society for the Protection of Animals), Animal Welfare Institute (AWI), Coalition for Consumer Information on Cosmetics (CCIC), and European Coalition to End Animal Experiments (ECEAE) must form alternative animal councils that promote the use of animal alternatives in cosmetic testing globally to ensure and advance animal welfare.

- The validation of alternative methods is the major challenge for cosmetic toxicity and safety testing. It may be worthwhile to develop more and more substantial incentives to encourage the government, business, and academia to participate in the validation process after creating an alternative test approach. This can be done, for instance, by designating a portion of public funds for research programs to only be used for the validation of in vitro and other alternative tests that are intended to replace animal testing.
- There is an urgent need to form country-specific federal agencies (for instance, the Indian Centre for Validation of Alternative Methods (InCVAM) in India) to institutionalize alternative testing methods for fostering cooperative relationships among domestic and foreign organizations for reviewing and validating proposed alternatives. This body will be intended to keenly respond to the current global trends by introducing and promoting alternative test methods developed by various organizations in the country. This body must provide policy support for developing and accepting alternative test methods that replace animal testing. It should also provide education and training regarding alternative test methods. These national agencies must also join the International Cooperation on Alternative Test Methods (ICATM), which includes other members like the Interagency Coordinating Committee on the Validation of Alternative Methods (ICCVAM), the European Union Reference Laboratory (EURL), the European Centre for the Validation of Alternative Methods (ECVAM), the Japanese Center for the Validation of Alternative Methods (JaCVAM), the Korean Centre for the Validation of Alternative Methods (KoCVAM), and Health Canada.

9.8 Conclusion

The fast-growing, highly competitive, and science-driven cosmetics sector contributes significantly to the social and economic well-being of national and regional economies worldwide (Singh et al. 2018). Due to the numerous potential applications of nanoparticles and their enhanced characteristics, there is an increasing rush to incorporate them into cosmetic preparations, and the cosmetic market is already overrun with "nano-enhanced" formulations. Nanocosmetics may offer many advantages, but one must recognize the risks associated with some nanomaterials. The nanomaterial risk assessment must be done item by item, employing relevant data. Furthermore, several regulatory agencies worldwide, each with their own set of laws and regulations, control cosmetic/nanocosmetic products. The international trade of cosmetics on the global market is substantially affected by numerous legislative measures undertaken by various nations. For decades, the cosmetic industry has been working to achieve international regulatory harmonization in cosmetic development and safety assessment for promoting global trade and animal welfare. To ensure global coverage, regulatory agencies must collaborate internationally in exchanging information about cosmetic ingredients, safety evaluation profiles, and their effects on human health. Harmonizing regulatory regulations has numerous advantages, including maintaining a favorable marketing environment, fostering productivity and competition, and minimizing unnecessary clinical testing duplication.

The welfare of animals is a subject that is as crucial as human welfare. In recent years, there has been a minor but considerable shift away from whole-animal testing toward in vitro and non-animal approaches, possibly as a result of advancements in biological techniques, ethical grounds, and in reaction to political and economic pressures. Numerous alternatives to using animals have been proposed; these alternatives must be effectively implemented. Several in vitro and computational models have been developed for the safety assessment of cosmetics and their ingredients, and some of these models have also been included in the test guidelines. Although most alternatives have not yet been fully validated, they have the potential to replace animal testing in the screening of cosmetics shortly. Alternative models have significant drawbacks, such as the inability to assess systemic toxicity and pharmacokinetic profiles and the difficulty of establishing complete physiological organ–organ interactions. As a result, the cosmetic must be tested in several contexts. In this regard, the OECD standards explicitly recommend using integrated technologies based on the AOP framework to produce more accurate results, prevent under- or overestimation of a particular cosmetic's toxicity, and improve understanding of the underlying mechanisms (Nabarretti et al. 2022).

Alternatives implementation in cosmetic testing is hindered by several barriers, including scientific constraints; traditional barriers; a lack of funding, strict laws, global harmonization, rigid regulations; bureaucratic barriers; and validation of alternatives. The obstacles to alternative implementation can be overcome by using integrated strategies based on AOP data, offering adequate funding and financial incentives for developing, validating, and using alternatives, and fostering international harmonization. Thus, research concerning the implementation of animal alternatives for the safety assessment of cosmetics and their ingredients is still a growing field that needs global cooperation between regulators, research institutes, universities, and industry. The efficient implementation of alternatives to animal testing in the cosmetic sector calls for extensive efforts to address numerous unmet needs for achieving policy changes, regulatory approval, and investment in innovation.

References

AG-Department of Health and Aged Care (2019) Ban on the use of animal test data for cosmetics

Ahmad U et al (2018) Strategies in development and delivery of nanotechnology based cosmetic products. Drug Res 68(10):545–552

Allocca M, Zola S, Bellosta P (2018) The fruit fly, Drosophila melanogaster: modeling of human diseases (part II). In: Perveen FK (ed) Drosophila melanogaster, pp 131–156

Alriquet M et al (2015) Characterization of SLC transporters in human skin. ADMET DMPK 3(1): 34–44

Barbero AM, Frasch HF (2016) Effect of frozen human epidermis storage duration and cryoprotectant on barrier function using two model compounds. Skin Pharmacol Physiol 29(1):31–40

Barthe M et al (2021) Safety testing of cosmetic products: overview of established methods and new approach methodologies (NAMs). Cosmetics 8(2):50
Bernauer U et al (2019) SCCS guidance on the safety assessment of nanomaterials in cosmetics-SCCS/1611/1
Boyd WA et al (2016) Developmental effects of the ToxCast™ phase I and phase II chemicals in Caenorhabditis elegans and corresponding responses in zebrafish, rats, and rabbits. Environ Health Perspect 124(5):586–593
Canada.ca(2017) Notification of cosmetics
Carrouel F et al (2020) Nanoparticles as anti-microbial, anti-inflammatory, and remineralizing agents in oral care cosmetics: a review of the current situation. Nano 10(1):140
Caruso C (2017) The beauty of "mini brains". Sci Am Mind 28(3):76–76
Castillo A, de la Guardia Y (2017) Spineless solutions: the potential of invertebrate animal models for advancing science in the developing world. EMBO Rep 18(11):1885–1888
Centre for Science and Environment (n.d.) Regulations for cosmetics
Clerbaux LA et al (2019) Membrane transporter data to support kinetically-informed chemical risk assessment using non-animal methods: scientific and regulatory perspectives. Environ Int 126: 659–671
Clift MJ, Doak SH (2021) Advanced in vitro models for replacement of animal experiments. Small 17:2101474
CliniExperts (2016) Cosmetic regulation in India
Commission of the European communities (2009) Commission regulation (EC) No 761/2009 of 23 July 2009 amending, for the purpose of its adaptation to technical progress, regulation (EC) no 440/2008 laying down test methods pursuant to regulation (EC) no 1907/2006 of the European parliament and of the council on the registration, evaluation, authorisation and restriction of chemicals (REACH). Method B.46 of the annex to 440/2008/EC. Off J Eur Communities:1–94
Commission of the European communities (2017) Commission Regulation (EC) No 761/2009 of 14 February 2017 Amending, for the purpose of its adaptation to technical Progress, regulation (EC) No 440/2008 laying down Test methods pursuant to regulation (EC) No 1907/2006 of the European Parliament and of the council on the registration, evaluation, authorisation and restriction of chemicals (REACH): method B.59 of the annex to 440/2008/EC. Off J Eur Communities 112:1–402
Crevedo (2022) Cosmetics registration process in Japan
Cronin M (2017) Non-animal approaches, the way forward. Report on a European Commission scientific conference held on 6–7 December 2016 at The Egg, Brussels, Belgium
Cruelty Free International (2016) Korea announces ban on animal testing for cosmetics
Doke SK, Dhawale SC (2015) Alternatives to animal testing: a review. Saudi Pharm J 23(3): 223–229
EC (2021) Report from the commission to the European parliament, the council, the European economic and social committee and the committee of the regions in accordance with Article 117(4) of REACH and Article 46(2) of CLP, and a review of certain elements of REACH in line with Articles 75(2), 138(2), 138(3) and 138(6) of REACH. Brussels, Belgium: 5.2.2013 COM, 2013(49 final)
ECVAM Scientific Advisory Committee (ESAC) (1998) ESAC statement on scientific validity of EpiSkin (SkinEthic)for corrosivity testing. ESAC, Ispra
Effiong DE et al (2019) Nanotechnology in cosmetics: basics, current trends and safety concerns—a review. Adv Nanopart 9(1):1–22
ESAC (2007) ESAC statement on scientific validity of EpiSkin and EpiDerm for skin irritation testing
European Commission (2015) Computers replace bunnies in Europe's cosmetics testing labs. The EU research and innovation magazine
European Commission, Joint Research Centre, Institute for Health and Consumer Protection (IHC) (2012) EURL ECVAM recommendation on the Direct Peptide Reactivity Assay (DPRA) for skin sensitisation testing

European Parliament and Commission of the European communities (2000) EU directive 2000/33/EU for the 21st amendment of annex v of the EU directive 86/906/EEC for classification and labelling of hazardous chemicals: test guideline B-40 skin corrosivity in vitro method and test guideline B-41 phototoxicity in vitro 3T3 NRU phototoxicity test. Off J Eur Communities L136: 90–107

European Union (2009a) Regulation (EC) no 1223/2009 of the European parliament and of the council. Off J Eur Union 342:59

European Union (2009b) Regulation (EC) no 1223/2009 of the European parliament and of the council of 30 November 2009 on cosmetic products. Off J Eur Union 342:59

FDA (2014) Guidance for industry safety of nanomaterials in cosmetic products

Ferreira M et al (2022) Overview of cosmetic regulatory frameworks around the world. Cosmetics 9(4):72

Fioravanzo E et al (2012) Role of in silico genotoxicity tools in the regulatory assessment of pharmaceutical impurities. SAR QSAR Environ Res 23(3–4):257–277

Fytianos G, Rahdar A, Kyzas GZ (2020) Nanomaterials in cosmetics: recent updates. Nanomaterials (Basel) 10(5):979

Halappanavar S et al (2020) Adverse outcome pathways as a tool for the design of testing strategies to support the safety assessment of emerging advanced materials at the nanoscale. Part Fibre Toxicol 17(1):1–24

Humane Society International (2021) Brazilian supreme court confirms Rio de Janeiro state ban on animal testing for cosmetics

Hunt PR, The C (2017) Elegans model in toxicity testing. J Appl Toxicol 37(1):50–59

Japan Ministry of Health (2014) Japanese law translation. Act on securing quality, efficacy and safety of products including pharmaceuticals and medical devices

Katz LM, Dewan K, Bronaugh RL (2015) Nanotechnology in cosmetics. Food Chem Toxicol 85: 127–137

Katz LM et al (2022) Regulation of cosmetics in the United States. Dermatol Clin 40(3):307–318

Khezri K, Saeedi M, Dizaj SM (2018) Application of nanoparticles in percutaneous delivery of active ingredients in cosmetic preparations. Biomed Pharmacother 106:1499–1505

Kim K, Park H, Lim KM (2015) Phototoxicity: its mechanism and animal alternative test methods. Toxicol Res 31(2):97–104

Kooijman M (2013) Why animal studies are still being used in drug development. Altern Lab Anim 41(6):P79–P81

Kumud M, Sanju N (2018) Nanotechnology driven cosmetic products: commercial and regulatory milestones. Appl Drug Res Clin Trials Regul Aff 5(2):112–121

Lachowicz J et al (2021) Zebrafish as an animal model for testing agents with antidepressant potential. Life 11(8):792

Lee M, Hwang JH, Lim KM (2017) Alternatives to in vivo Draize rabbit eye and skin irritation tests with a focus on 3D reconstructed human cornea-like epithelium and epidermis models. Toxicol Res 33(3):191–203

Liventsev V, Härmä A, Petkovic M (2021) Towards effective patient simulators. Front Artif Intell 4:195

Mademtzoglou D et al (2011) Applying the drosophila wing spot test to assess the genotoxic impact of 10 essential oil constituents used as flavouring agents or cosmetic ingredients. Flavour Fragr J 26(6):447–451

Melo A et al (2015) The role of nanomaterials in cosmetics: national and international legislative aspects. Química Nova 38:599–603

Nabarretti BH et al (2022) A review of alternative methods to the use of animals in safety evaluation of cosmetics. Einstein (São Paulo) 20:eRB5578

Nanda S (2018) Nanocosmeceuticals: retrospect, precepts and prospects. Pharma Times 50:73–78

Natsch A et al (2015) Predicting skin sensitizer potency based on in vitro data from KeratinoSens and kinetic peptide binding: global versus domain-based assessment. Toxicol Sci 143(2): 319–332

Nobile V (2016) Guidelines on cosmetic efficacy testing on humans. Ethical, technical, and regulatory requirements in the main cosmetics markets. J Cosmo Trichol 2:107
OECD (2004a) 428: skin absorption: in vitro method. OECD guidelines for the testing of chemicals, Section. 4: 1–8
OECD (2004b) Guidance document for the conduct of skin absorption studies. In OECD series on testing and assessment. ENV/JM/MONO, 2: 1–31
OECD (2011) Test no. 156, guidance notes for the estimation of dermal absorption values, OECD series on testing and assessment; OECD: Paris, France. ENV/JM/MONO, 36
OECD (2019) Test no. 492: reconstructed human Cornea-like Epithelium (RhCE) test method for identifying chemicals not requiring classification and labelling for eye irritation or serious eye damage. In: OECD guidelines for the testing of chemicals
Pastrana H, Avila A, Tsai CS (2018) Nanomaterials in cosmetic products: The challenges with regard to current legal frameworks and consumer exposure. NanoEthics 12(2):123–137
Pereira C et al (2012) Contribution of yeast models to neurodegeneration research. J Biomed Biotechnol 2012:941232
Peters D, Choi JS (2020) Status of cosmetics regulations in Korea. IRCL 3:73
Pomela M (2015) Cosmetics regulation in Brazil. The Brazil Business
Prinsen MK et al (2017) The isolated chicken eye test to replace the Draize test in rabbits. Regul Toxicol Pharmacol 85:132–149
Raj S et al (2012) Nanotechnology in cosmetics: opportunities and challenges. J Pharm Bioallied Sci 4(3):186
RedOrangePeach (2022) Animal testing in China
Revia RA, Wagner BA, Zhang M (2019) A portable electrospinner for nanofiber synthesis and its application for cosmetic treatment of alopecia. Nano 9(9):1317
Rigano L, Lionetti N (2016) Nanobiomaterials in galenic formulations and cosmetics. In: Nanobiomaterials in galenic formulations and cosmetics. Elsevier, Amsterdam, pp 121–148
SCCS (2012) Guidance on the safety assessment of nanomaterials in cosmetics. European Union
Shiseido (n.d.) What is nano particles?
Singh BM, Jain A, Mishra A (2018) Cosmetic regulations in India vs. globally and challenges in harmonization. Int J Pharm Sci Drug Res 10(3):150–157
Skincare (n.d.) Report: these countries have banned cosmetics animal testing
Sreedhar D et al (2020) Ban of cosmetic testing on animals: a brief overview. Int J Cur Res Rev 12(14):113
Su Z et al (2020) Final publication of the "regulations on the supervision and Administration of Cosmetics" and new Prospectives of cosmetic science in China. Cosmetics 7(4):98
Taylor K (2019) Recent developments in alternatives to animal testing. In: Herrmann K, Jayne K (eds) Animal experimentation: working towards a paradigm change, pp 585–609
The Human Society of the United States (n.d.) Cosmetics animal testing FAQ
The Times of India (2014a) Govt. bans cosmetic companies from testing on animals
The Times of India (2014b) India bans import of cosmetics tested on animals
Toronto Humane Society (2021) Cosmetic testing—a cruel and deadly practice
US-FDA (2018) Understanding over-the-counter medicines
US-FDA (2021a) Cosmetic products & ingredients
US-FDA(2021b) Animal testing & cosmetics
US-FDA (2022) Is it a cosmetic, a drug, or both? (or is it soap?)
US-FDA (n.d.) Regulation, I.C.O.C, About ICCR
Vermeire T et al (2013) OSIRIS, a quest for proof of principle for integrated testing strategies of chemicals for four human health endpoints. Regul Toxicol Pharmacol 67(2):136–145
Vonk RA et al (2015) Legal barriers for the use of alternatives to animal testing: do current EU regulations and guidelines for regulatory acceptance of medicinal products pose legal barriers?

Integrated Approaches to Testing and Assessment (IATA) for Safety of Cosmetics and Personal Care Products

10

Sunil Kumar Patel, Prakriti Gaur, Aditi Pandey, Akhilesh Kumar Yadav, Rama Shankar Sahu, and Bhartendu Nath Mishra

Abstract

Cosmetics and personal care products induce skin sensitization, and allergies are very common skin-related issues in people worldwide. Therefore, dermal safety assessment is a mandatory requirement before marketing these products. After the ban on animal testing for cosmetics, the safety assessment of these products is now very challenging. Integrated approaches to testing and assessment are the best option and can be used for the prediction of the toxicity or safety of chemicals based on the integration of multiple pieces of information generated via non-testing and testing methods for regulatory purposes. Moreover, along with traditional tools (in vivo and in vitro), IATA is widely including high-throughput screening, and computational approaches as new approach methods (NAM) for data generation, interpretation, and integration. Identifying the AOP for skin allergy also aids in the development of IATA for regulatory

S. K. Patel · P. Gaur
Photobiology Division, CSIR-Indian Institute of Toxicology Research, Lucknow, Uttar Pradesh, India

A. Pandey
Graduate Institute of Nanomedicine and Medical Engineering, Taipei Medical University, Taipei, Taiwan

A. K. Yadav
Regulatory Toxicology, CSIR-Indian Institute of Toxicology Research, Lucknow, Uttar Pradesh, India

R. S. Sahu
Package Reliability at Central Quality, Micron, Taichung, Taiwan

B. N. Mishra (✉)
Department of Biotechnology, Institute of Engineering & Technology, Lucknow, Uttar Pradesh, India
e-mail: bnmishra@ietlucknow.ac.in

A. B. Pant et al. (eds.), *Skin 3-D Models and Cosmetics Toxicity*, https://doi.org/10.1007/978-981-99-2804-0_10

decision-making. This chapter describes the role of IATA in the safety and toxicity prediction of cosmetics and personal care products.

Keywords

IATA · In chemico · New approach methods and adverse outcome pathway

10.1 Introduction

10.1.1 Cosmetics and Personal Care Products

Personal care products and cosmetics use are very common everywhere in the globe. It has a long list of chemicals such as soaps, hair dyes, emulsifiers, fragrances, ultraviolet absorbers, and preservatives (Hamilton and de Gannes 2011). Cosmetics and personal care products contain various ingredients such as colorants, fragrances, and preservatives to make them attractive and safe for consumers. However, these ingredients are a major cause of skin sensitization and skin allergy (Tan et al. 2014). Generally, dermal and ocular safety testing of cosmetics and personal care products is sufficient for commercial use. Therefore, before getting regulatory approval skin irritation, corrosion, phototoxicity, skin sensitization, eye irritation, corrosion, and skin absorption testing are compulsory (Fig. 10.1). If any chemical has deep penetration and is reaching systemic circulation, then systemic toxicity evaluation is important to avoid any organ-specific toxicity.

Carcinogenicity, genotoxicity, developmental toxicity, reproductive toxicity, and endocrine disruption testing are necessary to prevent systemic toxicity of cosmetics and personal care products and their constituents (Fig. 10.2).

The animal-based study was the gold standard for skin sensitization assessment of cosmetics and personal care products. However, after the ban on animal use for cosmetics and personal care product safety in 2013 by European Union and in 2014 by the Indian government, it was very difficult to assess the safety of these products. To overcome this, different alternative models were developed and approved by OECD to assess the skin sensitization/skin allergy of these products. The individual method was not sufficient to draw any regulatory decision. Therefore, integrated approaches are most appropriate in testing and safety assessment of cosmetics and personal care products.

10.1.1.1 IATA

IATA is used to combine various existing statistics and properties including physical and chemical properties, non-testing (QSAR), and testing methods (in vivo and in vitro and in chemico) based on information for regulatory decision (Browne et al. 2017, 2020; Sakuratani et al. 2018). Chemical regulation authorities are confronted due to intensive testing approaches including expensive, utilization of a large number of chemicals, time-consuming, and use of live animals to evaluate all chemicals in the development of personal healthcare products (Tollefsen et al.

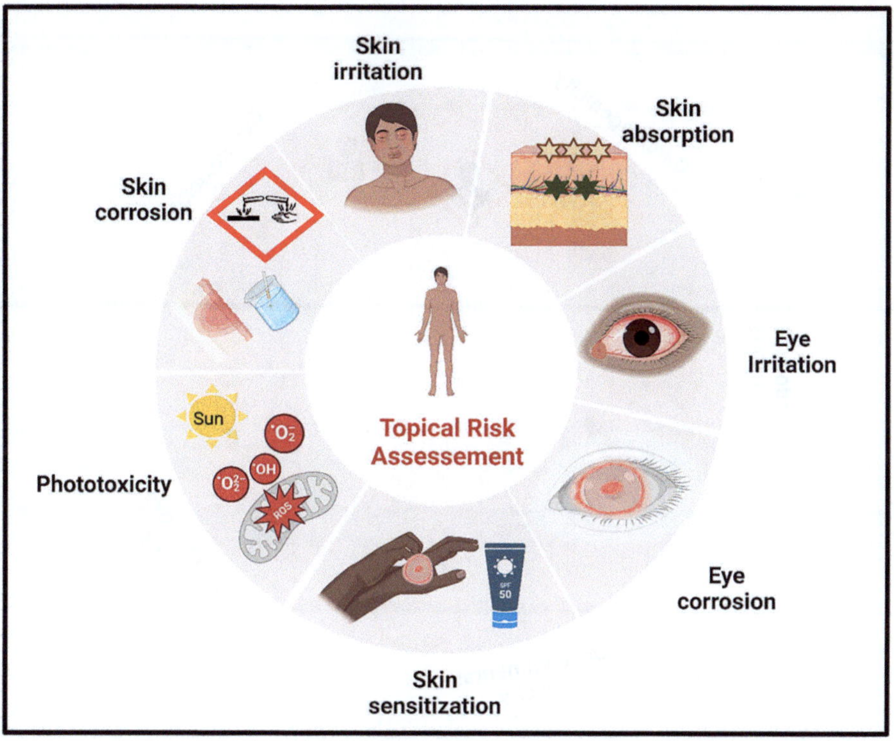

Fig. 10.1 Cutaneous and ocular risk assessment

2014). The need for robust and effective strategies for the evaluation of threats was imposed by the chemicals in humans via different routes including dermal, inhalation, and systemic exposure (Abd et al. 2016; Wills et al. 2016; Shah et al. 2018; Conway et al. 2020). Though the number of methods are existing for safety assessment, the single method for the prediction of toxicity/safety of chemical is not sufficient for regulatory decision (Pfuhler et al. 2020, 2021; Tollefsen et al. 2014). Therefore, considering the need for a combined approach via the use of IATA is more reliable for the prediction of toxicity of any chemical. Additionally, to evaluate proper risk assessment, there are progressive approaches to substituting methods for animal use in toxicology and refining to incorporate new approach method (NAM) (Brannen et al. 2016; OECD 2016a, b, 2017). Applying NAM as a solution to toxicological endpoints is included in IATA, e.g., defined approaches for testing and assessment and integrated testing strategies (Casati 2018; Eskes 2019). With acceleration, the artificial intelligence is imparting to incorporate and mix various streams (Fig. 10.3).

10.1.1.2 Non-Testing Methods

In silico approaches (QSAR, read across) are used to assess the safety/toxicity of cosmetics and personal care products.

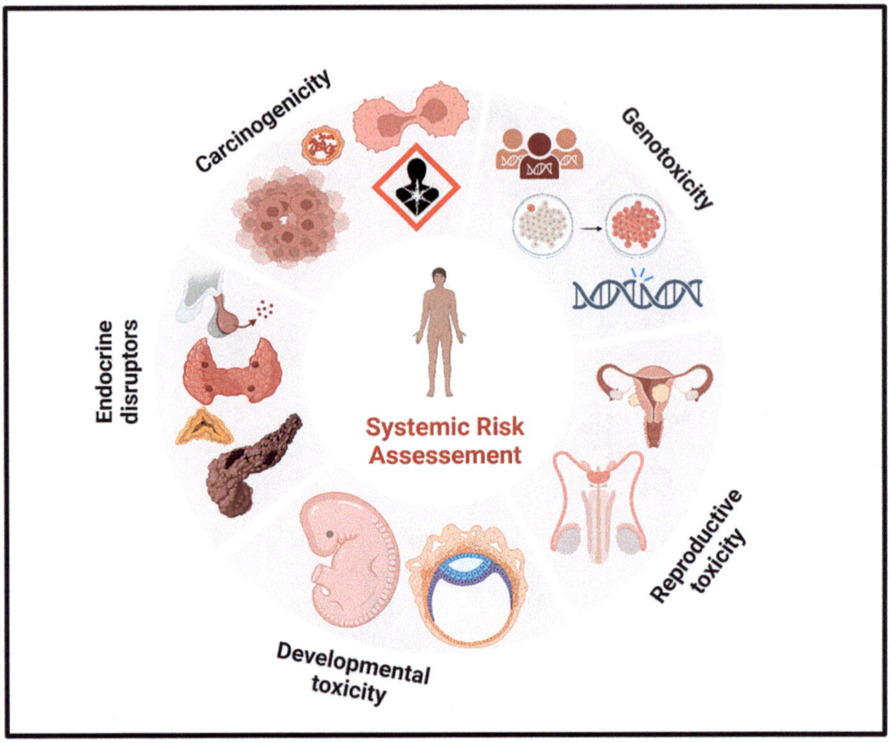

Fig. 10.2 Systemic risk assessment

10.1.1.3 Testing Methods

There are a variety of testing methods that can serve as important components of an IATA, including in vivo, in vitro, and in chemo experiments. IATA needs to move away from relying on in vivo/animal-based data and test components such as toxicogenomic and high-content/high-throughput screening (HC/HT) to address one or more adverse outcomes.

10.2 New Approach Methods

Various alternative methods such as omics technology (genomics, proteomics, and metabolomics), in silico models, and other advanced biotechnological and computational models can be considered as new approach methods. These new approach methods support the IATA in the exact prediction of the toxicity of chemicals.

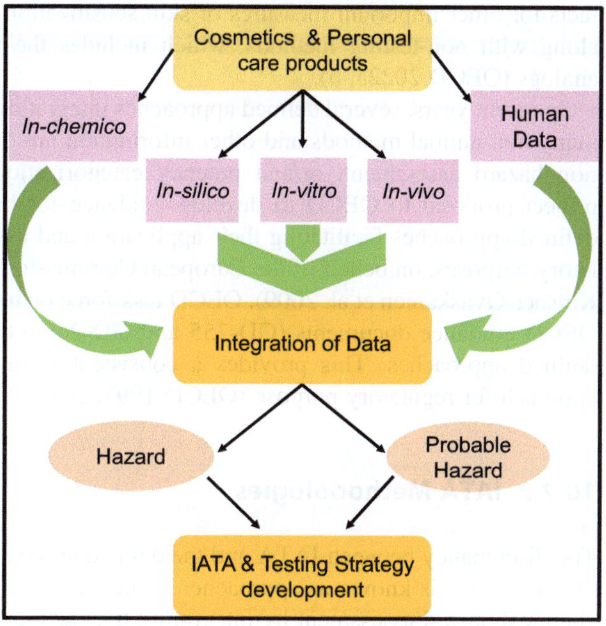

Fig. 10.3 Integrated risk assessment approaches

10.2.1 IATA in Skin Sensitization

Skin sensitization has been a regulatory endpoint, needed in various chemical sectors like industrial chemicals, cosmetics, and pesticides, being the center of concerted efforts for replacing animal testing over the years (Henning et al. 2009; Casati 2018 ; de Ávila et al. 2019; Kandarova and Hayden 2021). Various guinea–pig assay was migrated to reduced and distinguished LLNA, to demonstrate the dominance of in vitro and in silico methods (Basketter 2016). EURL organization for alternatives to animal models for testing has implemented strategies for skin sensitization. It has been of key importance in the assurance of the translation of the customary non-animal methods for skin sensitization into the internationally approved test guidelines (European Parliament and the Council of the European Union 2009; Rauscher et al. 2012; Chapman 2015). Consequently, from 2015 to 2017, in vitro and in chemico models and mechanisms are approved by the OECD under the first three crucial events of adverse outcome pathway for skin sensitization (Kandárová et al. 2006; Hecker et al. 2011; OECD 2022a, b). The current test methods cannot fulfill all regulatory requirements regarding skin sensitization potential and chemical potency in comparison with those that are provided by regulatory animal tests. LLNA (OECD TG 429) (OECD 2010a, b) or non-radioactive variants, like LLNA: BrdU-ELISA (OECD TG 442B) (OECD 2018), and LLNA: DA (OECD TG 442A). Considering this reason, data from the direct peptide reactivity assay, the ARE-Nrf2 luciferase assay, and the three protocols of dendritic cell activation (h-CLAT, interleukin-8, and Luc assay) must be considered for IATA. Moreover, combining other relevant information like physical–chemical properties, facts for other important measures of skin sensitization adverse outcome pathways along with non-testing methods, which includes the read across from chemical analogs (OECD 2022a, b).

In recent years, several defined approaches integrating information and facts from many non-animal methods and other information are developed for skin sensitization hazard assessment, or/and potency categorization. EURL ECVAM gave a project proposal to OECD to develop guidance for harmonized reporting of the defined approaches facilitating their application and assessment in IATA for regulatory purposes, on behalf of the European Commission (Hartung et al. 2004, 2013; Kinsner-Ovaskainen et al. 2009). OECD task force on hazard assessment reported in OECD guidance documents (GD-255 and GD 256, OECD 2016a, b, c) describes defined approaches. This provides a consistent format for describing a defined approach for regulatory purpose (OECD 1997, 2015, 2016a, b).

10.2.2 IATA Methodologies

The discrepancy between IATA and the defined approach is a key concept from the OECD. IATA is known as approaches, which are used in the prediction of risk or chemical hazard assessment by integrating the existing information and new information generated by the testing strategies. IATA is an iterative approach for answering a question in a defined regulatory context. The total evaluation process within the IATA mainly depends on weight of evidence, essentially implying an expert judgment for evaluating different pieces of information (OECD 2019, 2020, 2021a, b).

The intended design of IATA renders adaptability for particular regional requirements or regulatory statutes. The defined approaches can be considered a comparable alternative for the in vivo data if a similar kind of information is considered within the decision context of the IATA. As per the EURL ECVAM workshop, collaborated with ICATM, it was aimed to enrich the regulatory considered and adopted individual test methods, with the acceptance of defined approaches (although recognized for requirement differences in several sectors and jurisdictions) as an alternative for non-animal methodologies to skin sensitization evaluation by the chemicals used in several areas (Blaauboer et al. 1999; Clothier et al. 1999; Spielmann et al. 2006; Daniel et al. 2018; Strickland et al. 2019). This followed a consensus to maximize regulatory acceptance of data in defined approaches, for which international harmonization and standardization were necessary. This could be accomplished by developing an evaluation outline that allows an independent assessment of the defined approaches, wherein the reproducibility, relevance/predictive capacity, providing sufficient and equivalent information, mechanistically and biologically relevant, transparently described comparable to the reference animal test. Moreover, independent assessment by third parties and conflicting results and uncertainty in vivo data and defined approaches should also be considered, with predictions in the context of IATA (OECD 2016a, b).

10.3 International Cooperation in the Development of Alternative Test Methods

ICATM was promoting the use of DAs in the field of skin sensitization, and a framework was (Hoffmann 2015; Dumont et al. 2016; Roberts et al. 2016) defined based on the aforementioned criteria. Recent publications that evaluated the variability in the recorded animal datasets could be added to a meta-analysis (Kleinstreuer et al. 2018). Furthermore, the human data will be assembled and LLNA enactment against human data will be matched based on chemical data and variables (Basketter et al. 2014; Bell et al. 2017). The decision of definitions and a clear understanding of defined approaches and IATA were made to create information that brought all who were involved in the related area to a similar understanding. Many stakeholders were identified as important for progress in the field of DAs for skin sensitization assessment. Several scientists from research institutions, industry, and NGOs experienced with in vitro and in silico methods were also noted to form the building blocks of DAs. The US EPA, and other partners, has initiated many data-sharing pilots for encouraging industry stakeholders and other testing laboratories for easy access to non-animal methods and in vitro methods, respectively, for the areas related to toxicology and the broader scientific community, by the availability of such internal resources (https://www.aahp-abhp.org/node/1224). CAAT commissioned report suggests utilizing the existing data and monitoring future testing in toxicology by integrated testing strategies (Jaworska and Hoffmann 2010). OECD workshop and previous work (Jaworska et al. 2010) outlined the conceptual requirements as transparent, consistent, and hypothesis-driven (Jaworska et al. 2010).

10.4 Conclusion

IATA has the ability to reduce the usage of animal testing for safety studies of cosmetics and personal care products by utilizing testing and non-testing methods with integration with new approach methods.

References

Abd E, Yousef SA, Pastore MN, Telaprolu K, Mohammed YH, Namjoshi S et al (2016) Skin models for the testing of transdermal drugs. Clin Pharm 8:163

Basketter D (2016) Skin sensitisation, adverse outcome pathways and alternatives. Altern Lab Anim 44(5):431–436

Basketter DA, Alépée N, Ashikaga T, Barroso J, Gilmour N, Goebel C, Hibatallah J, Hoffmann S, Kern P, Martinozzi-Teissier S, Maxwell G, Reisinger K, Sakaguchi H, Schepky A, Tailhardat M, Templier M (2014) Categorization of chemicals according to their relative human skin sensitizing potency. Dermatitis 25(1):11–21

Bell SM, Phillips J, Sedykh A, Tandon A, Sprankle C, Morefield SQ, Shapiro A, Allen D, Shah R, Maull EA, Casey WM, Kleinstreuer NC (2017) An integrated chemical environment to support 21st-century toxicology. Environ Health Perspect 125(5):054501

Blaauboer BJ, Barratt MD, Houston JB (1999) The integrated use of alternative methods in toxicological risk evaluation: ECVAM integrated testing strategies task force report 1. Altern Lab Anim 27(2):229–237

Brannen KC, Chapin RE, Jacobs AC, Green ML (2016) Alternative models of developmental and reproductive toxicity in pharmaceutical risk assessment and the 3Rs. ILAR J 57(2):144–156

Browne P, Noyes PD, Casey WM, Dix DJ (2017) Application of adverse outcome pathways to US EPA's endocrine disruptor screening program. Environ Health Perspect 125(9):096001

Browne P, Van Der Wal L, Gourmelon A (2020) OECD approaches and considerations for regulatory evaluation of endocrine disruptors. Mol Cell Endocrinol 504:110675

Casati S (2018) Integrated approaches to testing and assessment. Basic Clin Pharmacol Toxicol 123 (S5):51–55

Chapman PM (2015) Including or excluding toxicity test data for development of a geometric mean. Environ Toxicol Chem 34(8):1691–1692

Clothier R, Willshaw A, Cox H, Garle M, Bowler H, Combes R (1999) The use of human keratinocytes in the EU/COLIPA international in vitro Phototoxicity test validation study and the ECVAM/COLIPA study on UV filter chemicals. Altern Lab Anim 27(2):247–259

Conway GE, Shah UK, Llewellyn S, Cervena T, Evans SJ, Al Ali AS, Jenkins GJ, Clift MJD, Doak SH (2020) Adaptation of the in vitro micronucleus assay for genotoxicity testing using 3D liver models supporting longer-term exposure durations. Mutagenesis 35(4):319–330

Daniel AB, Strickland J, Allen D, Casati S, Zuang V, Barroso J, Whelan M, Régimbald-Krnel MJ, Kojima H, Nishikawa A, Park H-K, Lee JK, Kim TS, Delgado I, Rios L, Yang Y, Wang G, Kleinstreuer N (2018) International regulatory requirements for skin sensitization testing. Regul Toxicol Pharmacol 95:52–65

de Ávila RI, Lindstedt M, Valadares MC (2019) The 21st century movement within the area of skin sensitization assessment: from the animal context towards current human-relevant in vitro solutions. Regul Toxicol Pharmacol 108:104445

Dumont C, Barroso J, Matys I, Worth A, Casati S (2016) Analysis of the local lymph node assay (LLNA) variability for assessing the prediction of skin sensitisation potential and potency of chemicals with non-animal approaches. Toxicol in Vitro 34:220–228

Eskes C (2019) The usefulness of integrated strategy approaches in replacing animal experimentation. Ann Ist Super Sanita 55(4):400–404

European Parliament and the council of the European Union (2009) Regulation (EC) No 1223/2009 of the European parliament and of the council. 342: 59

Hamilton T, de Gannes GC (2011) Allergic contact dermatitis to preservatives and fragrances in cosmetics. Dermatitis 14:16

Hartung T, Bremer S, Casati S, Coecke S, Corvi R, Fortaner S et al (2004) A modular approach to the ECVAM principles on test validity. Altern Lab Anim 32(5):467–472

Hartung T, Luechtefeld T, Maertens A, Kleensang A (2013) Integrated testing strategies for safety assessments. ALTEX 30(1):3–18

Hecker M, Hollert H, Cooper R, Vinggaard AM, Akahori Y, Murphy M, Nellemann C, Higley E, Newsted J, Laskey J, Buckalew A, Grund S, Maletz S, Giesy J, Timm G (2011) The OECD validation program of the H295R steroidogenesis assay: phase 3. Final inter-laboratory validation study. Environ Sci Pollut Res Int 18(3):503–515

Henning A, Schaefer UF, Neumann D (2009) Potential pitfalls in skin permeation experiments: influence of experimental factors and subsequent data evaluation. Eur J Pharm Biopharm 72(2): 324–331

Hoffmann S (2015) LLNA variability: an essential ingredient for a comprehensive assessment of non-animal skin sensitization test methods and strategies. ALTEX 32(4):379–383

Jaworska J, Hoffmann S (2010) Integrated testing strategy (ITS)–opportunities to better use existing data and guide future testing in toxicology. ALTEX 27(4):231–242

Jaworska J, Gabbert S, Aldenberg T (2010) Towards optimization of chemical testing under REACH: a Bayesian network approach to integrated testing strategies. Regul Toxicol Pharmacol 57(2):157–167

Kandarova H, Hayden PJ (2021) Standardised reconstructed skin models in toxicology and pharmacology: state of the art and future development. In: Schäfer-Korting M, Maria-Engler SS, Landsiedel R (eds) Organotypic models in drug development. Springer, Cham, pp 57–71

Kandárová H, Liebsch M, Spielmann H, Genschow E, Schmidt E, Traue D, Guest R, Whittingham A, Warren N, Gamer AO, Remmele M, Kaufmann T, Wittmer E, De Wever B, Rosdy M (2006) Assessment of the human epidermis model SkinEthic RHE for in vitro skin corrosion testing of chemicals according to new OECD TG 431. Toxicol in Vitro 20(5):547–559

Kinsner-Ovaskainen A, Akkan Z, Casati S, Coecke S, Corvi R, Negro GD, De Bruijn J, De Silva O, Gribaldo L, Griesinger C, Jaworska J, Kreysa J, Maxwell G, McNamee P, Price A, Prieto P, Schubert R, Tosti L, Worth A, Zuang V (2009) Overcoming barriers to validation of non-animal partial replacement methods/integrated testing strategies: the report of an EPAA–ECVAM workshop. Altern Lab Anim 37(4):437–444

Kleinstreuer NC, Hoffmann S, Alépée N, Allen D, Ashikaga T, Casey W et al (2018) Non-animal methods to predict skin sensitization (II): an assessment of defined approaches. Crit Rev Toxicol 48(5):359–374

OECD (1997). Test No 473: in vitro mammalian chromosome aberration test

OECD (2010a) Test no. 429: skin sensitisation

OECD (2010b) Test no. 442A: skin sensitization

OECD (2015) Test no. 404: acute dermal irritation/corrosion

OECD (2016a) Guidance document on the reporting of defined approaches to be used within integrated approaches to testing and assessment, organization for economic co-operation and development, Paris

OECD (2016b) Test no. 476: in vitro mammalian cell gene mutation tests using the Hprt and xprt genes

OECD (2017) Guidance document for the use of adverse outcome pathways in developing integrated approaches to testing and assessment (IATA). OECD Publishing

OECD (2018) Test no. 442B: skin sensitization

OECD (2019) Second edition–guidance document on integrated approaches to testing and assessment (IATA) for serious eye damage and eye irritation

OECD (2020) Test no. 437: bovine corneal opacity and permeability test method for identifying i) chemicals inducing serious eye damage and ii) chemicals not requiring classification for eye irritation or serious eye damage

OECD (2021a) Test no. 405: acute eye irritation/corrosion

OECD (2021b) Test no. 439: in vitro skin irritation: reconstructed human epidermis test method

OECD (2022a) Test no. 442E: in vitro skin sensitisation

OECD (2022b) Test no. 456: H295R steroidogenesis assay

Pfuhler S, van Benthem J, Curren R, Doak SH, Dusinska M, Hayashi M, Heflich RH, Kidd D, Kirkland D, Luan Y, Ouedraogo G, Reisinger K, Sofuni T, van Acker F, Yang Y, Corvi R (2020) Use of in vitro 3D tissue models in genotoxicity testing: strategic fit, validation status and way forward. Report of the working group from the 7th international workshop on genotoxicity testing (IWGT). Mutat Res Genet Toxicol Environ Mutagen 850–851:503135

Pfuhler S, Downs TR, Hewitt NJ, Hoffmann S, Mun GC, Ouedraogo G, Roy S, Curren RD, Aardema MJ (2021) Validation of the 3D reconstructed human skin micronucleus (RSMN) assay: an animal-free alternative for following-up positive results from standard in vitro genotoxicity assays. Mutagenesis 36(1):1–17

Rauscher H, Sokull-Klüttgen B, Stamm HJN (2012) The European commission's recommendation on the definition of nanomaterial makes an impact. Nanotoxicology 7(7):1195–1197

Roberts DW, Api AM, Aptula AO (2016) Chemical applicability domain of the local lymph node assay (LLNA) for skin sensitisation potency. Part 2. The biological variability of the murine local lymph node assay (LLNA) for skin sensitisation. Regul Toxicol Pharmacol 80:255–259

Sakuratani Y, Horie M, Leinala E (2018) Integrated approaches to testing and assessment: OECD activities on the development and use of adverse outcome pathways and case studies. Basic Clin Pharmacol Toxicol 123(S5):20–28

Shah UK, de Oliveira Mallia J, Singh N, Chapman KE, Doak SH, Jenkins GJS (2018) A three-dimensional in vitro HepG2 cells liver spheroid model for genotoxicity studies. Mutat Res Genet Toxicol Environ Mutagen 825:51–58

Spielmann H, Seiler A, Bremer S, Hareng L, Hartung T, Ahr H et al (2006) The practical application of three validated in vitro embryotoxicity tests: the report and recommendations of an ECVAM/ZEBET workshop (ECVAM workshop 57). Altern Lab Anim 34(5):527–538

Strickland J, Daniel AB, Allen D, Aguila C, Ahir S, Bancos S, Craig E, Germolec D, Ghosh C, Hudson NL, Jacobs A, Lehmann DM, Matheson J, Reinke EN, Sadrieh N, Vukmanovic S, Kleinstreuer N (2019) Skin sensitization testing needs and data uses by US regulatory and research agencies. Arch Toxicol 93(2):273–291

Tan CH, Rasool S, Johnston GA (2014) Contact dermatitis: allergic and irritant. Clin Dermatol 32(1):116–124

Tollefsen KE, Scholz S, Cronin MT, Edwards SW, de Knecht J, Crofton K, Garcia-Reyero N, Hartung T, Worth A, Patlewicz G (2014) Applying adverse outcome pathways (AOPs) to support integrated approaches to testing and assessment (IATA). Regul Toxicol Pharmacol 70(3):629–640

Wills JW, Hondow N, Thomas AD, Chapman KE, Fish D, Maffeis TG, Penny MW, Brown RA, Jenkins GJS, Brown AP, White PA, Doak SH (2016) Genetic toxicity assessment of engineered nanoparticles using a 3D in vitro skin model (EpiDerm™). Part Fibre Toxicol 13(1):50

Approaches for In Silico Validation of Safety (Toxicity) Data for Cosmetics

11

Tanmayee Basu, Rashi Chugh, Ranjit Singh Gujjar, and Atul Kumar Upadhyay

Abstract

It has been proven that computational approaches can be used to find endpoints that can help with a cosmetic safety assessment. Thousands of mice, guinea pigs, rats, and rabbits die each year due to torturous experiments. This paradigm shift has enabled the highest number of regulations of chemical safety assessments while also mandating the use of alternate methodologies, such as in silico approaches, whenever applicable, to evaluate different products for individual users from the US and Europe and other countries worldwide. Some people believe that animal testing is a reliable and quick approach to ensure that items are safe for human consumption as it helps to find the movement of the compound through the biological membrane and its action through it. There is also a practical realization well within the toxicity testing discipline that alternative techniques would not supersede in vivo models on a resembling scale. SEURAT-I was indeed a flagship project creating the academic and developing foundations necessary to develop strategies to supplement conventional repeated

T. Basu · R. Chugh
Department of Biotechnology, Thapar Institute of Engineering & Technology, Patiala, Punjab, India

R. S. Gujjar
Division of Crop Improvement, Indian Institute of Sugarcane Research, Lucknow, Uttar Pradesh, India

A. K. Upadhyay (✉)
Department of Biotechnology, Thapar Institute of Engineering & Technology, Patiala, Punjab, India

Division of Crop Improvement, Indian Institute of Sugarcane Research, Lucknow, Uttar Pradesh, India
e-mail: atul.upadhyay@thapar.edu

A. B. Pant et al. (eds.), *Skin 3-D Models and Cosmetics Toxicity*,
https://doi.org/10.1007/978-981-99-2804-0_11

dose systemic toxicity testing consumer monitoring with QSAR methods, read across frameworks, TTC approach, or other omics or other computational techniques. Alternative methods of testing and validating the toxicity of cosmetic products to animals must be incorporated into cosmetic industries to promote business ethics.

Keywords

Safety · Computational approaches · Cosmetics · Bio-toxicity

11.1 Introduction

The increased influences the demand and popularity of cosmetics worldwide in skin-related disparities and the need for people to look good. Globally, the cosmetic industry is enormous, reaching a value of US$ 357.5 billion in the year 2021, with an expected value to reach US$ 508.3 billion by the year 2027 (Business Wire 2022). Skin and personal care product goods, hair products, antiperspirants, scents, and cosmetics and beauty products are among some of the product kinds with segments of the market. The functional compounds are combinations of synthetic chemical compounds with overall health benefits.

When associated with consumers, the primary benchmark is safety and toxicity-free products. Thus, product testing becomes prime for any manufacturer, ensuring the quality and safety of each ingredient used, and the cosmetic product is the manufacturer's or distributor's legal responsibility. Toxicological studies also become a part of testing for the manufacturer (US-FDA 2022). A product formulator plays a crucial role in the cosmetic industry in identifying the right ingredients for the perfect blend in any personal care product. A consultant must carry out routine screening tests. It is also essential to analyze and evaluate the stability and the toxicity of the cosmetics/personal care product formulations prior to consumer use, as these come in direct contact with our skin for a significant period (Tanner 2022). Animals have been used in research to evaluate the suitability of cosmetic industry for humans.

Countless mice, small rodents, rodents, and rabbits are slaughtered annually as a consequence of such cruel investigations (Villalobos et al. 2014). In many cases, they are not given any anesthetic at all. Tests for skin and eye irritation, allergies, poisoning, and other ailments might be conducted, damage to the genome, birth abnormalities, and cancer consequences, to name a few (Rise for Animals 2022). However, animal testing is a contentious issue in both the pharmaceutical and cosmetic industries. Some people believe that animal testing is a reliable and quick approach to ensure that items are safe for human consumption.

In contrast, others argue that it is unnecessary because other testing methods are available (White 2022). According to research, customers are interested in sustainability (Sheehan and Lee 2014). Accordingly, animal testing in the cosmetics industry has always been a polarizing topic. It is crucial in the development and safety of cosmetics while also infringing on experimental animals' survival rights. Hence, animal experimentation is immoral in cosmetology R&D and manufacturing that is because the outcomes do not really aid population well-being and the approach results in animal suffering and killings (Kabene and Baadel 2019). However, several alternatives are available, and use of such animals to test cosmetics is extremely limited.

Alternatives to animals must be incorporated into cosmetic industries to promote business ethics. Companies can use scientific barrier evaluation to discover alternatives to animal test subjects and learn how to use animals correctly in medical and cosmetics tests. A few approaches that can anticipate acts of animal remorselessness by beauty care product organizations incorporate advanced consideration of human cells or tissues, computer modeling strategies, and tests on willing volunteers. Companies must join animal-free tests to diminish the hazard of creature enduring and, as a result, progress their trade morals (Doke and Dhawale 2015). A series of toxicity tests determine a cosmetic ingredient's hazardous potential and is part of the hazard identification process. Toxicological data relevant to humans has traditionally been collected by studying the toxicological profiles of chemicals on animals, preferably utilizing the same exposure route as in people. Toxicological studies are frequently conducted via the oral route, with extrapolation to the cutaneous route required (Vinardell and Mitjans 2017). The employment of an array of computational algorithms to assess toxicity based on the chemical structure of the substances is a crucial aspect of the strategy for developing alternatives to detect the hazard of cosmetic ingredients (and several other types of chemicals). Computational techniques can include a reliable inventory of structures, toxicological information, and data databases to produce safe exposure limits, models, and algorithms. Relevant assays considering toxicity pathways, examined in high-throughput screening assays, may eventually be added to these.

The development of animal-free toxicity testing methodologies, also known as alternative tests, has become a hot topic in toxicological science, resulting in a paradigm change in traditional animal-based toxicity evaluations (Garthoff 2005; Turley et al. 2019; Gironde et al. 2020). Cosmetics made through animal research, including cosmetic materials or products, were banned by the European Union in 2013 (European Commission 2009). As a result, new methods for ensuring the protection of cosmetic products other than animal research became inevitable. As little more than an outcome, the novel toxicity analysis technologies turned its attention to a mechanism-based technique, with the intent of deeper grasp into the pathways that lead to unfavorable biochemical processes in order to better safeguard human health and the environment (Hatherell et al. 2020; Fischer et al. 2020).

Among some of the approaches for an altruistic testing alternative for laboratory safety-level evaluation are (1) in vitro methods, (2) in silico methods, (3) read-across framework, and (4) in chemical techniques (Madden et al. 2020b; Bassan et al. 2021). The techniques can be used to measure risk and internal exposure. Computational methods cover many techniques and concepts and a wide range of endpoints. It has been proven that computational approaches can be used to find endpoints that can help with a cosmetic safety assessment. This chapter aims to overview several dry laboratory techniques for safety evaluation. The assessment of potential danger to a list of ingredients in a product is likely to be the first stage in the safety review of a cosmetic product. There are a variety of resources and approaches that can be utilized to evaluate cosmetic product ingredients.

11.2 Estimating Ingredients of Cosmetic Products: Models and Regulations

A plethora of beliefs is often made concerning exposure to ingredients of cosmetics. Scientific committee on consumes safety (SCCS) gives standardized value to the exposure of ingredients of cosmetic products frequency of product application, quantity of product applied, retentiveness, and the different ways of uptake of the product; it can be via oral administration, inhalation, or by dermal route (SCCS 2016, 2018; Madden et al. 2020b). The SCCS provides solutions for dealing with various product sensitivities while taking into account various administration approaches. The Creme RIFM model (https://www.cremeglobal.com/creme-rifm/) is another technique that covers use data from over 36,000 users from the US and European populations. This model allows you to generate the value of aggregate exposure in order to evaluate scents in compounds. It has been updated and expanded since its first release to incorporate additional cosmetics, hygiene products, and hair care (Bernauer et al. 2021; Safford et al. 2017). If the data become accessible, this method might be used to a wider spectrum of cosmetics components. The technique of probabilistic aggregate exposure modeling has been devised for fragrances and vitamins, which are arising from cosmetic product usage, nutrition, and nutraceuticals (Safford et al. 2015; Comiskey et al. 2017).

In silico models like RIFM databases are also helpful in evaluating the frequency of product utilization combination of different products used simultaneously during a day. This will evaluate different products for individual users from the USA and Europe (Tozer et al. 2019). Further data are also available on human exposure from Human Biomonitoring studies. In our day-to-day life, people are using so many unknown chemicals, remaining unaware of the effects of those chemicals—human biomonitoring tool's objective is to measure the exposure of toxic substances to people by evaluating metabolites of human samples such as blood or urine. Human biomonitoring can only integrate toxicity-level assessment information until the initial stage, but it provides valuable data for future use.

11.2.1 The Cosmetics Regulation of the European Union (EU) (EC/1223/2009)

In 2003, the EU finally agreed to ban all sorts of animal testing in its historical Seventh Amendment to the Cosmetic Directive (Directive 76/768/EEC) from September 11, 2004 (European Commission 2004, 2009, 2018). The European Commission (EC) also made sure that after that date, the commercialization (i.e., products import and selling) in the market that had been tested on animals outside of Europe was to be outlawed (Taylor and Rego Alvarez 2020; European Commission 2009, 2010); however, an extension for the total prohibition on the marketing of such products was allowed until March 11, 2013 (EC). In 2009, the Cosmetic Directive was rewritten as a regulation, although all of these rules remained (Regulation 1223/2009) (EC (European Commission) 2009). Cosmetics Regulation (EC) No. 1223/2009, CLP Regulation (EC No. 1272/2008), and REACH Regulation (EC No. 1907/2006) pertain to all cosmetic commodities in the EU.

In 2005, the European Partnership for Alternative Approaches to Animal Testing (EPAA) was made in coopetition venture uniting the EC, European industry trade groups, and commercial organizations to encourage the formulation and deployment of substitute regulatory evaluation methods (European Commission 2001, 2018). In 2009, the EC and Cosmetics Europe each invested 25 million Euros in the establishment of replacements for animals for long-term toxicological analysis in a program entitled SEURAT-1, in response to the imminent 2013 deadline (see www.seurat-1.eu) (Taylor and Rego Alvarez 2020); the details of this project are discussed in the later section of this chapter.

11.2.2 Organisation for Economic Cooperation and Development (OECD)

The Organisation for Economic Cooperation and Development (OECD) (www.oecd.org) is a global membership organization composed of EU and non-EU nations. One of its responsibilities is to assist participating countries in establishing and standardizing ways to evaluate the risk to public health and the environment, such as environmental exposure assessment procedures. Existing safety evaluations have focused on experiments conducted following the Test Guidelines (TG) of the OECD, which provides a degree of confidence in the returns generated. Currently, TG for in silico methods is none but needs to adapt to substitute animal experimentation with non-test approaches (Taylor and Rego Alvarez 2020); however, the OECD and numerous government entities have created a variety of publications, especially relevant to (Q) SARs, that provide guidelines about using and presenting in silico techniques. The OECD is receiving cooperation from national policymakers and researchers from North America, Europe, and Asia to popularize these tools.

11.3 Next-Generation Risk Assessment (NGRA)

The term "Next Generation Risk Assessment" (NGRA) refers to a hypothesis-driven, a risk assessment technique based on contact that incorporates *in silico*, in vitro, and *in chemico* strategies to aid in animal-free ethical decision-making (Dent et al. 2018, 2021; Rogiers et al. 2020), with a perception to incorporate additional data types within safety selection. A fundamental was published by the US National Academies of Sciences (NAS) in the year 2007 with the title "*Toxicity*

Testing in the 21st Century, A Vision and a Strategy" (NAS 2007; Krewski et al. 2010a, b; National Research Council 2007) followed by a report titled "*Exposure Science in the 21st Century* in the year 2012 (NAS 2012) and an interpretive structure of the former transcripts in the year 2017 namely '*21st Century Science to Improve Risk-Related Evaluations*'" (NAS 2017). With an emphasis on exposure concerns, this report explores the achievements and risk assessment issues are associated with analyzing and combining various forms (and quantities) of data. Instead of relying on a safety assessment of documented diseases in animals, this study asserts that concentrations that trigger modifications in cellular signaling pathways that contribute to detrimental consequences should be understood. This paper presented a desirable and possible vision, given recent developments in molecular methods, bioinformatics, and systems biology (Rogiers et al. 2020; USEPA 2014). In Europe, a unique European Chemicals Agency (ECHA) Topical Scientific Workshop on the use of data and information from new approach methodologies (NAMs) was organized in April 2016, outlining their potential and present constraints to enhance regulatory compliance and choices relating to the evaluation of chemical compounds (ECHA 2012a, b). In the year 2017, the ECHA also published a "Read Across Assessment Framework (RAAF)" to inculcate the application of read-across data in non-animal testing models (Patlewicz et al. 2018; Kuseva et al. 2019).

The International Cooperation on Cosmetics Regulation (ICCR) established nine principles for the NGRA of cosmetic ingredients in 2018, offering a viable path forward for animal-free safety decision-making. The ICCR is an international network of cosmetics regulatory bodies from Brazil, Canada, the EU, Japan, and the United States that works voluntarily. The ICCR was established in 2007 to create a transnational framework for maintaining and enabling the most significant degree of global consumer protection by fostering regulatory convergence and lowering trade barriers (Dent et al. 2018). Pace with the rapid expansion of toxicity hazard identification and risk evaluation science and the potential even by NAMs as detailed in the NAS and ECHA studies, ICCR realized a pivotal shift in the cosmetics safety review is achievable. As a response, the ICCR convened a partnership steering committee consisting of specialists from every regulating body and an industry to concur on and emphasize the essentials for incorporating NAMs into an integrated approach for assessment process of cosmetic constituents (or "Next Generation" Risk Assessment).

There are nine principles corresponding to the risk assessment's ultimate aim, how it should be carried out, and how it can be published (Fig. 11.1) (Amaral et al. 2018; Dent et al. 2018). In July 2019, a workshop was conducted to review how well the nine ICCR principles are now being actively implemented in NGRA clinical studies being undertaken in various organizations and to investigate how the approach used may enhance safety results in vulnerability assessment utilizing NAMs. The goals and accomplishments of the workshop are described in the publication by Dent et al. 2021, which are as follows:

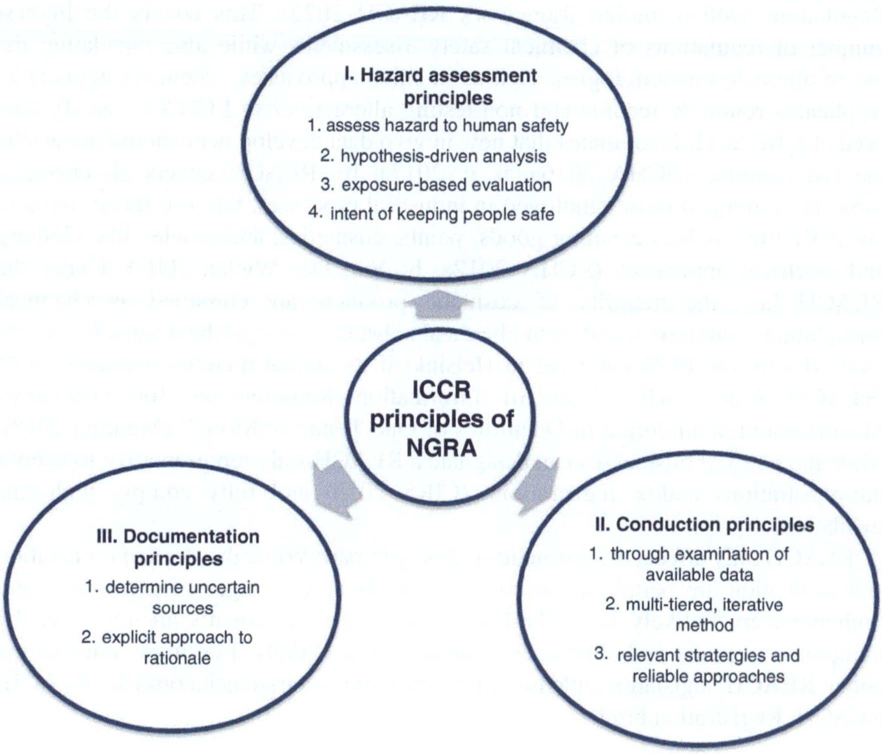

Fig. 11.1 Nine ICCR principles of NGRA that govern the adoption of new approaches in the risk analysis of cosmetic chemicals were discussed by Dent et al. (2018)

- To see whether the NGRA for cosmetic compounds can safeguard human well-being as conventional (animal-based) safety testing.
- Review some NGRA cases for cosmetic components, agree on what worked effectively, and highlight discrepancies.
- To agree on the subsequent actions that must be taken to make NGRA a regular occurrence for the hazard analysis of cosmetic compounds.

11.3.1 The Reach Chemicals Regulation (EC/1907/2006)

REACH seeks to promote human health and environmental protection by identifying chemical compounds' essential characteristics more accurately and earlier (REACH 2012). The ECHA is in charge of implementing REACH (Registration, Evaluation, Authorisation, and Restriction of Chemicals) in the EU. The amendment of EU chemical legislation in 2006 facilitated the emergence of alternative approaches (European Commission 2006; Taylor and Rego Alvarez 2020), and REACH went into effect on June 1, 2007, replacing a vast number of European Directives and

Regulations with a unified framework REACH 2022). This covers the highest number of regulations of chemical safety assessments while also mandating the use of alternate methodologies, such as in silico approaches, whenever applicable. Applicants routinely recommend non-testing alternatives to ECHA to satisfy data needs for REACH. It stipulates that new in vivo data development should always be the last recourse (ECHA 2016a, b, c, 2017a, b). REACH covers all chemical substances, not just those employed in industrial processes, but also those found in our daily life, such as cleaning goods, paints, cosmetics, and articles like clothing and electrical appliances (ECHA 2012a, b; Van Der Wielen 2007). Under the REACH law, the majority of cosmetic products are classified as chemical formulations (mixtures), and each chemical substance or ingredient must be priorly indexed with the ECHA located in Helsinki if its annual quantity exceeds 1 tons (REACH Annex XII. “Standard Information Requirements for Substances Manufactured or Imported in Quantities of One Tonne or More”) (Merenyi 2018), while the non-EU businesses can designate a REACH-only representative to submit pre-registrations and/or registrations (CIRS 2013) and fully comply with this regulation.

REACH only affects the cosmetics industry in part: While the steps of registration and evaluation are pertaining to cosmetic products, the stages of permission and limitations are unlikely to apply because cosmetic ingredients are regulated by numerous agencies and directives (Pouillot et al. 2009). For more information about REACH legislation, please go to http://www.cirs-reach.com/EU_REACH/REACH_Registration.html.

11.3.2 SEURAT-I Project

“Safety Evaluation Ultimately Replacing Animal Testing (SEURAT)”-I (http://www.seurat-1.eu) was a flagship project with the collaboration of 70 European public-private research joint projects (equally sponsored and funded) led by the EC’s Framework Programme 7 Health Programme (https://ec.europa.eu/research/fp7/) administered by DG Research and Innovation and Cosmetics Europe (https://www.cosmeticseurope.eu/) to eliminate animal testing of chemical compounds and ensure the highest degree of consumer safety (Gocht et al. 2015; Berggren et al. 2017). The report *Toxicity Testing in the Twenty-First Century: A Vision and a Strategy* by the National Research Council of the United States (National Research Council 2007) was a massive inspiration for the program. SEURAT-I was one of the most extraordinary EU ventures on radical solutions yet undertaken. A scientific strategy was implemented around the driving premise of using a toxicological mode-of-action approach to defining how any chemical could harm public health (Boobis et al. 2008; Ankley et al. 2010; Krewski et al. 2010a, b; Gocht et al. 2015) and applies it to the development of complementing conceptual, computational (in silico), and laboratory (in vitro) model allows for the identification of numerical transit points, which is required for safety evaluations (Sturla et al. 2014). The actual objective was to make ab initio conclusions based on comprehensive knowledge of

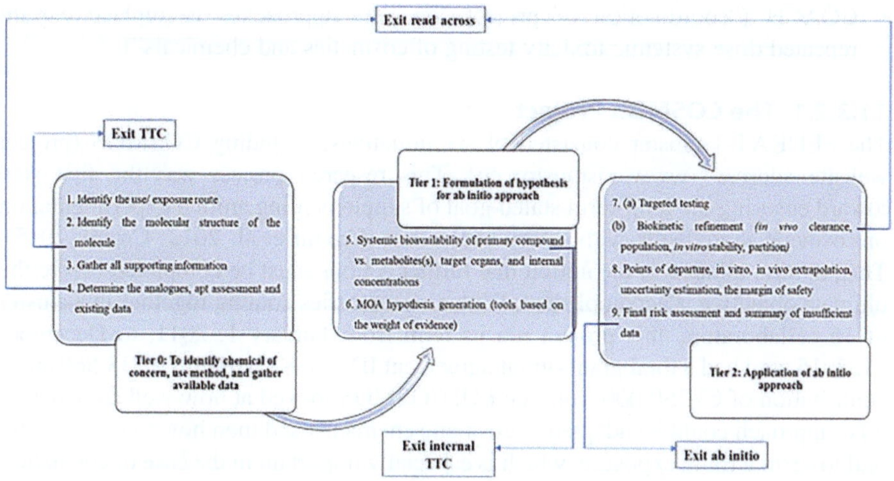

Fig. 11.2 Process flow of evaluating the efficacy of cosmetic compounds rather than using animal models for NGRA as adopted from Berggren et al. (2017) and Dent et al. (2018). (Copyright from Elsevier, first published by Berggren et al. in Computational Toxicology, 4, 2017)

toxicology pathways (Boobis et al. 2008) and to provide a standardized risk management plan approach for recurring exposure toxicity to forecast a no adverse effect level (NOAEL) of a cosmetic-relevant chemical under a given exposure circumstance (Daston et al. 2014; Thomas et al. 2013). The primary pipeline for chemical risk evaluation is built on the SEURAT-1 theoretical model but expanded, intending to provide a tool to help the evaluator through the many steps to be considered and decision-making (Berggren et al. 2017; Yang et al. 2017). We can use Thresholds of Toxicological Concern (TTC) or read-across techniques with this procedure (Schultz et al. 2015; Williams et al. 2016). SEURAT-I was indeed creating the academic and developing foundations necessary to develop strategies to supplement conventional repeated dose systemic toxicity testing consumer monitoring (Fig. 11.2).

The seven cluster projects (central data management and maintenance projects, as well as a coordination and support projects belonging to five research initiatives) under the SEURAT-I initiative include the following:

- Scr&Tox (stem cells for relevant, efficient extended and normalized toxicology)
- HeMiBio (hepatic microfluidic bioreactor)
- DETECTIVE (detection of endpoints and biomarkers of repeated dose toxicity using in vitro systems)
- COSMOS (integrated in silico models for the prediction of human repeated dose toxicity of cosmetics to optimize safety)
- Notox (predicting long-term toxic effects using computer models based on systems characterization of organotypic cultures)
- ToxBank (supporting integrated data analysis and servicing of alternative testing methods in toxicology)

- COACH ("coordination of projects on new approaches to replace current repeated dose systemic toxicity testing of cosmetics and chemicals")

11.3.2.1 The COSMOS Project

The SEURAT-1 cluster consisted of six initiatives, including COSMOS (project website address—www.cosmostox.eu). This research project was the first step toward ensuring the long-term stated goal of supplementing animal experimentation of cosmetic ingredients with safety evaluation (Cronin et al. 2012; Cronin 2015). This, in turn, alluded to the notion that further actions must be completed before the ultimate objective is accomplished. With nine countries coming together in a cluster of 15 collaborators, this project ran its term from January 1, 2011, to December 31, 2015, and had a total grant (grant agreement ID: 266835) of €6,79,733,560 and a contribution of €3,350,000 from the EU. COSMOS looked at how well the existing TTC approach could be adapted to cosmetic chemicals and then how to extend from oral to dermal route exposure, which is especially important in the case of cosmetics. The COSMOS initiative was a one-of-a-kind collaboration that addressed the cosmetic industry in terms of comprehensive screening demands without using animals (Cronin et al. 2012; Cronin 2015).

COSMOS' principal goal was to create accessible and open-source technologies and procedures for estimating the long-term detrimental consequences of cosmetic chemicals on consumers (Yang et al. 2021). The study produced implications and regulations to expand the usability and final authorities of the present TTC method for cosmetic components. On September 9, 2015, the COSMOS Symposium on Computational Tools for Safety Assessment was convened in Liverpool, United Kingdom. The one-day session provided an overview of the EU COSMOS Project's accomplishments and impact.

The International Life Sciences Institute, Europe (https://ilsi.eu/eu-projects/past-projects/cosmos/) was one of the partners of the COSMOS project and contributed as two experts groups for the TTC approach; their observations were published in the research work of Williams et al. 2016 and Yang et al. 2017. Further general information is available at the following URLs:

- COSMOS Database|http://www.cosmostox.eu/what/COSMOSdb/
- COSMOS Space|http://cosmosspace.cosmostox.eu
- COSMOS KNIME Web Portal|http://www.cosmostox.eu/what/knime/

11.4 Intuitive and In Silico Methodologies for Impact Prediction

If there is insufficient evidence and the threshold of toxicological concern (TTC) is incapable of predicting risk, computational and in silico approaches can be used to estimate cosmetic chemicals for future hazards. There are numerous methodologies that may be used to analyze the potential serendipity of cosmetics components using in silico computational approaches. Hence, it can give information about the safety level of different ingredients in a product. Nowadays, upsurge use of computational approaches is due to the replacement and reduction in the rate of animal testing; along with this benefit in silico approach is a cost-efficient and rapid process of toxicity assessment.

Statistical researchers use a combination of data mining algorithms to discover relationships among chemical composition and function. These models rely on data that can generate computer algorithms without the need for specialized knowledge (Tintó-Moliner and Martin 2020). A negative prediction is more accurate than a positive prediction, even though it does not rely on direct mechanistic insight. Hybrid techniques combine practical information with statistically based principles to address each flaw.

Toxicology prediction research utilizing AI has recently become popular (Wu and Wang 2018; Ciallella and Zhu 2019). Artificial intelligence (AI) is an in silico system that "adapts" the chemical composition and hazard effects of chemicals. Because animal studies are restricted, this methodology can be used to assess the safety of cosmetic compounds. AI techniques such as artificial neural networks (ANNs) and machine learning are extensively trained to determine chemical skin irritability and cytotoxicity (Hirota et al. 2015, 2018; Wilm et al. 2019).

11.4.1 Quantitative Structure–Activity Relationships (QSARs) for Dermal Absorption

QSAR is a significant technique in the field of bioinformatics. The primary goal of QSAR is to establish a statistical link between molecule characteristics and dynamics. Although machine learning techniques outperform other approaches in terms of prediction rates, they lack interpretability (Potts and Guy 1992). Most of those are professionally developed models, which, ideally, will add biochemistry, kinematic, and distal pharmacology, and suitable empirical methods toward the discussion. A cosmetic corporation will rarely engage in research unless they have a specific interest. They rely on third parties or current QSARs to construct their own to do so. Most industries, including the cosmetic industry, use QSAR to evaluate products' toxicity levels (ECHA 2016a, b, c), e.g., carcinogenicity and skin sensitization. For quantitative measurement of chemicals, first, it needs to be modeled to evaluate endpoints like ADME parameter calculation, lethal dose, and half-maximal effective concentration; second, it is required to generate descriptors based on the chemical structure of compounds to generate a model. Most of the time, interpretable descriptors are favored for the generation of QSAR. Generally used descriptors used those related to portioning tissue: blood portioning coefficient (log P). This shows the relative nature of compounds, like their hydrophobicity and lipophilicity. It helps to find the movement of the compound through the biological membrane and its action through it (Madden et al. 2020b).

At last, QSAR needs a statistical approach to link descriptor with activity (safety level or any other factor of interest (Madden et al. 2020b). Many statistic approaches were proposed, spanning from simple linear progression to multiple regression analysis, depending on whether a two or more distinct classifiers are intended. The

Potts and Guy skin permeability mathematical formulation is shown below, wherein Kp signifies the dermal coefficient of permeability.

Log Kp = 0:71 log P0:0061 M Wt 6:3

$N = 93$; R2 = 0.67

R2 is the "correlation coefficient" in this case, and it demonstrates the variation in Kp represented by the descriptors LogP.

The value of r describes the correlations; whether it comes out to be positive or negative correlations, a value above 0.7 for correlation coefficients shows that it is good to use. If the r-value comes close to 1, it shows very unrealistic behavior for finding biological activity (Potts and Guy 1992).

QSAR models are very approachable methods for evaluating cosmetic product ingredients for skin permeability. The data on which QSAR is based are accessible through resources such as EDETOX (https://research.ncl.ac.uk/edetox/theedetoxdatabase/) and HuskinDB (https://huskindb.drug-design.de/data/) (Hewitt et al. 2020). In a recent study, a consistent technique was used to analyze the permeability of 56 substances pertinent to cosmeceuticals across and around human skin, and it had a high degree of reproducibility (Hewitt et al. 2020). RIFM proposed another in silico approach-based model for skin absorption, primarily for epidermal rapid screening for perfumes, with a permeation value ranging between 10% and 80% premised upon Jmax (Laroche et al. 2018). Neither of these algorithms can yield definitive estimates of makeup ingredient structural accessibility following topical contact. They must be maneuvered to discern substances that have an increasing or decreasing potential for systemic bioavailability; cutaneous permeation is insufficient (Table 11.1).

11.4.1.1 Structural Rules Capturing Structure–Activity Relationships

Structural alerts are one of the simple and easy ways to assess the toxicity of compounds. Structural alerts are also known as toxic fragments. In 1985, John Ashby's concept of structural alert for structural analysis of chemical carcinogen compounds (Ashby 1985). Many structural features are responsible for the toxic properties of compounds that give rise to structural alerts like mutagenicity, skin sensitization, and organ toxicity. If any other compound shows, the same structural alert indicates the risk potential to show some effects. For example, aromatic amine and an α, β-unsaturated aldehyde are electrophiles capable of reacting with the nucleophilic site within DNA and protein, respectively, leading to skin sensitization (Madden et al. 2020b). The presence of a functional group in these compounds is responsible for eliciting toxicity or any other potential hazard (Madden et al. 2020b). The relationship between molecular structure and activity of compound can easily derive structural alert and can be used to evaluate the potential risk. Statistical analysis and interpretability are two computational approaches for finding structural alerts. Most of the methods are based on a systematic analysis approach to find some substructures that occur very frequently in toxic compounds compared to non-toxic ones. On the other side, the machine learning approach is more accepted due to algorithms for pattern detection of compounds (Cherkasov et al. 2014). SAR has also been included in a number of prognostic toxicology applications and browser

Table 11.1 A non-exhaustive collection of freely accessible professional QSAR platforms for toxicology prognosis, including carcinogenicity and genotoxicity (Kim et al. 2021)

Name of software	URL	Features
1. Ambit (IDEAconsult Ltd.)	https://ambitlri.ideaconsult.net/tool2	• For toxicity and metabolism, knowledge-based expert systems are used. AMBIT incorporates a number of in silico estimation techniques (such as Toxtree)
2. Danish QSAR predictions database (DK EPA)	http://qsar.food.dtu.dk	• Estimates primarily predicated on over 200 (Q)SARs through both public and private sources, encompassing genotoxicity and carcinogenicity throughout male and female rats and mice in vivo and in vitro
3. LAZAR (in silico toxicology, GmbH)	https://lazar.in-silico.de/predict	• Models for mutagenicity and carcinogenicity are included in this statistics-built software
4. OECD QSAR Toolbox	http://toolbox.oasis-lmc.org/	• Contains "profilers" for genotoxicity and carcinogenicity and experimental observation databases
5. Oncologic, United States Environmental Protection Agency (US EPA)	https://www.epa.gov/reviewingnewchemicals-under-toxic-substancescontrolact-tsca	• Carcinogenicity estimates based on knowledge
6. TEST (US EPA)	https://www.epa.gov/chemical-research/toxicity-estimation-software-tool-test	• Models from many external sources are included in the software, which predicts various endpoints, including Ames mutagenicity
7. Toxtree (EU JRC—IDEAconsult Ltd.)	https://eurl-ecvam.jrc.ec.europa.eu/ https://apps.ideaconsult.net/data/ui/Toxtree	• SARs are presented for cytotoxic effects, carcinogenic effects, and in vivo chromosomal aberrations test
8. VEGA (Istituto Di Ricerche Farmacologiche Mario Negri)	https://www.vegahub.eu	• In silico models and tools for assessing various endpoints, such as mutagenicity and carcinogenicity

services, as seen in Table 11.1. Toxtree is one of the user-friendly open resource software that helps find the toxic risk of compounds by using the decision tree approach. Using structural information of compounds, chemicals are kept in different toxicity classes. A toxicologist may also utilize OCHEM to anticipate the physiological characteristics of substances. The benefit of using structural alert is that results are very transparent and easily accessible, reducing testing of products on animals.

11.4.2 PBK Model

Because of ethical and legal considerations, non-animal methodologies are increasingly being conducted to estimate the sustainability of chemicals for commercial use. We show how, in the absence of additional animal evidence, a physiologically based kinetic (PBK) framework for something like the cosmetic UV blocker constituent homovalvate was constructed and validated to support its safety (Dent et al. 2021). Prior to the EU animal testing ban in 2013, the intravenous (IV) rat PBK theory was established and verified utilizing legacy in vivo data using PK-Sim® (Bessems et al. 2017). These models evaluate the parameters of chemical absorption, distribution, metabolism, and excretion (ADME) (Gellatly and Sewell 2019). The models can take into account varying modes of transmission, taxa, ages, ethnicity, sexuality, illness condition, and other characteristics. These models' purpose is to anticipate an acceptable exposure measure—a dosage parameter that is intrinsically connected to the detrimental response, such as the highest concentration, which might be attained within that tissue. The primary goal of this model seems to determine an acceptable dosage meter for assessing cytotoxic consequences. The PBK model requires appropriate knowledge for deployment in the cosmetic sector. Previously, models were created using ordinary differential equations (ODE) or MATLAB (Cronin et al. 2022). To effectively employ PBK systems, the assessment team must develop meaningful and quite well judgments concerning this same model's structure for such specific topic being discussed (e.g., which divisions are crucial and suitable exposure paradigm), as well as the legitimacy of the input parameters (exploratory or computed attributes) and the model's susceptibility toward the variables adopted (Madden et al. 2020a).

11.4.3 Grouping and Read Across

Read across is one of the conceptually simple processes for evaluating chemical safety or toxicity level. It really is the act of estimating terminal data through one or perhaps more document (origin) compounds, which are already believed for being analogous using endpoint data from one or more data-poor (target) chemicals (Berggren et al. 2015; Madden et al. 2020b). Chemicals are grouped based on shared properties they share with other groups, and information interpreted from one member of the group is used to infer from other members of the group (ECHA 2017a, b). Read across the main objective is to find similarities between the chemicals. It can be based on carbon chain length, chemical fingerprinting, mechanism of action, or specific functional groups in chemical structure (Berggren et al. 2015). ToxMatch (from IDEAconsult) and the Compound Similarity toolset (from ChemMine Tools) are two examples of software that may be used to evaluate structural similarity in compounds. Analog selection should result in an accurate read-across prognosis for in vivo responsiveness (Madden et al. 2020b). The resemblance in chemical composition, but rather more vitally, the resemblance in behavior, is factored into the equation. ADME profile (i.e., pharmacokinetics and (toxic)

Table 11.2 A list of some of the best open-source read-across utilities

Name of the tool	Features
1. AMBIT IDEAconsult Ltd. (Bulgaria) http://cefic-lri.org/lri_toolbox/ambit/	• A Web-based stand-alone tool • User-dependent qualitative approach • Manual biological similarity selection • Chemical input in the form of—name, identifiers, SMILES, InChI • Output report in the form of—s docx or xlsx, data matrix as xlsx
2. OECD Toolbox LMC, Bourgas (Bulgaria) www.qsartoolbox.org	• A stand-alone tool working on a client/server basis • Both qualitative and quantitative approaches • Presence of both manual + automatic filters for similarity search • Accepted input formats—CAS, name, SMILES, structure drawing, MOL, SDF • Output formats—IUCLID format, pdf and RTF files of prediction report, text files of data, image files of plots, etc. • Visualize data as 2D standard plots
3. CBR Fourches Lab at North Carolina State University (USA) http://www.fourcheslaboratory.com/software	• Standalone tool • Automatic biological similarity selection • Qualitative approach • Accepted formats of input are Molfile, descriptors as txt • Visualize data as a radial plot of neighbors
4. ToxRead Istituto Di Ricerche Farmacologiche Mario Negri (Italy) https://www.vegahub.eu/portfolio-item/toxread/	• Stand-alone tool • A qualitative approach to check mutagenicity while quantitative for bioconcentration factor • Automatic filters for similarity selection • Input chemical format is SMILES • Visualization of data as interactive neighbor plot • Output in the form of an image file of the plot
5. CIIPro Zhu Research Group at Rutgers University (USA) http://ciipro.rutgers.edu/	• A Web-based tool • Manual + automatic filters for similar selection • Uses the VEGA similarity algorithm • Accepted chemical formats are PubChem CID, CAS, IUPAC, SMILES, and InChI • Data visualization as activity plots

activity are toxicokinetic) (Cronin et al. 2022). Alexander-White et al., in the year 2022, based on the EU SEURAT-I project and the ICCR principles, established a pragmatic and systemic 10-step framework to illustrate how read across can be employed NAM in the absence of TTC will aid in consumer safety evaluation (Alexander-White et al. 2022) (Table 11.2).

11.4.4 The Threshold of Toxicological Concern (TTC) Approach

TTC is a statistical likelihood technique toward assessing chemical toxicity in the lack of chemical-based toxicology studies. This implies establishing a universal absolute threshold for all substances under whom there is minimal substantial risk to individual well-being. In chemical-specific toxicity evidence, SCCS considers the TTC approach a suitable supporting tool for evaluating the safety of cosmetic compounds with known chemical structures (European Commission 2018; Worth et al. 2012; Williams et al. 2016; Yang et al. 2017). TTC values are applied using a decision framework that checks the composition of ingredients step by step. TTC's underlying database has been critical in determining robust and accurate thresholds, which requires in-depth analysis and assessment of acceptable toxicological data (Cronin et al. 2022). One of the greatest applications of the TTC approach was the basis of the EU COSMOS project (Yang et al. 2017) as it is a plausible solution to many safety risk management difficulties (Ellison et al. 2019); and it is also a part of the ab initio approach of NGRA (Daston et al. 2014; Gocht et al. 2015) as explained in the former sections of this chapter. Topical sensitivity evaluation is critical in the TTC method for cosmetic chemicals. Internal contact with cosmetic chemicals should be used in risk assessments, including the TTC approach (Kim et al. 2021). The Munro database and COSMOS dataset have been created using NOAELs (Munro et al. 1996) of chemicals obtained by oral exposure with a 100% permeability hypothesis. Williams et al. elucidated that the application of risk evaluation criteria premised on repeated dosage data of cosmetology constituents is the application of TTC (Williams et al. 2016).

11.5 The Relevance of In Silico Technologies in Adverse Outcome Pathways (AOPs)

During the last decade, the use of AOPs has indeed been established as a mechanistic utilitarian technique with vast applications in the disciplines of toxicology and mitigation strategies of chemical compounds, and their usage in the cosmetics sector is publicly recognized and highly documented as well (National Research Council 2007; Tollefsen et al. 2014; Burden et al. 2015; Vinken et al. 2020). The AOP notion indicates a robust structure that allows insights from in silico models, bioinformatics, in vitro experiments, high-throughput screening, *omics* technologies, and biological systems to be deeply implemented and unanswered questions addressed (Madden et al. 2020b). Recently, the OECD has extensively encouraged the establishment of AOPs (OECD 2012a, b; Yamada et al. 2020); the OECD-AOP initiative (www.oecd.org/chemicalsafety/testing/adverse-outcome-pathways-molecular-screening-and-toxicogenomics.htm) is actively building a variety of AOPs for a myriad of sophisticated toxicological outcomes (OECD 2012a, b; Madden et al. 2020b; Yamada et al. 2020) after the complete ban of animal testing as inflicted by the European ordinance (EC (European Commission) 2009). The goal of AOPs is to describe and collect current understanding of the biologically viable and experimentally validated grounds for forecasting basal toxicity from mechanical evidence (OECD 2013). The AOP Knowledge Base (AOP-KB; https://aopkb.org) (Sachana 2018; Wittwehr et al. 2015), associated with its wiki (https://aopwiki.org/), and its documentation is arranged in a structured, navigable, and direct way, following a set of criteria and guidelines (OECD 2016a, b) that make it easier to assess eligibility for specific governance needs (Wittwehr et al. 2016).

The OECD organized a workshop on "*Using Mechanistic Information in Forming Chemical Categories*" in December 2010 in Washington, DC, USA, in lieu of the situation of the *Use of Adverse Outcome Pathways in the Development of Categories* (OECD 2011a, b; Sakuratani et al. 2018). The AOP for skin sensitization was created in 2011 and 2012 (OECD 2011a, b; Schultz et al. 2016) as a reflection of the workshop's accomplishment and at the request of OECD member nations (Sakuratani et al. 2018; Schultz et al. 2016). An AOP depicts current insights into the interactions among two reference points, the molecular initiating event (MIE) (Burden et al. 2015; Ankley et al. 2010) and an adverse outcome (AO), interlinked by a sequence of key events (KEs), and whenever feasible the relations between the KEs (KERs) (Schultz et al. 2016; Delrue et al. 2016; OECD 2013).

High-throughput in vitro techniques can be used to signal hazardous potential since AOPs represent the sequence of essential processes that lead to adversity at multiple levels of biological organization (Villeneuve et al. 2014; Villeneuve 2015; Vinken et al. 2020). There is also a practical realization well within the toxicity testing discipline that alternative techniques would not supersede in vivo models on a resembling scale. Hence, AOPs will indeed be utilized to feed and lead a multidisciplinary approach (Wittwehr et al. 2016) to verification and validation; AOPs might thus serve as a link between non-animal methodologies (Burden et al. 2015; Knapen et al. 2018) and systems toxicology, thereby improving the domain of non-animal safety evaluation (Schultz et al. 2016). To assist their implementation in regulatory decision-making, there is a need for an empirical foundation to understand the outcomes of innovative test techniques and related prediction models (Tollefsen et al. 2014; Sakuratani et al. 2018). A paradigm of this type might have three key components: the AOP, non-animal test techniques, and in silico methodologies addressing essential parts of the AOP, as well as their related modeling techniques for an appropriate policy framework (Delrue et al. 2016; Yamada et al. 2020; Hecker and LaLone 2019; Wittwehr et al. 2016; Villeneuve 2015). A tangible solution to such proactive diagnostics, hypothesis-driven Integrated Approaches to Testing and Assessment (IATA), has been advocated (Tollefsen et al. 2014; OECD 2017a, b; Madden et al. 2020b).

11.5.1 The Way Forward

Blending incredibly challenging biological systems with large-scale methodologies provides a greater understanding of the complexities of the biological response to cosmetic compounds, improving the possibility of predicting clinical reactions in vivo and finding novel substitutes for animal experimentations (Zimbardi 2018). In this regard, the *omics* technology has emerged as sophisticated technology enabling trying to analyze whole genetic or molecular, or metabolite fingerprints (Lee et al. 2020a, b; Pirih and Kunej 2017), integrating analyses to enhance the evaluation and monitor the toxicity testing of cosmetic compounds (Lee et al.

2020a, b; He and Jia 2021), and offering valuable means of assessing the hazard and efficacy tests that cannot be evaluated on animals (Kim et al. 2021; van Delft et al. 2014). Cosmetics and the personal care industry are also broadening their horizons by investigating the potential of deep learning artificial intelligence (AI) and machine learning (ML) to aid in toxicity testing and product selection (Nambiar 2021; Kim and Lee 2021) for artificially creating algorithms that automatically extract facts and figures from multivariate data and analyze it even further (SciForce 2019). To generate state-of-the-art models through numerous ways, such as logistic regression, linear support vector machine (SVM), artificial neural networks (ANNs), and decision tree classifiers (Umer et al. 2020; Ma et al. 2021). Even successful businesses like that Coty and L'Oréal appear to be going into AI (de Jesus 2020) through virtual mirrors and Alexa skills, which might suggest how AI will inevitably change the panorama of the cosmetics and personal care sector over the next couple of years (Ma et al. 2021).

11.6 Conclusions

Animal cruelty has always been the dispute concerning animal experimentation and the need for searching for alternative testing methods. It has indeed been a constant conflict, but at the very minimum, a stagnation that researchers and scientists need to take a strong stance and make their moral judgments. In vitro and in silico approaches are gaining momentum as technology develops and shifts the battleground slightly. Various computational techniques are available for assessing safety levels, estimating exposure, and hazard identification for any cosmetic products. The organizational dimension has lately managed to include a repercussion: Several controlled trials are simply too expensive, take too long, or generate inaccurate results (Meigs 2018). High advances are that lots of high-quality databases are available for toxicology evaluation. They provide well-curated information, which reduces the use of animals in research to evaluate the safety-level assessment of cosmetic products. Recent advances in technology and understanding of mechanistic approaches, primarily through AOP, have helped prove a more insightful side of computational and in silico approaches for predicting the toxicity level of a product. Computational techniques range from structural rules to various databases and models read-across approaches to fill data gaps using closely related chemical structures and properties of various compounds. NGRA implies a combination practice of in silico and in vitro methods for animal-free testing of products and incorporating a new type of database for safety-level assessment. Overall, a range of computational approaches increasing confidence with well-curated results and leading research toward a very ethical pathway for society's benefit, such as hazard and safety assessment of a plethora of products and will continue the trend started by SCCS.

The debate over alternative testing has historically been considered primarily science-based. Furthermore, it also demands a reassessment of fundamental components of where and how regulatory toxicity studies are now done. However, it also raises concerns about a complicated legislative framework that is not structured or equipped to reform swiftly. Overall, establishing a somewhat more fundamental perspective to regulating toxicity testing is a contemporary "Artemis," an extraction point that, if traversed, would quickly relegate several traditional procedures to an obsolete.

References

Alexander-White C, Bury D, Cronin M, Dent M, Hack E, Hewitt N et al (2022) A 10-step framework for the use of read-across (RAX) in next generation risk assessment (NGRA) for cosmetics safety assessment. Regul Toxicol Pharmacol 129:105094

Amaral R, Ansell J, Boisleve F, Cubberley R, Dent M, Hatao M, Weiss C (2018) Report for international cooperation on cosmetics regulation regulators and industry joint working group (JWG): integrated strategies for safety assessment of cosmetic ingredients: Part 2

Ankley G, Bennett R, Erickson R, Hoff D, Hornung M, Johnson R et al (2010) Adverse outcome pathways: a conceptual framework to support ecotoxicology research and risk assessment. Environ Toxicol Chem 29(3):730–741

Ashby J (1985) Fundamental structural alerts to potential carcinogenicity or non-carcinogenicity. Environ Mutagen 7(6):919–921

Bassan A, Alves VM, Amberg A, Anger LT, Auerbach S, Beilke L et al (2021) In silico approaches in organ toxicity hazard assessment: current status and future needs in predicting liver toxicity. Comput Toxicol 20:100187

Berggren E, Amcoff P, Benigni R, Blackburn K, Carney E, Cronin M et al (2015) Chemical safety assessment using read-across: assessing the use of novel testing methods to strengthen the evidence base for decision making. Environ Health Perspect 123(12):1232–1240

Berggren E, White A, Ouedraogo G, Paini A, Richarz A, Bois F et al (2017) Ab initio chemical safety assessment: a workflow based on exposure considerations and non-animal methods. Comput Toxicol 4:31–44

Bernauer U, Bodin L, Chaudhry Q, Coenraads PJ, Dusinska M, Ezendam J et al (2021) The SCCS notes of guidance for the testing of cosmetic ingredients and their safety evaluation, 11th revision, 30–31 march 2021, SCCS/1628/21. Regul Toxicol Pharmacol 127:105052

Bessems J, Paini A, Gajewska M, Worth A (2017) The margin of internal exposure (MOIE) concept for dermal risk assessment based on oral toxicity data–a case study with caffeine. Toxicology 392:119–129

Boobis A, Doe J, Heinrich-Hirsch B, Bette Meek ME, Munn S, Ruchirawat M et al (2008) IPCS framework for analyzing the relevance of a noncancer mode of action for humans. Crit Rev Toxicol 38(2):87–96

Burden N, Sewell F, Andersen M, Boobis A, Chipman J, Cronin M et al (2015) Adverse outcome pathways can drive non-animal approaches for safety assessment. J Appl Toxicol 35(9): 971–975

Business Wire (2022) Global cosmetics market trends, share, size, growth, opportunities and forecast 2022–2027

Cherkasov A, Muratov EN, Fourches D, Varnek A, Baskin II, Cronin M et al (2014) QSAR modeling: where have you been? Where are you going to? J Med Chem 57(12):4977–5010

Ciallella HL, Zhu H (2019) Advancing computational toxicology in the big data era by artificial intelligence: data-driven and mechanism-driven modelling for chemical toxicity. Chem Res Toxicol 32(4):536–547

CIRS (2013) REACH and cosmetics

Comiskey D, Api AM, Barrett C, Ellis G, McNamara C, O'Mahony C et al (2017) Integrating habits and practices data for soaps, cosmetics and air care products into an existing aggregate exposure model. Regul Toxicol Pharmacol 88:144–156

Cronin M (2015) CORDIS|European Commission|integrated in silico models for the prediction of human repeated dose toxicity of cosmetics to optimise safety
Cronin M, Madden J, Richarz A (2012) The COSMOS project: a Foundation for the Future of computational modelling of repeat dose toxicity|AltTox.org
Cronin M, Enoch S, Madden J, Rathman J, Richarz A, Yang C (2022) A review of in silico toxicology approaches to support the safety assessment of cosmetics-related materials. Comput Toxicol 21:100213
Daston G, Knight D, Schwarz M, Gocht T, Thomas R, Mahony C, Whelan M (2014) SEURAT: safety evaluation ultimately replacing animal testing—recommendations for future research in the field of predictive toxicology. Arch Toxicol 89(1):15–23
de Jesus A (2020) Artificial intelligence for beauty and cosmetics-current applications. Emerj, Artificial Intelligence Research
Delrue N, Sachana M, Sakuratani Y, Gourmelon A, Leinala E, Diderich R (2016) The adverse outcome pathway concept: a basis for developing regulatory decision-making tools. Altern Lab Anim 44(5):417–429
Dent M, Amaral R, Da Silva P, Ansell J, Boisleve F, Hatao M et al (2018) Principles underpinning the use of new methodologies in the risk assessment of cosmetic ingredients. Comput Toxicol 7: 20–26
Dent M, Vaillancourt E, Thomas R, Carmichael P, Ouedraogo G, Kojima H et al (2021) Paving the way for application of next generation risk assessment to safety decision-making for cosmetic ingredients. Regul Toxicol Pharmacol 125:105026
Doke S, Dhawale S (2015) Alternatives to animal testing: a review. Saudi Pharm J 23(3):223–229
EC (European Commission) (2009). Regulation No 1223/2009 of The European Parliament and of the Council on cosmetic products
ECHA (2012a) Guidance on information requirements and chemical safety assessment chapter R.19: uncertainty analysis. European Chemicals Agency, Helsinki
ECHA (2012b) Understanding REACH
ECHA (2016a) Evaluation under REACH Progress report–executive summary and recommendations to registrants
ECHA (2016b) New approach methodologies in regulatory science proceedings of a scientific workshop. Proceedings of a scientific workshop, Helsinki
ECHA (2016c) Practical guide–how to use and report (Q)SARs. Report no. 1831–6727
ECHA (2017a) The use of alternatives to testing on animals for the REACH regulation
ECHA (2017b) Read-across assessment framework (RAAF)
Ellison C, Blackburn K, Carmichael P, Clewell H, Cronin M, Desprez B et al (2019) Challenges in working towards an internal threshold of toxicological concern (iTTC) for use in the safety assessment of cosmetics: discussions from the cosmetics Europe iTTC working group workshop. Regul Toxicol Pharmacol 103:63–72
European Commission (2001) White paper: strategy for a future chemicals policy
European Commission (2004) Alternative testing methods: European research to replace and reduce animal tests
European Commission (2006) REACH regulation (EC)No1907/2006. Regulation concerning the registration, evaluation, authorisation and restriction of chemicals (REACH), establishing ECHA. Off J Eur Communities L396:1–849
European Commission (2009) Report from the commission to the European parliament and the council report on the development. Validation and legal acceptance of alternative methods to animal tests in the field of cosmetics
European Commission (2010) Directive 2010/63/EU of the European Parliament and of the Council of September 22nd 2010 on the protection of animals used for scientific purposes. Off J Eur Communities 276:33–79
European Commission (2018) 10th revision of the SCCS notes of guidance for the testing of cosmetic ingredients and their safety evaluation
Fischer I, Milton C, Wallace H (2020) Toxicity testing is evolving! Toxicol Res 9(2):67–80

Garthoff B (2005) Alternatives to animal experimentation: the regulatory background. Toxicol Appl Pharmacol 207(2):388–392

Gellatly N, Sewell F (2019) Regulatory acceptance of in silico approaches for the safety assessment of cosmetic-related substances. Comput Toxicol 11:82–89

Gironde C, Dufour C, Furger C (2020) Use of LUCS (light-up cell system) as an alternative live cell method to predict human acute oral toxicity. Toxicol Rep 7:403–412

Gocht T, Berggren E, Ahr H, Cotgreave I, Cronin M, Daston G et al (2015) The SEURAT-1 approach towards animal free human safety assessment. ALTEX 32(1):9–24

Hatherell S, Baltazar MT, Reynolds J, Carmichael PL, Dent M, Li H, Ryder S, White A, Walker P, Middleton AM (2020) Identifying and characterizing stress pathways of concern for consumer safety in next-generation risk assessment. Toxicol Sci 176(1):11–33

He J, Jia Y (2021) Application of omics technologies in dermatological research and skin management. J Cosmet Dermatol 21(2):451–460

Hecker M, LaLone C (2019) Adverse outcome pathways: moving from a scientific concept to an internationally accepted framework. Environ Toxicol Chem 38(6):1152–1163

Hewitt NJ, Grégoire S, Cubberley R, Duplan H, Eilstein J, Ellison C et al (2020) Measurement of the penetration of 56 cosmetic relevant chemicals into and through human skin using a standardized protocol. J Appl Toxicol 40(3):403–415

Hirota M, Fukui S, Okamoto K, Kurotani S, Imai N, Fujishiro M, Kyotani D, Kato Y, Kasahara T, Fujita M (2015) Evaluation of combinations of in vitro sensitization test descriptors for the artificial neural network-based risk assessment model of skin sensitization. J Appl Toxicol 35(11):1333–1347

Hirota M, Ashikaga T, Kouzuki H (2018) Development of an artificial neural network model for risk assessment of skin sensitization using human cell line activation test, direct peptide reactivity assay, Keratinosensâ and in silico structure alert parameter. J Appl Toxicol 38(4): 514–526

Kabene S, Baadel S (2019) Bioethics: a look at animal testing in medicine and cosmetics in the UK. J Med Ethics Hist Med 12:15

Kim S, Lee S (2021) Deep learning approach for cosmetic product detection and classification. Comput Mater Contin 69(1):713–725

Kim K, Kwack S, Lee J, Kacew S, Lee B (2021) Current opinion on risk assessment of cosmetics. J Toxicol Environ Health Part B 24(4):137–161

Knapen D, Angrish M, Fortin M, Katsiadaki I, Leonard M, Margiotta-Casaluci L et al (2018) Adverse outcome pathway networks I: development and applications. Environ Toxicol Chem 37(6):1723–1733

Krewski D, Andersen ME, Tyshenko MG, Krishnan K, Hartung T, Boekelheide K, Wambaugh JF, Jones D, Whelan M, Thomas R et al (2010a) Toxicity testing in the 21st century: Progress in the past decade and future perspectives. Arch Toxicol 94:1–58

Krewski D, Acosta D, Anderson M, Anderson H, Bailar J, Boekelheide K et al (2010b) Toxicity testing in the 21st century: a vision and a strategy. J Toxicol Environ Health Part B 13(2–4): 51–138

Kuseva C, Schultz T, Yordanova D, Tankova K, Kutsarova S, Pavlov T et al (2019) The implementation of RAAF in the OECD QSAR toolbox. Regul Toxicol Pharmacol 105:51–61

Laroche C, Aggarwal M, Bender H, Benndorf P, Birk B, Crozier J et al (2018) Finding synergies for 3Rs–toxicokinetics and read-across: report from an EPAA partners' forum. Regul Toxicol Pharmacol 99:5–21

Lee B, Choi M, Shin I, Kim J, Choi Z, Kim K et al (2020a) Risk communication for labeling all ingredients in consumer products. J Toxic Environ Health A 83(13–14):509–524

Lee J, Kim H, Kang K, Jeong H, Song M, Tae I et al (2020b) Integration of transcriptomics, proteomics and metabolomics identifies biomarkers for pulmonary injury by polyhexamethylene guanidine phosphate (PHMG-p), a humidifier disinfectant, in rats. Arch Toxicol 94(3):887–909

Ma X, Zhang F, Wei H, Xu L (2021) Deep learning method for makeup style transfer: a survey. Cogn Robot 1:182–187

Madden J, Tan Y, Blaauboer B, Paini A (2020a) Development and application of physiologically-based kinetic (PBK) models. Comput Toxicol 13:100116

Madden JC, Enoch SJ, Paini A, Cronin MT (2020b) A review of in silico tools as alternatives to animal testing: principles, resources and applications. Altern Lab Anim 48(4):146–172

Meigs L (2018) Animal testing and its alternatives–the most important omics is economics. ALTEX 35:275–305

Merenyi S (2018) REACH: regulation(EC) no 1907/2006. 152–158

Munro I, Ford R, Kennepohl E, Sprenger J (1996) Correlation of structural class with no-observed-effect levels: a proposal for establishing a threshold of concern. Food Chem Toxicol 34(9): 829–867

Nambiar K (2021) 3 applications of AI in the cosmetic industry

NAS (2007) Toxicity testing in the twenty-first century: a vision and a strategy. National Academies Press, Washington DC

NAS (2012) Exposure science in the 21st century: a vision and a strategy. National Academies Press, Washington DC

NAS (2017) Using 21st century science to improve risk-related evaluations. National Academies Press, Washington, DC

National Research Council (2007) Toxicity testing in the 21st century: a vision and a strategy. National Academies Press, Washington, DC

OECD (2011a) Series on testing and assessment no. 138: report of the workshop on using mechanistic information in forming chemical categories (176pp)

OECD (2011b) Series on testing and assessment no. 139: report of the expert consultation on scientific and regulatory evaluation of organic chemistry based structural alerts for the identification of protein-binding chemicals (100pp)

OECD (2012a) OECD series on testing and assessment no 168: the adverse outcome pathway for skin sensitisation initiated by covalent binding to proteins. Part 2: use of the AOP to develop chemical categories and integrated assessment and testing approaches.(46pp)

OECD (2012b) OECD series on testing and assessment no. 168: the adverse outcome pathway for skin sensitisation initiated by covalent binding to proteins. Part 1: scientific evidence (59pp)

OECD (2013) Guidance document on developing and assessing adverse outcome pathways. Environment, health and safety publications. Series on testing and assessment, no. 184, Vol. 6

OECD (2016a) Guidance document on the reporting of defined approaches to be used within integrated approaches to testing and assessment. Series on testing and assessment no. 255

OECD (2016b) Users' handbook supplement to the guidance document for developing and assessing adverse outcome pathways. Environment, health and safety publications, series on testing and assessment no. 233

OECD (2017a) Guidance document on the reporting of defined approaches and individual information sources to be used within integrated approaches to testing and assessment (IATA) for skin sensitisation. Series on testing and assessment no. 256

OECD (2017b) Adverse outcome pathways, molecular screening and toxicogenomics

Patlewicz G, Cronin M, Helman G, Lambert J, Lizarraga L, Shah I (2018) Navigating through the minefield of read-across frameworks: a commentary perspective. Comput Toxicol 6:39–54

Pirih N, Kunej T (2017) Toward a taxonomy for multi-omics science? Terminology development for whole genome study approaches by omics technology and hierarchy. OMICS 21(1):1–16

Potts RO, Guy RH (1992) Predicting skin permeability. Pharm Res 9(5):663–669

Pouillot A, Polla B, Polla A (2009) REACH: impact on the US cosmetics industry? J Cosmet Dermatol 8(1):3–7

REACH (2012) Chemicals–environment—European Commission

REACH (2022) Health and safety authority

Rise for Animals (2022) Seeking justice for animals in labs

Rogiers V, Benfenati E, Bernauer U, Bodin L, Carmichael P, Chaudhry Q et al (2020) The way forward for assessing the human health safety of cosmetics in the EU–workshop proceedings. Toxicology 436:152–421

Sachana M (2018) Approaching the development of alternative testing methods using information available in the adverse outcome pathway knowledgebase (AOP-KB). Toxicol Lett 295:S41

Safford B, Api AM, Barratt C, Comiskey D, Daly EJ, Ellis G et al (2015) Use of an aggregate exposure model to estimate consumer exposure to fragrance ingredients in personal care and cosmetic products. Regul Toxicol Pharmacol 72(3):673–682

Safford B, Api AM, Barratt C, Comiskey D, Ellis G, McNamara C et al (2017) Application of the expanded Creme RIFM consumer exposure model to fragrance ingredients in cosmetic, personal care and air care products. Regul Toxicol Pharmacol 86:148–156

Sakuratani Y, Horie M, Leinala E (2018) Integrated approaches to testing and assessment: OECD activities on the development and use of adverse outcome pathways and case studies. Basic Clin Pharmacol Toxicol 123:20–28

SCCS (2016) SCCS notes of guidance for the testing of cosmetic ingredients and their safety evaluation 9th revision. Revision of April 25th 2016

SCCS (2018) SCCS notes of guidance for the testing of cosmetic ingredients and their safety evaluation 10th revision, 24–25

Schultz T, Amcoff P, Berggren E, Gautier F, Klaric M, Knight D et al (2015) A strategy for structuring and reporting a read-across prediction of toxicity. Regul Toxicol Pharmacol 72:586–601

Schultz T, Dimitrova G, Dimitrov S, Mekenyan O (2016) The adverse outcome pathway for skin sensitisation: moving closer to replacing animal testing. Altern Lab Anim 44(5):453–460

SciForce (2019) Machine learning: changing the beauty industry

Sheehan KB, Lee J (2014) What's cruel about cruelty-free: an exploration of consumers, moral heuristics, and public policy. J Anim Ethics 4(2):1

Sturla S, Boobis A, FitzGerald R, Hoeng J, Kavlock R, Schirmer K et al (2014) Systems toxicology: from basic research to risk assessment. Chem Res Toxicol 27(3):314–329

Tanner B (2022) Cosmetic testing–microchem laboratory

Taylor K, Rego Alvarez L (2020) Regulatory drivers in the last 20 years towards the use of in silico techniques as replacements to animal testing for cosmetic-related substances. Comput Toxicol 13:100–112

Thomas R, Philbert M, Auerbach S, Wetmore B, Devito M, Cote I et al (2013) Incorporating new technologies into toxicity testing and risk assessment: moving from 21st century vision to a data-driven framework. Toxicol Sci 136(1):4–18

Tintó-Moliner A, Martin M (2020) Quantitative weight of evidence method for combining predictions of quantitative structure-activity relationship models. SAR QSAR Environ Res 31(4):261–279

Tollefsen K, Scholz S, Cronin M, Edwards S, de Knecht J, Crofton K et al (2014) Applying adverse outcome pathways (AOPs) to support integrated approaches to testing and assessment (IATA). Regul Toxicol Pharmacol 70(3):629–640

Tozer S, O'Mahony C, Hannah J, O'Brien J, Kelly S, Kosemund-Meynen K, Alexander-White C (2019) Aggregate exposure modelling of vitamin a from cosmetic products, diet and food supplements. Food Chem Toxicol 131:110–549

Turley A, Isaacs K, Wetmore B, Karmaus A, Embry M, Krishan M (2019) Incorporating new approach methodologies in toxicity testing and exposure assessment for tiered risk assessment using the RISK21 approach: case studies on food contact chemicals. Food Chem Toxicol 134: 110–819

Umer S, Mohanta P, Rout R, Pandey H (2020) Machine learning method for cosmetic product recognition: a visual searching approach. Multimed Tools Appl 80(28–29):34997–35023

USEPA (2014) Next generation risk assessment: incorporation of recent advances in molecular, computational, and systems biology. National Center for environmental assessment, Office of Research and Development. U.S. Environmental Protection Agency, Washington DC

US-FDA (2022) Product testing of cosmetics

van Delft J, Mathijs K, Polman J, Coonen M, Szalowska E, Verheyen G et al (2014) Hepatotoxicity screening on in vitro models and the role of omics. In: Kleinjans J (ed) Toxicogenomics-based cellular models. Academic Press, Cambridge, pp 193–212

Van Der Wielen A (2007) REACH: next step to a sound chemicals management. J Expo Sci Environ Epidemiol 17(1):S2–S6

Villalobos A, Tenneriello B, Koppisch E, Fagan J (2014) Cosmetic testing on animals

Villeneuve D (2015) In response: the path forward for the adverse outcome pathway framework-a regulatory perspective. Environ Toxicol Chem 34(9):1938–1940

Villeneuve D, Crump D, Garcia-Reyero N, Hecker M, Hutchinson T, LaLone C et al (2014) Adverse outcome pathway (AOP) development I: strategies and principles. Toxicol Sci 142(2):312–320

Vinardell M, Mitjans M (2017) Alternative methods to animal testing for the safety evaluation of cosmetic ingredients: an overview. Cosmetics 4(3):30

Vinken M, Kramer N, Allen T, Hoffmans Y, Thatcher N, Levorato S et al (2020) The use of adverse outcome pathways in the safety evaluation of food additives. Arch Toxicol 94(3):959–966

White E (2022) An analysis of animal testing in beauty products and its ethicality

Williams F, Rothe H, Barrett G, Chiodini A, Whyte J, Cronin M et al (2016) Assessing the safety of cosmetic chemicals: consideration of a flux decision tree to predict dermally delivered systemic dose for comparison with oral TTC. Regul Toxicol Pharmacol 76:174–186

Wilm A, Stork C, Bauer C, Schepky A, Kühnl J, Kirchmair J (2019) Skin doctor: machine learning models for skin sensitization prediction that provide estimates and indicators of prediction reliability. Int J Mol Sci 20(19):33–48

Wittwehr C, Munn S, Landesmann B, Whelan M (2015) Adverse outcome pathways knowledge base (AOP-KB). Toxicol Lett 238(2):S309

Wittwehr C, Aladjov H, Ankley G, Byrne H, de Knecht J, Heinzle E et al (2016) How adverse outcome pathways can aid the development and use of computational prediction models for regulatory toxicology. Toxicol Sci 155(2):326–336

Worth A, Cronin M, Enoch S, Fioravanzo E, Fuart-Gatnik M, Pavan M, Yang C (2012) Applicability of the threshold of toxicological concern (TTC) approach to cosmetics–preliminary analysis. Eur Union 10:5059

Wu Y, Wang G (2018) Machine learning based toxicity prediction: from chemical structural description to transcriptome analysis. Int J Mol Sci 19(8):23–58

Yamada T, Ashikaga T, Kojima H, Hirose A (2020) Challenge for adverse outcome pathway (AOP)-based chemical safety assessment. Yakugaku Zasshi 140(4):481–484

Yang C, Barlow S, Muldoon Jacobs K, Vitcheva V, Boobis A, Felter S et al (2017) Thresholds of toxicological concern for cosmetics-related substances: new database, thresholds, and enrichment of chemical space. Food Chem Toxicol 109:170–193

Yang C, Cronin M, Arvidson K, Bienfait B, Enoch S, Heldreth B et al (2021) COSMOS next generation–a public knowledge base leveraging chemical and biological data to support the regulatory assessment of chemicals. Comput Toxicol 19:100–175

Zimbardi D (2018) Gene expression screening to identify activation profiles in skin after exposition to cosmetic ingredients. In: 2nd pan-American for conference alternative methods. Natura innovation, Cajamar, SP

Role of Omics Approach in the Toxicity/Safety Study of Cosmetics

12

Prakriti Gaur, Priyanka Rana, Diksha Pathania, Saumya Shukla, Swastika Suvirya, Shakti Vinay Shukla, and Deepti Chopra

Abstract

The use of cosmetic items has increased significantly across all age groups globally without knowing their hazardous impact in long-term exposure. Additionally, these cosmetics have some ingredients, especially preservatives, colorant, and fragrances that promote inflammation and allergy in users' skin under exposure of UVR/sunlight. Therefore, identification and development of predictive molecular signatures for cosmetics toxicity via use of omics approaches are important for safety of consumers. OMICS includes biological techniques that aid in the detection of potential adverse outcome pathways (AOPs) of any

P. Gaur · D. Chopra (✉)
Photobiology Laboratory, Systems Toxicology and Health Risk Assessment Group, CSIR-Indian Institute of Toxicology Research (CSIR-IITR), Lucknow, Uttar Pradesh, India

P. Rana · D. Pathania
Photobiology Laboratory, Systems Toxicology and Health Risk Assessment Group, CSIR-Indian Institute of Toxicology Research (CSIR-IITR), Lucknow, Uttar Pradesh, India

Academy of Scientific and Innovative Research (AcSIR), CSIR-IITR Campus, Lucknow, Uttar Pradesh, India

S. Shukla
Photobiology Laboratory, Systems Toxicology and Health Risk Assessment Group, CSIR-Indian Institute of Toxicology Research (CSIR-IITR), Lucknow, Uttar Pradesh, India

Department of Biochemistry, School of Dental Sciences, Babu Banarasi Das University, Lucknow, Uttar Pradesh, India

S. Suvirya
Department of Dermatology, Venereology and Leprosy, King George's Medical University, Lucknow, Uttar Pradesh, India

S. V. Shukla
Fragrance & Flavour Development Centre, Kannauj, Uttar Pradesh, India

A. B. Pant et al. (eds.), *Skin 3-D Models and Cosmetics Toxicity*,
https://doi.org/10.1007/978-981-99-2804-0_12

toxicological response. These AOPs elucidate underlying molecular principles and affirm key molecular events that take place at various levels of biological organizations. This chapter will demonstrate how the omics method has been integrated into the toxicological evaluation of cosmetics and help in development of safe and quality cosmetic products.

Keywords

Cosmetics · Cosmetic toxicity · UVR · Omics

12.1 Introduction

Skin is an important organ that is mainly exposed with cosmetics. Their use triggers a chain of chemical reactions that alter the expression of several genes and protein synthesis, significantly altering cellular responses and resulting in skin allergies, phototoxicity, aging, and skin cancer (Khan and Alam 2019). Alternative test models and methods after the animal use banned for cosmetics toxicity testing by EU in 2013 and Indian in 2014 will be desperately needed for the detection of harmful effects upon the usage of cosmetic compounds.

Molecular alterations at various levels of biological organization have been quantified as a result of improvements in traditional toxicological testing. The strategy that was based on the conventional method of animal testing was superseded by the new paradigm of "system toxicology." The "omics" method used by this system technology encompasses genomics, transcriptomics, proteomics, and metabolomics (Fig. 12.1).

Omics is a vast system of biological methods, which assists in the identification of potential adverse outcome pathways (AOPs) (Wang et al. 2021), which confirms significant molecular events occurring at different levels of biological organizations and thus aids in the elucidation of underlying molecular principles. Omics techniques offer mechanistic analysis for the discovery of important targets, indicators, and toxicity pathways in toxicological evaluation (Gouveia et al. 2019). With the aid of these techniques, it is possible to identify adaptive reactions to low toxicant concentrations that do not cause toxicity but instead subject cells to oxidative stress, which is known to be a damaging action mechanism. Proteomics, the study of proteins at the systems level, and metabolomics, the study of cellular metabolic processes, are some of the omics approaches that are being used in the study of various biological responses that involve thousands of genes, proteins, and metabolites, respectively. Genomics and transcriptomics currently refer to the study of alterations in gene expression at the genome-wide level (Reay and Cairns 2021).

In order to evaluate the pertinent outcomes or endpoints that would aid in the development of customized skin care products, this article focuses on the function of omics technology as a holistic approach, its application, and its significance in the toxicity assessment of the cosmetics.

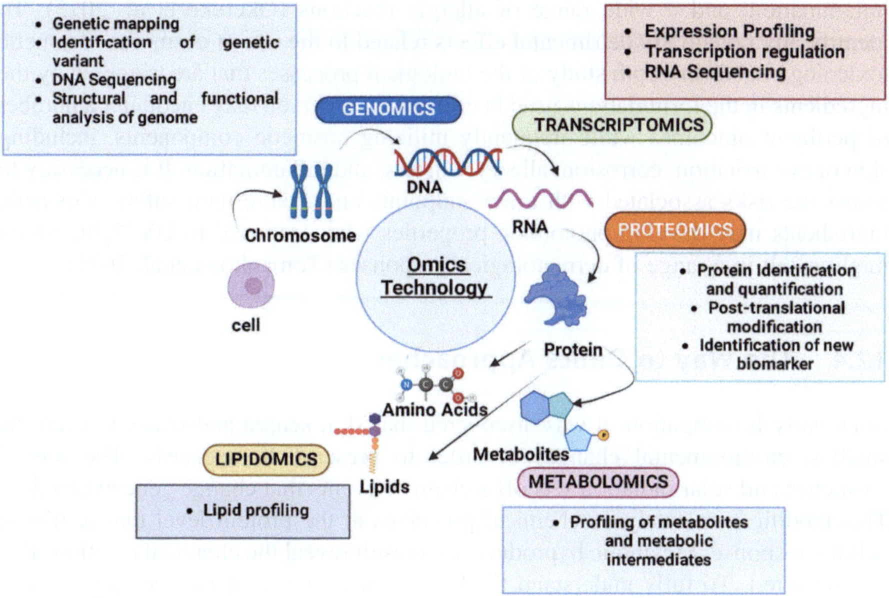

Fig. 12.1 Overview of omics approaches

12.2 Cosmetic Toxicology

Our skin is exposed to a variety of chemicals that are found in many cosmetic items almost every day. As a result, people of all ages continue to experience a steady increase in demand for cosmetic items worldwide. Cosmetic preservatives and fragrances have been connected to a number of serious health problems, including endocrine disruption, cancer, mutation, and reproductive damage (Mishra and Rahi 2022). Many heavy metals are utilized in cosmetics, and even in very little amounts, they can impair the body's vital organs. These negative effects of metals in cosmetics have been linked to a wide range of problems, including organ failure, cancer, respiratory disorders, and intellectual disability (Paithankar et al. 2021).

12.3 Need to Study Cosmetic Toxicity

Safety and toxicological testing is one of the most significant and crucial tests in the field of cosmetic toxicology. In order to conduct numerous toxicity and safety tests on the various cosmetic products and their effects on the skin when applied topically, safety and toxicological assessment of the raw chemicals used in cosmetics is necessary. The study of cosmetic toxicology enables the prediction of potential hazards connected to the use of cosmetic products, risks that, if a person is exposed to them, may result in undesirable effects such as skin redness, a burning sensation, inflammation, and a wide range of allergic reactions (Okereke et al. 2015). To identify any potentially detrimental effects related to the use of cosmetics, cosmetic toxicology needs in-depth study of the biological processes that are triggered by the ingredients in the formulations used in cosmetics. A person may encounter a number of pertinent outcomes while frequently utilizing cosmetic components, including skin or eye irritation, corrosion, allergy, blisters, and inflammation. It is necessary to assess the risks associated with these endpoints and ensure their safety. Cosmetic ingredients may develop phototoxic properties when exposed to UV light, which might result in a range of dermatological responses (Tomankova et al. 2011).

12.4 The Way to Omics Approaches

In an early investigation, it was discovered that skin senses and reacts to even the smallest environmental changes in order to preserve homeostasis. The use of cosmetics and solar radiation sets off a chain of events that change gene expression. This modification catalyzes chemical processes at the protein level that results in cellular response. Metabolic byproducts as a result reveal the chemical reactions that had occurred. To fully understand the biological response, it is necessary to view these complex events holistically, which involve many different genes, proteins, and metabolites. Only a few methods were available to researchers in the previous decade for the analysis of these substances. However, advances in technology have transformed the study of skin's responses to the environment.

Omics approaches have evolved in the present day with the development of high-throughput technology for the analysis of diseases and online datasets of various biological samples (Table 12.1). With the study of connections between biochemical, molecular, and environmental factors, omics studies—which comprise genomes, transcriptomics, proteomics, metabolomics, and epigenomics—provide comprehensive knowledge of system biology (Manzoni et al. 2018).

12.4.1 Genomics

There are two ways that genomics can be used to investigate the toxicity of cosmetics. First, it can support the persistence and detectability of chemicals in humans. Second, it can be used to seek new toxicity pathways. qRT-PCR, DNA chips, Serial Analysis of Gene Expression (SAGE), genome-wide association study (GWAS), and CRISPR/Cas are examples of genomic technology (Jinek et al. 2012). Following the use of cosmetic goods, prolonged exposure to UV-R causes photosensitization, which results in burning, inflammatory, and painful sensations. As a result, a person develops various dermatological problems like psoriasis, acne, and dermatitis. The gene expression profile changes with time. The identification of genes involved in skin disorders should provide a better mechanistic understanding given recent advances in genomics. A small variation known as an SNP is connected to a certain disease through GWAS, allowing researchers to investigate and identify

Table 12.1 Omics approaches with research objects and different methods of analysis (He and Jia 2022)

Sr. No	Omics approaches	Research objects	Methods of analysis
1.	Genomics	DNA	1. q-RT-PCR 2. GWAS(genome-wide association studies) 3. SAGE(serial analysis of gene expression) 4. CRISPR/Cas(clustered regularly interspaced short palindromic repeats)
2.	Transcriptomics	RNA	1. Gene chip 2. EST (expressed sequence tags) 3. SAGE (serial analysis of gene expression) 4. RNA-Seq
3.	Proteomics	Proteins	1. 2D gel electrophoresis 2. Mass spectrometry (mass spectrometry) 3. Capillary electrophoresis 4. Yeast 2-hybrid system
4.	Metabolomics	Metabolites with relative molecular mass less than 1000	1. NMR (nuclear magnetic resonance) 2. MS (mass spectrometry) 3. LC-MS (liquid chromatography–mass spectrometry)
5.	Lipidomics	Lipid	1. ESI-MS (electrospray ionization–mass spectrometry) 2. DESI-MS (desorption electrospray ionization–mass spectrometry) 3. UPLC-QTOF-MS (ultra-high-performance liquid chromatography–quadrupole time-of-flight mass spectrometry)
6.	Microbiomics	Microorganisms	1. 16S-rRNA Seq 2. Metagenomic data analysis

the main gene responsible for the disease's development (Ober and Yao 2011). This might be a useful approach to investigate population genetic variants that contribute to skin disorders. A researcher may be able to accurately anticipate treatment techniques in precision skincare or cosmetic products using data from GWAS investigations. In a single experiment, gene chips or microarrays detect the expression levels of almost all human genes. They can be used to examine how the skin responds to treatments like topical application of personal care products and to external factors like sun exposure. They have been used, for example, to identify molecular distinctions between the skin of different ages (young skin and older skin) (Fig. 12.2) (Robinson et al. 2008). This information might be used to develop newer skin beneficial products and treatments that will eventually improve skin health.

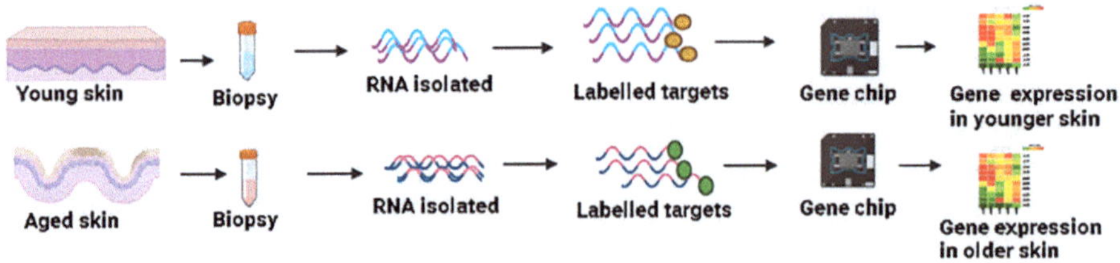

Fig. 12.2 Schematic representation of genomics workflow

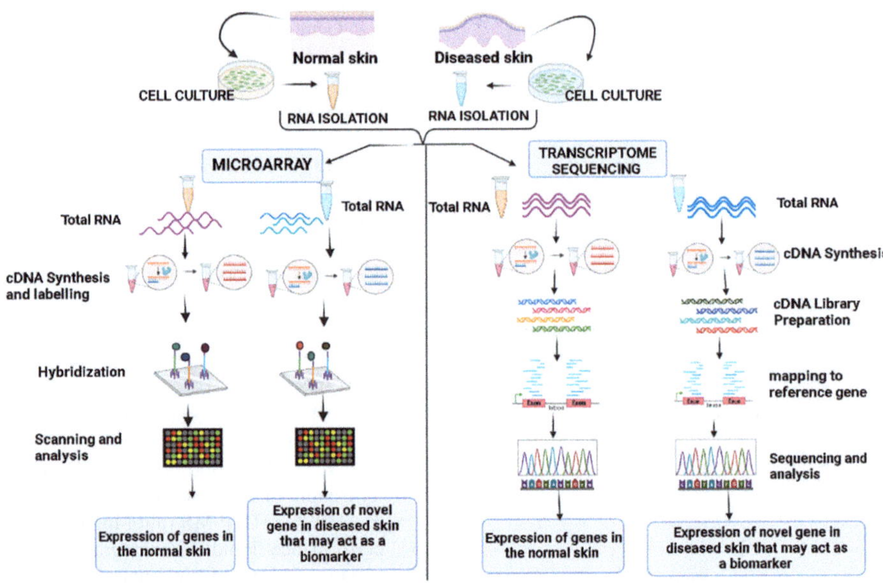

Fig. 12.3 Schematic representation of transcriptomic workflow

12.4.2 Transcriptomics

The use of a molecular method called transcriptomics, which examines the expression levels of genes with well-known biological activity, can help us to better understand how toxicological mechanisms work (Cui and Paules 2010). Better sequencing-based technologies are now available, and they promise to teach us more about how skin cells function (Fig. 12.3). Technologies like RNA sequencing (RNA-Seq) and digital gene expression profiling (DGE) leverage new advancements in next-generation sequencing to assess the genes being expressed in a sample in a more accurate and thorough manner (Kimball et al. 2012).

In RNA-Seq/DGE, the quantity of a particular transcript is calculated without taking into account the intensity of the fluorescent signal. Instead, it concentrates on counting the number of times each gene occurs while sequencing several genes that are expressed in a sample. Additionally, RNA-Seq/DGE gives a more accurate overview of all the RNAs in a sample, including mRNAs, microRNAs, and other ncRNAs (noncoding RNA) species (Kimball et al. 2012).

Primitive monolayer keratinocytes produced from damaged and non-psoriatic tissue slices were studied using RNA-seq (Swindell et al. 2017). Additionally, a whole skin biopsy from the same individual was examined. Compared to the transcriptome of the entire skin, there is a greater difference between the damaged and normal psoriasis keratinocytes. Significant gene overlap around disease-related SNPs may indicate that the keratinocytes of people with psoriasis are less differentiated. Skin aging is a result of both genetic and environmental factors. Sun-protected epidermis, sun-damaged skin before the ear, and sun-protected skin behind the ear were all subjected to a sequence comparison analysis of gene expression (Urschitz et al. 2002). Genes expressed in human skin and genes with differential expression in response to UV exposure were found using SAGE. In skin that had been exposed to the sun, before the ear, 19 distinct labels were diminished (at least four times lower), whereas 15 labels were increased. Numerous genes whose transcription levels change in response to sunlight are expressed by epidermal keratinocytes that might act as a novel biomarker.

12.4.3 Proteomics

Proteomics is a study of proteins that quantifies and qualitatively analyses the numerous types of protein content in a cell, including protein–protein interactions, protein–ligand interactions, and post-translational modifications (Jensen 2006). The proteomic method aids in the protein characterization of skin biology (Fig. 12.4). Atopic skin disease develops and worsens as a result of numerous factors. In contrast to a control group, the skin proteome of an individual with atopic dermatitis (AD) showed higher and more significant upregulation than blood protein, demonstrating inflammation and cardiovascular features in the skin proteome of the AD patient and their interaction with the blood proteome and skin genome (Pavel et al. 2020).

The quantity of cytokines and structural proteins can be measured using a clinical sampling technique (noninvasive) like tape stripping to isolate and collect protein from skin surface in order to learn more about skin irritation and inflammation and the condition of the skin barrier in healthy and damaged skin (like dandruff) (Keurentjes et al. 2021). Using particular antibody-based ELISA techniques, many proteins can be measured and evaluated simultaneously (Kerr et al. 2011).

12.4.4 Metabolomics

Analysis of the biochemical, physiological, and chemical changes in a biological system is provided by metabolomics. A wide range of fields use the analysis of metabolites to determine safety. In the past few years, metabolomics has become a novel method for evaluating the safety profile of chemical compounds in regulatory

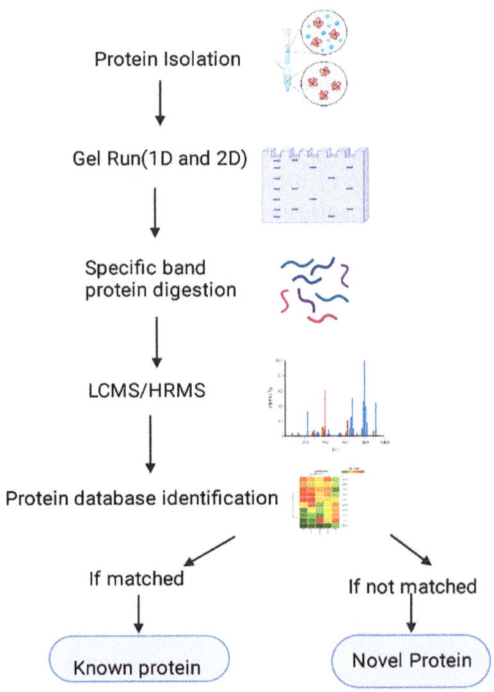

Fig. 12.4 Schematic representation of proteomic workflow

toxicology. Metabolomics is similar to analyzing disrupted metabolic pathways in terms of toxicity (Ramirez et al. 2013). It simplifies the process of determining potentially harmful substances and their target. The metabolomic method also elucidated a compound's mode of action and its impact on the target organ. The endogenous compounds that are altered during cellular metabolism include nucleotides, amino acids, steroids, phospholipids, carbohydrates, and their derivatives (Patti et al. 2012). These metabolites are byproducts of proteins, mRNA, and genes and have the ability to control the expression and function of other biomolecules (Ramirez et al. 2013). Metabolites provide a clear reflection of the metabolic processes taking place in a system, allowing for analysis of the role that particular biochemical pathway plays in the production of a given metabolite profile and the effects of perturbations to those pathways (Fig. 12.5). Metabolomics facilitates detection since any slight alteration in gene or protein expression directly reflects metabolite changes.

Continuous exposure to UV-R causes erythema, wrinkles, and loss of skin moisture (Amaro-Ortiz et al. 2014). Mice treated with green tea catechin and those exposed to UV-B had their skin metabolites examined using the MS method. The findings demonstrated that UV-B exposure in ECGT-treated mice reduced alterations in metabolites including ceramide, amino acid, and lysophospholipid, but purine bases, lactoses, and ascorbic acid were most affected (Jung et al. 2015). These modifications in skin metabolite can also be utilized as a biomarker to determine how a specific substance with a hazardous potential affects the skin,

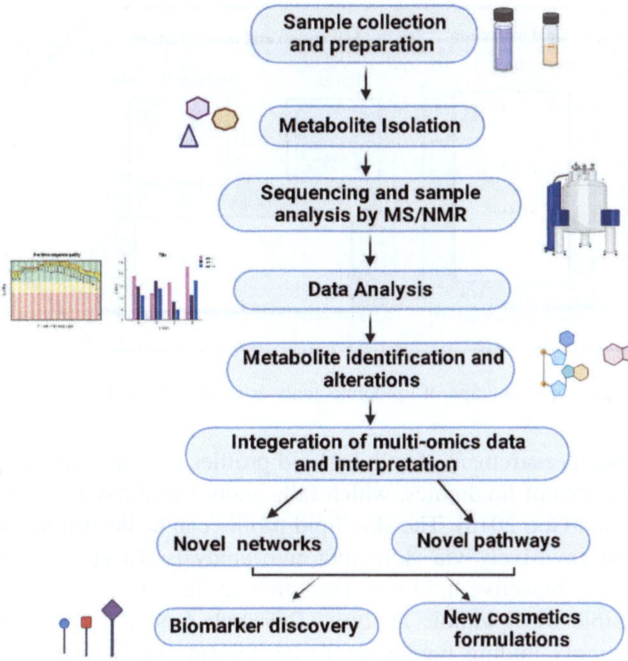

Fig. 12.5 Schematic representation of metabolomics workflow

which could offer a theoretical framework for formulations and creams for the cosmetic industry.

When metabolic pathways are dysregulated in specific diseases, such as psoriasis, GC-MS is employed to find potential biomarkers that are associated with psoriatic patients as opposed to healthy people (Kang et al. 2017). Understanding the diseases connected to the various skin allergies better may aid in the discovery of possible biomarkers. Skin metabolites to some extent reflect the state of the skin. Analyzing different skin metabolites through metabolomics can help us understand different skin illnesses better while also improving formulations for skin care products.

12.4.5 Lipidomics

The outermost skin layer, or epidermis, is a lipid-rich area that provides structural support and prevents chemical access. Lipids, particularly ceramides, acyl ceramides, cholesterol, cholesterol esters, and non-esterified fatty acids (NEFA, also known as free fatty acids), are the main constituents of the extracellular space of the epidermis. These lipids are arranged across many bilayers (Knox and O'Boyle 2021). For the integrity and functionality of the skin, an active lipid metabolism and fatty acid profile are essential (Kendall and Nicolaou 2013).

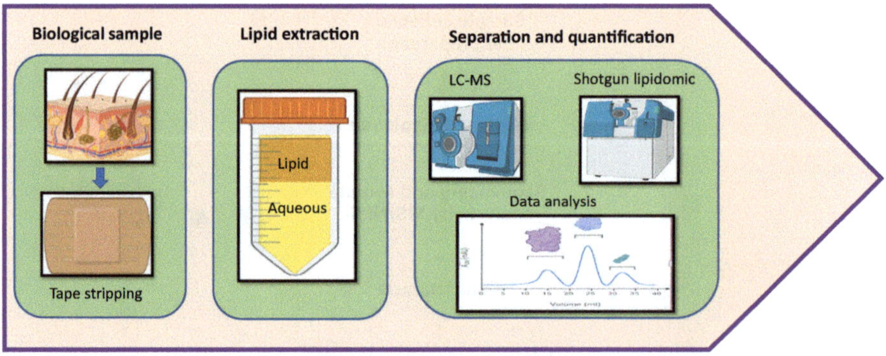

Fig. 12.6 Diagrammatic representation of lipidomic analysis in skin research

Identification and measurement of cellular lipid profiles in biological samples are the focus of the subfield of lipidomics, which falls under metabolomics (Ahluwalia et al. 2022; Lydic and Goo 2018). The skin lipid profile can be determined quickly, accurately, and non-invasively via skin lipidomic analysis (Li et al. 2016). The identification of many bioactive lipid mediators that are involved in immune function is made possible by lipidomic analysis. Research funded by the (skincare) business is increasingly finding lipidomics to be helpful, and it may be used to evaluate the effectiveness of both cosmetic formulations and skincare active ingredients. Oils, fats, waxes, and lipid antioxidants including carotenoids, retinoids, and tocopherols are used as active ingredients in cosmetic and personal care formulations, and lipidomic analysis is necessary for both quality assurance and efficacy testing (Ahmad and Ahsan 2020). A biological specimen may be subjected to lipidomic profiling directly or after being extracted with an organic solvent (Yang and Han 2016). Shotgun or LC-based lipidomics were typically the two kinds of sophisticated mass spectrometry techniques (MS) used for lipid profiling (Han and Ye 2021). Liquid chromatography–mass spectrometry (LC-MS or LC-MS/MS), MALDI-MS, ion mobility MS, high mass accuracy MS, and tandem MS were used for LC-based lipidomic analysis, whereas multidimensional MS, MALDI-MS, and DESI-MS were used for shotgun-based lipidomic investigations. The steps involved in lipidomic analysis of biological samples include sample collection, sample processing, data capture, and data processing (Fig. 12.6) (Hyötyläinen and Orešič 2015).

Healthy skin requires a particular lipid composition to maintain a barrier that offers defense and prevents excessive water loss, facilitates cell–cell communication, and controls epidermal homeostasis (Murakami et al. 2018). Lipidomic study improves our knowledge of how skin lipid composition can be altered by cosmetic products, which can lead to a number of dermatological issues. An important chronic inflammatory skin condition known as atopic dermatitis is frequently characterized by a compromised skin barrier and increased trans-epidermal water loss (TEWL) (Williamson et al. 2020). Changes in stratum corneum (SC) lipids have been extensively studied. Patients with atopic dermatitis (AD) had different SC lipid profiles, according to retrospective investigations. Furthermore, lipidomic study indicates that ceramide levels are significantly lower in AD patients compared to healthy people (Emmert et al. 2021). Through lipidomic research, ceramidase overexpression in psoriatic patients was discovered (Łuczaj et al. 2021). Smeden and his colleagues conducted a thorough LC-MS-based lipidomic investigation that identified significant changes in the SC lipid profile in NTS (Netherton syndrome) patients (van Smeden et al. 2020).

As a result, lipidomics offers fresh perspectives on how cosmetic ingredients affect skin lipid profiles or the skin microbiome–lipidome relationship. Based on these findings, there is also significant potential for customized cosmetics by segmenting consumer groups according to skin lipid composition.

12.4.6 Microbiomics

Microorganisms such as bacteria, viruses, yeast, fungus, and archaea all live on human skin. These skin-dwelling microorganisms make up the skin barrier, which facilitates the maintenance of healthy skin (Byrd et al. 2018). The skin microbiota, which makes up our top layer of skin, is continually in contact with outside factors including UV radiation, pollution, and cosmetic additives (Skowron et al. 2021). It is hypothesized that the use of synthetic chemical ingredients in modern cosmetics has an impact on skin bacteria (Wallen-Russell 2018). Application of cosmetic products to the skin, including soaps, shampoos, lotions, moisturizers, anti-aging, and hygiene items, might alter the lipid layer that protects the skin and affect the diversity of resident microflora (Pinto et al. 2021). The active ingredients in cosmetics may promote the growth of some microbial species or may hinder them. In the future, skin microbiomics may be used as a comprehensive strategy to assess whether cosmetic compounds are good for the skin or harm it.

The study of the entire microbiota, or bacterial population, is the focus of the discipline of microbiomics, which is constantly expanding. The goal of the field of microbiomics is to comprehend a specific microbial community's makeup and how it could shift over time or in response to a specified pressure. The composition of a particular microbial community is examined using high-throughput sequencing of the 16S rRNA gene (Kim et al. 2021).

Genomic DNA from the targeted sample was collected, and PCR was used to amplify the 16S rRNA gene (PCR). The amplified 16S rRNA is sequenced utilizing the standard sequencing methods. The collected sequence must be matched to widely available online databases in order to identify the bacterium (Zemb et al. 2020). Analysis has advanced from kingdom to strain level with the development of shotgun metagenomic technology in recent years since it simultaneously captures all genetic material in sample and provides sufficient resolution for identifying species and strains (Fig. 12.7) (Quince et al. 2021) Numerous dermatoses pathogenesis has been demonstrated to involve changes to the cutaneous microbiota, according to studies' reference. Higher microbial community biodiversity was a sign of healthier skin. In terms of microbial diversity, the synthetic and "natural" product categories

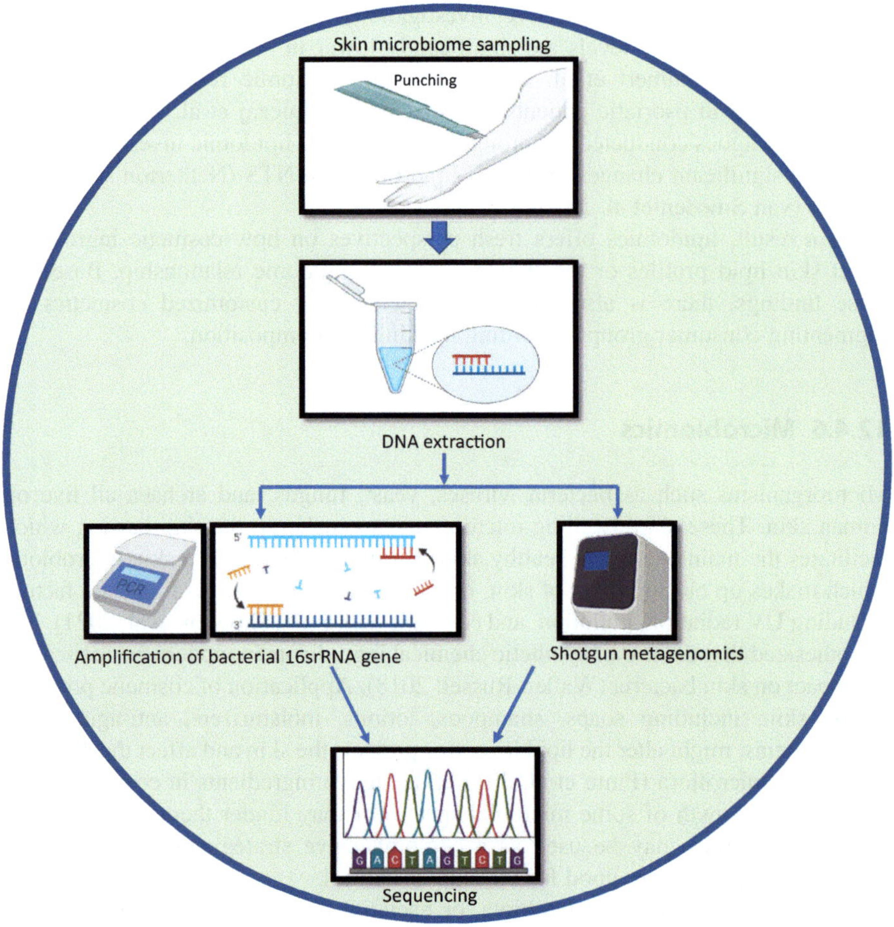

Fig. 12.7 Diagrammatic representation of skin microbiome research workflow

have shown the slowest growth during periods of 2 and 4 weeks. Face wash demonstrated the fastest average growth rate because it has no artificial components (Wallen-Russell 2018). It was further concluded that cosmetic product, preservatives, can remain active on the skin and, if used frequently, alter the local microbiota over time (Holland and Bojar 2002). In this essence, the pool of resistance genes basically increases, and these genes can spread to other microorganisms (transients) that are infectious agents and spread around the neighborhood through local microflora. One of the elements that stimulate the skin microbiota is N-acetylglucosamine, a component typically presents in skincare products and a precursor to hyaluronic acid (Skowron et al. 2021). Instead of getting rid of microorganisms, antiperspirants and foot powders increased the variety of microbial flora in armpits and between toes (Bouslimani et al. 2019). Lipids included in moisturizers promote the growth of lipophilic microorganisms including staphylococcus and propionium bacteria (Diaz and Ditre 2020). Companies are working to grow beneficial microorganisms to treat both more severe and less serious conditions like eczema and acne and minor ailments like dryness and wrinkles. Cosmetic manufacturers have begun looking into the relationship between a healthy microbiome and healthy skin (Reisch 2017). In conclusion, skin microbiota is important to the cosmetic industry. The microbiome is where the future of cosmetics is taking place, and how skin care products affect the microbial makeup of the skin is crucial. A precise skincare strategy should be developed using the microbiomic approach and relevant cosmetic ingredients based on microbiological and chemical evidence that are important players in host defense.

12.5 Conclusion

Omics technology is currently widely employed in all scientific disciplines, including biology, medicine, and nutrition. The development of omics techniques in the field of cosmetic toxicity will advance knowledge of the etiology and disorders associated with the skin. It is anticipated that the development of novel techniques for examining molecular mechanisms and related signaling pathways in the diagnosis, treatment, and creation of personalized precision medicine will lead to the production of more relevant scientific data in the identification and management of various skin diseases, their prognosis, and the identification of new biomarkers.

References

Ahluwalia K, Ebright B, Chow K, Dave P, Mead A, Poblete R, Louie SG, Asante I (2022) Lipidomics in understanding pathophysiology and pharmacologic effects in inflammatory diseases: considerations for drug development. Meta 12(4):333

Ahmad A, Ahsan H (2020) Lipid-based formulations in cosmeceuticals and biopharmaceuticals. Biomed Dermatol 4(1):1–10

Amaro-Ortiz A, Yan B, D'Orazio JA (2014) Ultraviolet radiation, aging and the skin: prevention of damage by topical cAMP manipulation. Molecules 19(5):6202–6219

Bouslimani A, da Silva R, Kosciolek T, Janssen S, Callewaert C, Amir A, Dorrestein K, Melnik AV, Zaramela LS, Kim JN, Humphrey G (2019) The impact of skin care products on skin chemistry and microbiome dynamics. BMC Biol 17(1):1–20

Byrd AL, Belkaid Y, Segre JA (2018) The human skin microbiome. Nat Rev Microbiol 16(3): 143–155

Cui Y, Paules RS (2010) Use of transcriptomics in understanding mechanisms of drug-induced toxicity. Pharmacogenomics 11(4):573–585

Diaz D, Ditre CM (2020) The effect of cleansers on the skin microbiome. Pract Dermatol:62–65

Emmert H, Baurecht H, Thielking F, Stölzl D, Rodriguez E, Harder I, Proksch E, Weidinger S (2021) Stratum corneum lipidomics analysis reveals altered ceramide profile in atopic dermatitis patients across body sites with correlated changes in skin microbiome. Exp Dermatol 30(10): 1398–1408

Gouveia D, Almunia C, Cogne Y, Pible O, Degli-Esposti D, Salvador A, Cristobal S, Sheehan D, Chaumot A, Geffard O, Armengaud J (2019) Ecotoxicoproteomics: a decade of progress in our understanding of anthropogenic impact on the environment. J Proteome 198:66–77

Han X, Ye H (2021) Overview of lipidomic analysis of triglyceride molecular species in biological lipid extracts. J Agric Food Chem 69(32):8895–8909
He J, Jia Y (2022) Application of omics technologies in dermatological research and skin management. J Cosmet Dermatol 21(2):451–460
Holland KT, Bojar RA (2002) Cosmetics. Am J Clin Dermatol 3(7):445–449
Hyötyläinen T, Orešič M (2015) Optimizing the lipidomics workflow for clinical studies—practical considerations. Anal Bioanal Chem 407(17):4973–4993
Jensen ON (2006) Interpreting the protein language using proteomics. Nat Rev Mol Cell Biol 7(6): 391–403
Jinek M, Chylinski K, Fonfara I, Hauer M, Doudna JA, Charpentier E (2012) A programmable dual-RNA–guided DNA endonuclease in adaptive bacterial immunity. Science 337(6096): 816–821
Jung ES, Park HM, Lee KE, Shin JH, Mun S, Kim JK, Lee SJ, Liu KH, Hwang JK, Lee CH (2015) A metabolomics approach shows that catechin-enriched green tea attenuates ultraviolet B-induced skin metabolite alterations in mice. Metabolomics 11(4):861–871
Kang H, Li X, Zhou Q, Quan C, Xue F, Zheng J, Yu Y (2017) Exploration of candidate biomarkers for human psoriasis based on gas chromatography-mass spectrometry serum metabolomics. Br J Dermatol 176(3):713–722
Kendall AC, Nicolaou A (2013) Bioactive lipid mediators in skin inflammation and immunity. Prog Lipid Res 52(1):141–164
Kerr K, Darcy T, Henry J, Mizoguchi H, Schwartz JR, Morrall S, Filloon T, Wimalasena R, Fadayel G, Mills KJ (2011) Epidermal changes associated with symptomatic resolution of dandruff: biomarkers of scalp health. Int J Dermatol 50(1):102–113
Keurentjes AJ, Jakasa I, Kezic S (2021) Research techniques made simple: stratum corneum tape stripping. J Investig Dermatol 141(5):1129–1133
Khan AD, Alam MN (2019) Cosmetics and their associated adverse effects: a review. J Appl Pharm Sci Res:1–6
Kim KB, Kwack SJ, Lee JY, Kacew S, Lee BM (2021) Current opinion on risk assessment of cosmetics. J Toxicol Environ Health Part B Crit Rev 24(4):137–161
Kimball AB, Grant RA, Wang F, Osborne R, Tiesman JP (2012) Beyond the blot: cutting edge tools for genomics, proteomics and metabolomics analyses and previous successes. Br J Dermatol 166:1–8
Knox S, O'Boyle NM (2021) Skin lipids in health and disease: a review. Chem Phys Lipids 236: 105055
Li S, Ganguli-Indra G, Indra AK (2016) Lipidomic analysis of epidermal lipids: a tool to predict progression of inflammatory skin disease in humans. Expert Rev Proteomics 13(5):451–456
Łuczaj W, Gęgotek A, Skrzydlewska E (2021) Analytical approaches to assess metabolic changes in psoriasis. J Pharm Biomed Anal 205:114359
Lydic TA, Goo YH (2018) Lipidomics unveils the complexity of the lipidome in metabolic diseases. Clin Transl Med 7(1):1–13
Manzoni C, Kia DA, Vandrovcova J, Hardy J, Wood NW, Lewis PA, Ferrari R (2018) Genome, transcriptome and proteome: the rise of omics data and their integration in biomedical sciences. Brief Bioinform 19(2):286–302
Mishra G, Rahi S (2022) Need of toxicity studies for cosmetic products and their approaches. Biol Sci 2(1):105–109
Murakami M, Yamamoto K, Taketomi Y (2018) Phospholipase A2 in skin biology: new insights from gene-manipulated mice and lipidomics. Inflamm Regen 38(1):1–10
Ober C, Yao TC (2011) The genetics of asthma and allergic disease: a 21st century perspective. Immunol Rev 242(1):10–30
Okereke JN, Udebuani AC, Ezeji EU, Obasi KO, Nnoli MC (2015) Possible health implications associated with cosmetics: a. Science 3(5–1):58–63

Paithankar JG, Saini S, Dwivedi S, Sharma A, Chowdhuri DK (2021) Heavy metal associated health hazards: an interplay of oxidative stress and signal transduction. Chemosphere 262: 128350

Patti GJ, Yanes O, Siuzdak G (2012) Metabolomics: the apogee of the omics trilogy. Nat Rev Mol Cell Biol 13(4):263–269

Pavel AB, Zhou L, Diaz A, Ungar B, Dan J, He H, Estrada YD, Xu H, Fernandes M, Renert-Yuval Y, Krueger JG (2020) The proteomic skin profile of moderate-to-severe atopic dermatitis patients shows an inflammatory signature. J Am Acad Dermatol 82(3):690–699

Pinto D, Ciardiello T, Franzoni M, Pasini F, Giuliani G, Rinaldi F (2021) Effect of commonly used cosmetic preservatives on skin resident microflora dynamics. Sci Rep 11(1):1–7

Quince C, Nurk S, Raguideau S, James R, Soyer OS, Summers JK, Limasset A, Eren AM, Chikhi R, Darling AE (2021) STRONG: metagenomics strain resolution on assembly graphs. Genome Biol 22(1):1–34

Ramirez T, Daneshian M, Kamp H, Bois FY, Clench MR, Coen M, Donley B, Fischer SM, Ekman DR, Fabian E, Guillou C (2013) Metabolomics in toxicology and preclinical research. ALTEX 30(2):209

Reay WR, Cairns MJ (2021) Advancing the use of genome-wide association studies for drug repurposing. Nat Rev Genet 22(10):658–671

Reisch MS (2017) Cosmetics: the next microbiome frontier. Mitsui Chemicals Catalysis Science Award 95: 30–34

Robinson M, Tiesman J, Binder R, Juhlin K (2008) Immune and inflammatory gene expression profiles of chronological skin aging and photoaging. J Am Acad Dermatol 58(2):AB34

Skowron K, Bauza-Kaszewska J, Kraszewska Z, Wiktorczyk-Kapischke N, Grudlewska-Buda K, Kwiecińska-Piróg J, Wałecka-Zacharska E, Radtke L, Gospodarek-Komkowska E (2021) Human skin microbiome: impact of intrinsic and extrinsic factors on skin microbiota. Microorganisms 9(3):543

Swindell WR, Sarkar MK, Liang Y, Xing X, Baliwag J, Elder JT, Johnston A, Ward NL, Gudjonsson JE (2017) RNA-seq identifies a diminished differentiation gene signature in primary monolayer keratinocytes grown from lesional and uninvolved psoriatic skin. Sci Rep 7(1):1–13

Tomankova K, Kejlova K, Binder S, Daskova A, Zapletalova J, Bendova H, Kolarova H, Jirova D (2011) In vitro cytotoxicity and phototoxicity study of cosmetics colorants. Toxicol in Vitro 25(6):1242–1250

Urschitz J, Urban Z, Granda C, Souza KA, Lupp C, Csiszar K, Boyd CD, Iobst S, Schilling K, Scott I (2002) A serial analysis of gene expression in sun-damaged human skin. J Investig Dermatol 119(1):3–13

van Smeden J, Al-Khakany H, Wang Y, Visscher D, Stephens N, Absalah S, Overkleeft HS, Aerts JM, Hovnanian A, Bouwstra JA (2020) Skin barrier lipid enzyme activity in Netherton patients is associated with protease activity and ceramide abnormalities [S]. J Lipid Res 61(6):859–869

Wallen-Russell C (2018) The role of every-day cosmetics in altering the skin microbiome: a study using biodiversity. Cosmetics 6(1):2

Wang X, Li F, Chen J, Ji C, Wu H (2021) Integration of computational toxicology, toxicogenomics data mining, and omics techniques to unveil toxicity pathways. ACS Sustain Chem Eng 9(11): 4130–4138

Williamson S, Merritt J, De Benedetto A (2020) Atopic dermatitis in the elderly: a review of clinical and pathophysiological hallmarks. Br J Dermatol 182(1):47–54

Yang K, Han X (2016) Lipidomics: techniques, applications, and outcomes related to biomedical sciences. Trends Biochem Sci 41(11):954–969

Zemb O, Achard CS, Hamelin J, De Almeida ML, Gabinaud B, Cauquil L, Verschuren LM, Godon JJ (2020) Absolute quantitation of microbes using 16S rRNA gene metabarcoding: a rapid normalization of relative abundances by quantitative PCR targeting a 16S rRNA gene spike-in standard. Microbiologyopen 9(3):e977

Application of 3D Skin Model Systems in Assessing the Safety/Toxicity of Fragrances and Flavors, Hair Dyes, Preservatives, and Colorants

13

Priyanka Rana, Sunil Kumar Patel, Saumya Shukla, Shakti Vinay Shukla, Divya Dubey, Deepti Chopra, and Malini Kotak

Abstract

Since cosmetics are essential for human existence, it is crucial to guarantee their efficacy and safety. The current chapter focuses on cosmetic ingredients like preservatives, fragrances, and colorants and the involvement of 3D skin model systems in their safety evaluation. At the beginning of the chapter, we briefly understand different types of 3D skin models currently available and the advantages of using these systems over current in vitro assays when it comes to toxicity testing for cosmetic products. Because the primary route of exposure for

P. Rana · S. K. Patel
Photobiology Laboratory, Systems Toxicology and Health Risk Assessment Group, CSIR-Indian Institute of Toxicology Research, Lucknow, Uttar Pradesh, India

Academy of Scientific and Innovative Research, AcSIR Headquarters, CSIR-HRDC Campus, Ghaziabad, Uttar Pradesh, India

S. Shukla · D. Chopra
Photobiology Laboratory, Systems Toxicology and Health Risk Assessment Group, CSIR-Indian Institute of Toxicology Research, Lucknow, Uttar Pradesh, India

Department of Biochemistry, School of Dental Sciences, Babu Banarasi Das University, Lucknow, Uttar Pradesh, India

S. V. Shukla
Fragrance & Flavour Development Centre, Ministry of MSME, Govt. of India, Kannauj, Uttar Pradesh, India

D. Dubey
Department of Molecular Biosciences, The Wenner-Gren Institute, Stockholm University, Stockholm, Sweden

M. Kotak (✉)
ITC Life Sciences and Technology Center, ITC Ltd., Bangalore, Karnataka, India
e-mail: malini.kotak@itc.in

A. B. Pant et al. (eds.), *Skin 3-D Models and Cosmetics Toxicity*,
https://doi.org/10.1007/978-981-99-2804-0_13

cosmetic products is dermal, we then go on to give general regulatory guidelines for dermal toxicity testing and recommendation for 3D skin models in dermal testing. The chapter then delves into details of each ingredient mentioned above and reports the usage of 3D skin models in the safety testing of these ingredients.

Keywords

3D skin models · Fragrance · Flavors · Phototoxicity · Cosmetics · Hair dye

13.1 Introduction

Since cosmetics are essential for human existence, it is crucial to guarantee their efficacy and safety. The frequency of cosmetic goods' unintended side effects is rising with the use of these products daily. Skin irritation reactions that are mild to severe in severity are the most common side effects. On the other side, sensitive people, including susceptible groups like youngsters, can experience severe skin reactions from cosmetics. The most often used compounds in cosmetics include a variety of preservatives, perfumes, antioxidants, vehicles, ultraviolet absorbers, humectants, emollients, emulsifiers, acrylates, hair dyes, and nail polish components.

Previously, animal models were widely used to establish the in vivo safety of cosmetic products but starting in March 2009, the seventh amendment to the European Union (EU) Cosmetics Directive banned animal models for this purpose (Thakkar et al. 2022). As such, the use of animal models has several drawbacks: (1) differences in the adsorption, distribution, and metabolism of cosmetic products; (2) inability to accurately predict disease development owing to the short life span of animal models; (3) discrepancies in adverse effects of ingredients on animals and humans (Ng and Yeong 2019). These issues have spurred the development of numerous 3D in vitro skin models for testing, and unlike animal models that show species differences, 3D skin models are incredibly accurate in identifying irritation caused by cosmetics and substances (Xiao et al. 2020).

Several in vitro skin models have been successfully created at this point. Models for the epidermis include EpiSkin, LabCyte EPI-MODEL, EpiDerm, EpiKutis, epiCS, and SkinEthic RHE. Models for the whole skin include T-Skin, EpiDerm FT, Phenion FT, and FulKutis, and the melanin skin models include SkinEthic RHPE, MelanoDerm, epiCS-M, and MelaKutis (Ma et al. 2021). The safety of cosmetic products is established after evaluating the toxicological profile of individual ingredients and the final product. The critical endpoints for the toxicological profile are (1) eye and skin irritation, (2) skin sensitization, (3) skin penetration/absorption, and (4) genotoxicity. Toxicological studies are often performed to mimic in-use exposure to cosmetic products.

The Organization for Economic Cooperation and Development (OECD) has investigated and documented various skin toxicology tests. OECD Test Guidelines (TG) for cosmetic testing involving in vitro skin tissue models include (a) skin

Table 13.1 OECD guidelines for toxicology testing involving skin model systems

Toxicology test	OECD guideline	Description	In vitro testing model
Skin absorption	TG 428	Absorption of cosmetic/ingredient through passive diffusion	Excised human or animal skin
Skin corrosion	TG 431	Irreversible skin damage as a result of application of cosmetic/ingredient	EpiDerm™ EpiSkin™ SkinEthic™ RHE epiCS®
Skin irritation	TG 439	Irreversible skin damage as a result of application of cosmetic/ingredient	EpiDerm™ EpiSkin™ LabCyte EPI-model SkinEthic™ RHE
Skin sensitization	TG 442D	Allergic response as a result of application of cosmetic/ingredient	KeratinoSens™

Table 13.2 OECD guidelines for genotoxicity testing

In vitro tests	OECD guideline	Tests for
Bacterial reverse mutation test (Ames test)	TG 471	Gene mutations
Mammalian cell gene mutation test/ mouse lymphoma assay (MLA)	TG 490	Gene mutations
Mammalian chromosomal aberration test	TG 473	Irreversible skin damage as a result of application of cosmetic/ingredient
Mammalian cell micronucleus test	TG 487	Allergic response as a result of application of cosmetic/ingredient

absorption (OECD TG 428), (b) skin corrosion (OECD TG 431), (c) skin irritation (OECD TG 439), and (d) skin sensitization (OECD TG 442D). Details of these guidelines are given in Table 13.1.

Another critical parameter of toxicology testing is genotoxicity. Genotoxicity testing of a cosmetic/ingredient includes tests that can detect (1) mutagenicity at the gene level, (2) chromosome breakage and/or rearrangements, and (3) numerical chromosome change. OECD guidelines mention four in vitro tests that have been validated to test for these genotoxicity endpoints. These are detailed in Table 13.2.

From Table 13.2, it is clear that no individual test covers all three genotoxicity endpoints. Therefore, they are not used as stand-alone tests but combined in batteries to ensure adequate coverage of endpoints. However, these tests have a tendency to give high false-positive results. With in vivo testing, the cosmetic ingredients with positive tests in this standard in vitro genotoxicity assays would be developed further. To address this issue, skin tissues were combined with classical genotoxicity read-out parameters to create a 3D comet assay and reconstructed skin micronucleus (RSMN) test as non-animal follow-up assays to test genotoxicity of dermal exposure products (Reus et al. 2013; Curren et al. 2006). These 3D skin models, consisting of the well-differentiated multi-layered model of the human epidermis, offer more physiologically relevant results for dermally cosmetic products compared to standard in vitro micronucleus tests. The metabolizing capacity of these models is biologically more relevant to human skin and, therefore, more representative of dermal exposure risks compared to the limited metabolic capacity of induced exogenous rat liver S9 used in standard in vitro genotoxicity tests. Lastly, similar to human skin, these models have relatively less phase I activity and more phase II detoxification activity, making them more relevant for the genotoxicity assessment of dermal products (Reisinger et al. 2018).

The 3D Skin Comet assay, which can detect both chromosomal damage and DNA lesions leading to gene mutations, complements the RSMN test, which sees only chromosomal damage. The Scientific Committee on Consumer Safety (SCCS), an independent expert panel of the European Commission, recommends using both the 3D Skin Comet assay and the RSMN as a follow-up on suspected misleading positive results from the in vitro standard testing battery (SCCS/1532/14). The "Use of 3D Tissues in Genotoxicity Testing" working group (WG) met at the seventh IWGT meeting in Tokyo in November 2017 to discuss the progress of skin, airway, and liver tissue equivalents and how they may fit into a genotoxicity testing strategy. The IWGT WG concluded that the 3D Skin Comet and micronucleus assays are sufficiently validated to undergo an independent peer review of the validation study, followed by the development of individual OECD Test Guidelines (Pfuhler et al. 2020).

13.2 Fragrances and Flavors

Fragrances are a mixture of substances that contain strong-smelling organic compounds with distinctive, pleasant odors. A fragrance ingredient might also be a flavors ingredient. Flavors are processed differently for purity, labeled, and regulated differently than fragrances (Bauer et al. 2008). Fragrances are now used to both prettify the wearer and to add appealing scents to cosmetic products, giving rise to an internal feeling of beauty. Fragrances are found in most cosmetics, personal care products, air fresheners, aromatherapy products, laundry supplies, and cleaners (Steinemann 2009). Fragrances are derived from natural and synthetic sources. Natural fragrances like essential oils are often derived from organic sources such as plants and trees. Synthetic fragrances are formulated in laboratories. Synthetic scents mimic natural accords rather than sourcing them from forestry and animals. However, fragrances are made up of 10–300 different components (Johansen et al. 2020) that contain ingredients, plenty of which may be allergens that can cause skin allergic reactions such as urticaria (hives), eczema, phototoxicity, photoallergy, skin discoloration (dyschromia), eyelid dermatitis, irritant dermatitis, and contact dermatitis. However, under trade secrets, a company is not required to disclose the entire ingredient list (Zeliger 2007). Therefore, the ambiguous labeling of other fragrance ingredients as "perfume" or "fragrance" impedes diagnostic and preventive measures. Fragrance ingredients such as coumarin and oakmoss absolute have been reported as an inducer of photoallergic contact dermatitis (Arribas et al. 2013). Evernia prunastri, cinnamal, isoeugenol, eugenol, cinnamyl alcohol, benzyl alcohol, cinnamaldehyde, benzyl cinnamate, farnesol, hydroxycitronellal, and more than 160 fragrances have resulted in skin sensitivity to cosmetic products (Mahajan 2022). The prominent utilization of essential oils such as ylang-ylang oil and jasmine absolute also makes patients sensitive to fragrances (Frosch et al. 2002). Contact urticaria is commonly caused by cinnamal, cinnamyl alcohol, and Myroxylon pereirae, but menthol, vanillin, and benzaldehyde have also been reported (Tanaka et al. 2004). A study was conducted to investigate the prevalence of contact dermatitis caused by various indigenous cosmetics among Indian users and found that among positive patients, 14% of the cases were caused by fragrances/deodorants (Goyal et al. 2019). The photo patch test is routinely used to detect allergic responses to varied antigens such as fragrances, sunscreens, and drugs (Rai and Thomas 2016). In photo patch assessment, fragrance mix 1 (FM1), fragrance mix 2 (FM2), and Balsam of Peru are recognized markers for fragrance allergy (Reeder 2020), but still patch testing may overlook fragrance-induced allergic patients because of false-positive and false-negative results (Lazzarini et al. 2013). Therefore, a better approach is required to properly analyze the safety/toxicity of fragrance ingredients. In this search, the 3D skin model is promising. To evaluate the genotoxicity detection potential of the RSMN test, Thakkar et al. (2022) evaluated 22 fragrance materials in vitro and in vivo as per OECD guidelines. Of 22 fragrance materials, 18 were positive in the standard in vitro assay but negative in *the* in vivo assay. These 18 fragrance materials were also harmful in the RSMN test, indicating that the RSMN assay is an essential alternative to in vivo animal assay as its results were 100% in agreement with *the* in vivo assay. Although the number of fragrances tested in 3D tissue model-based assays is fewer, the outcomes are promising, so the dataset necessitates expanding these models to an OECD guideline phase to accompany cell culture-based models (Corsini and Galbiati 2019).

13.3 Preservatives

Cosmetics are produced in a nonsterile and hygienically controlled environment. However, the products can get contaminated during manufacturing or during consumer use. Cosmetics are frequently in contact with nonsterile human skin, especially when creams are in jars, making them vulnerable to contamination by bacteria. Cosmetics that contain water, oils, peptides, and carbohydrates are excellent growing environments for bacteria. As a result, cosmetic products require very high preservation to prevent microbial growth, cosmetic product spoilage, and skin illness. An ideal preservative should have wide-ranging activity against all microorganisms, be efficient at low concentrations, not affect flavor or color, not interact with any other ingredients, have a high-water solubility and a low oil solubility to stay in the water phase, be stable across the entire pH and temperature spectrum with an unlimited shelf life, be safe to use when concentrated and diluted, and be inexpensive. Cosmetic manufacturers use various approved preservative systems like organic acids, alcohols, phenols, aldehydes, formaldehyde donors, isothiazolinones, biguanides, quaternary ammonium compounds, nitrogen compounds, heavy metal derivatives, and inorganic compounds (Halla et al. 2018).

Though preservatives are essential in ensuring product safety, they can have unfavorable effects that may manifest immediately or years after prolonged cosmetic use. These side effects can cause everything from minor skin irritation to estrogenic activity, and more recently, it has been speculated that they may be able to cause human breast tumors (Polati et al. 2007). Preservatives like bronopol and clioquinol have phototoxic effects in vitro and may also result in photosensitization when applied topically (Placzek et al. 2005). Methylparaben causes DNA damage caused by oxidative stress and apoptosis via the mitochondria and ER. MP has adverse effects, and prolonged contact with human skin could worsen existing skin conditions (Dubey et al. 2017). With the help of type I photodynamic reaction, instability of various organelles, and DNA damage resulting in death, photosensitized triclosan generated ROS-mediated oxidative stress (Dubey et al. 2019). Cosmetic preservatives cause allergy in the afflicted area, such as the face, neck, hand, and armpits, according to the Norwegian Institute of Public Health, 2013. Additionally, it may result in health issues, including contact urticaria and itching accompanied by swelling and redness of the skin.

In addition to the side effects mentioned above, it has been reported that residual activity of cosmetic preservatives can alter the balance of skin microbiota, compromising the aerobiosis of skin, mucous membranes, and the scalp. Pinto et al. (2021) used Labskin 3D (Innovenn Ltd., USA) to test the effect of 11 commonly found cosmetic preservatives on three skin resident bacteria: *Propionibacterium acnes*, *Staphylococcus epidermidis*, and *Staphylococcus aureus*. They also evaluated the impact of these preservatives on the expression of histone deacetylase 3 (HDAC3), which plays a crucial role in the relationship between skin microbiota and inflammation. Using the 3D skin model, they were able to elucidate the effect of various combinations of these preservatives on skin resident microflora and recommend the correct choice and dosage of preservatives to preserve or restore homeostasis of the skin microbiome.

13.4 Colorants and Hair Dyes

Colors play a crucial role in the appeal and marketing of cosmetic products. This group of ingredients is most tightly regulated in the cosmetic industry. Cosmetic colorants can be divided into two broad categories: organic and inorganic. Synthetic dyes, lakes, and botanicals are three major types of organic color additives. Inorganic color additives consist of mineral compounds like iron oxide and zinc oxide. Both organic and inorganic color additives can be further classified as dye or pigment based on their solubility. Dyes are hydro- or oil-soluble and are mainly found in skin care products or toiletries. On the other hand, pigments are insoluble, remain in particulate form, and are found primarily in toothpaste or makeup (Guerra et al.

2018). In the cosmetic industry, synthetic dyes are preferred over natural dyes due to their low production costs and long-lasting properties.

Based on the use of colorants, cosmetic products can be classified into a few broad categories: (1) products with colorants to color the skin, e.g., makeup items like lipstick, and mascara; (2) products with colorants to color the product itself, e.g., body wash and shampoo; and (3) product with colorants to color the hair, e.g., hair dyes. Hair color products are reported to dominate the cosmetic dye market. Hair dyes are divided into permanent (oxidative), temporary, and semi-permanent, out of which permanent hair dyes are widely used. These dyes are reported to be solid allergens; therefore, these substances' safety is a significant concern (Guerra et al. 2017).

Hair dyes contain aromatic amines, which are found to be carcinogenic and mutagenic to animals and humans. The International Agency for Research on Cancer (IARC) classified aromatic amines as a probable carcinogen in hairdressers and barbers. One research study stated that out of 11 hair dyes tested, 8 were found to be a carcinogen of the urinary bladder (Chung 2016). Hair dyes are reported to cause non-Hodgkin lymphoma and leukemia (Towle et al. 2017). Ingredients of hair dyes such as p-methyl aminophenol and toluene-2,5-diamine also cause acute to immediate effects leading to hypersensitive reactions such as anaphylaxis or respiratory syndrome (He et al. 2022).

In summary, cosmetic colorants, which come in contact with humans through dermal exposure, are reported to cause mutagenic, genotoxic, and carcinogenic effects.

The 3D Skin Comet assay has been applied in multiple studies to evaluate the safety of hair dyes. Basic Brown 17, a hair dye ingredient, was genotoxic in the bacterial reverse mutation test. This was followed up with a 3D Skin Comet assay and in vitro mammalian cell gene mutation tests (according to OECD TGs 476 and 490). All three follow-up tests gave negative results, which were accepted in the weight-of-evidence approach by SCCS, and the dye was considered safe to use. Zanoni et al. (2014) developed a 3D epithelium skin equivalent using HaCaT cells to test the cytotoxicity and oxidative stress potential of Basic Red 51 (BR51) dye, a temporary hair dye. They observed that BR51 exposure induced cell injury as the cells underwent apoptosis and also induced reactive oxygen species. They concluded that consumer or professional exposure to BR51 is harmful to health.

13.5 Conclusion

Toxicology testing of cosmetic products, which involves the evaluation of ingredients and end products, is necessary to ensure the safety of consumers. Since the ban on animal testing of cosmetic ingredients/products, extensive progress has been made in developing and validating 3D skin models. These models have several benefits over standard in vitro tests and have been recommended by several international regulatory bodies to test cosmetic products' skin irritation, corrosion, and genotoxicity potential. These 3D skin models have proved to be powerful alternatives to animal testing to establish the safety of dermal cosmetic products and ingredients.

References

Arribas MP, Soro P, Silvestre JF (2013) Allergic contact dermatitis to fragrances: part 2. Actas Dermosifiliogr (Engl Ed) 104(1):29–37

Bauer K, Garbe D, Surburg H (2008) Common fragrance and flavour materials: preparation, properties and uses. Wiley, Hoboken, NJ

Chung KT (2016) Azo dyes and human health: a review. J Environ Sci Health C 34(4):233–261

Corsini E, Galbiati V (2019) Contact allergy to fragrances: in vitro opportunities for safety assessment. Cosmetics 6(1):3

Curren RD, Mun GC, Gibson DP, Aardema MJ (2006) Development of a method for assessing micronucleus induction in a 3D human skin model (epi Derm™). Mutat Res Genet Toxicol Environ Mutagen 607(2):192–204

Dubey D, Chopra D, Singh J, Srivastav AK, Kumari S, Verma A, Ray RS (2017) Photosensitized methyl paraben induces apoptosis via caspase-dependent pathway under ambient UVB exposure in human skin cells. Food Chem Toxicol 108:171–185

Dubey D, Srivastav AK, Singh J, Chopra D, Qureshi S, Kushwaha HN, Singh N, Ray RS (2019) Photoexcited triclosan induced DNA damage and oxidative stress via p 38 MAP kinase signalling involving type I radicals under sunlight/UVB exposure. Ecotoxicol Environ Saf 174:270–282

Frosch PJ, Johansen JD, Menné T, Pirker C, Rastogi SC, Andersen KE, Bruze M, Goossens A, Lepoittevin JP, White IR (2002) Further important sensitizers in patients sensitive to fragrances: II. Reactivity to essential oils. Contact Derm 47(5):279–287

Goyal S, Sajid N, Husain S (2019) Contact dermatitis due to local cosmetics: a study from northern India. Indian J Dermatol 64(6):461–464

Guerra E, Lamas JP, Llompart M, Garcia-Jares C (2017) Determination of oxidative hair dyes using miniaturized extraction techniques and gas chromatography-tandem mass spectrometry. Microchem J 132:308–318

Guerra E, Llompart M, Garcia-Jares C (2018) Analysis of dyes in cosmetics: challenges and recent developments. Cosmetics 5(3):47

Halla N, Fernandes IP, Heleno SA, Costa P, Boucherit-Otmani Z, Boucherit K, Rodrigues AE, Ferreira IC, Barreiro MF (2018) Cosmetics preservation: a review on present strategies. Molecules 23(7):1571

He L, Michailidou F, Gahlon HL, Zeng W (2022) Hair dye ingredients and potential health risks from exposure to hair dyeing. Chem Res Toxicol 35:901

Johansen JD, Uter W, Frosch P, Lepoittevin JP (2020) Contact allergy to fragrances. Contact Dermatitis:1–33

Lazzarini R, Duarte I, Ferreira AL (2013) Patch tests. An Bras Dermatol 88:879–888

Ma X, Wang F, Wang B (2021) Application of an in vitro reconstructed human skin on cosmetics in skin irritation tests. J Cosmet Dermatol 20(6):1933–1941

Mahajan VK (2022) Perfumes and associated allergens: a brief review. Cosmoderma 2:21

Ng WL, Yeong WY (2019) The future of skin toxicology testing–three-dimensional bioprinting meets microfluidics. Int J Bioprint 5(2.1):237

Pfuhler S, van Benthem J, Curren R, Doak SH, Dusinska M, Hayashi M, Heflich RH, Kidd D, Kirkland D, Luan Y, Ouedraogo G (2020) Use of in vitro 3D tissue models in genotoxicity testing: strategic fit, validation status and way forward. Report of the working group from the 7th international workshop on genotoxicity testing (IWGT). Mutat Res Genet Toxicol Environ Mutagen 850:503135

Pinto D, Ciardiello T, Franzoni M, Pasini F, Giuliani G, Rinaldi F (2021) Effect of commonly used cosmetic preservatives on skin resident microflora dynamics. Sci Rep 11(1):1–7

Placzek M, Krosta I, Gaube S, Eberlein-König B, Przybilla B (2005) Evaluation of phototoxic properties of antimicrobials used in topical preparations by a photohaemolysis test. Acta Derm Venereol 85(1):13–16

Polati S, Gosetti F, Gennaro MC (2007) Preservatives in cosmetics. Analytical methods. In: Salvador A, Chisvert A (eds) Analysis of cosmetic products. Elsevier, Amsterdam, pp 211–241

Rai R, Thomas M (2016) Photopatch and UV-irradiated patch testing in photosensitive dermatitis. Indian Dermatol Online J 7(1):12–16

Reeder MJ (2020) Allergic contact dermatitis to fragrances. Dermatol Clin 38(3):371–377

Reisinger K, Blatz V, Brinkmann J, Downs TR, Fischer A, Henkler F, Hoffmann S, Krul C, Liebsch M, Luch A, Pirow R (2018) Validation of the 3D skin comet assay using full thickness skin models: transferability and reproducibility. Mutat Res Genet Toxicol Environ Mutagen 827:27–41

Reus AA, Reisinger K, Downs TR, Carr GJ, Zeller A, Corvi R, Krul CA, Pfuhler S (2013) Comet assay in reconstructed 3D human epidermal skin models—investigation of intra-and inter-laboratory reproducibility with coded chemicals. Mutagenesis 28(6):709–720

Steinemann AC (2009) Fragranced consumer products and undisclosed ingredients. Environ Impact Assess Rev 29(1):32–38

Tanaka S, Matsumoto Y, Dlova N, Ostlere LS, Goldsmith PC, Rycroft RJG, Basketter DA, White IR, Banerjee P, Mcfadden JP (2004) Immediate contact reactions to fragrance mix constituents and Myroxylon pereirae resin. Contact Dermatitis 51(1):20–21

Thakkar Y, Moustakas H, Aardema M, Roy S, Pfuhler S, Api AM (2022) Use of the epi Derm TM 3D reconstructed skin micronucleus assay for fragrance materials. Mutagenesis 37(2):89–111

Towle KM, Grespin ME, Monnot AD (2017) Personal use of hair dyes and risk of leukemia: a systematic literature review and meta-analysis. Cancer Med 6(10):2471–2486

Xiao L, Liao F, Fan Y, Miwa N (2020) Enzyme-digested Colla Corii Asini (E'jiao) accelerates wound healing and prevents ultraviolet A-induced collagen synthesis decline and wrinkle formation in three-dimensional skin equivalents. Hum Cell 33(4):1056–1067

Zanoni TB, Tiago M, Faião-Flores F, de Moraes Barros SB, Bast A, Hageman G, de Oliveira DP, Maria-Engler SS (2014) Basic red 51, a permitted semi-permanent hair dye, is cytotoxic to human skin cells: studies in monolayer and 3D skin model using human keratinocytes (HaCaT). Toxicol Lett 227(2):139–149

Zeliger HI (2007) Cosmetics: toxicity and regulatory requirements in the US. In: Global regulatory issues for the cosmetics industry. William Andrew Publishing, Norwich, pp 63–70

Toxicokinetic and Toxicodynamic Studies of Cosmetics and Personal Care Products Using 3D Skin Models: Progress Made and Path Ahead

14

Amita Singh, Vipin Kumar, Anurag Kumar Gautam, Ashok Kumar Yadav, and M. R. Vijayakumar

Abstract

Generally, cosmetic products have very rare health hazards, but it cannot be stated that they are safer in use. Cosmetic products are used by all age group populations and the sensitive group of the human population like pregnant women, old people, and small kids. Therefore, the toxicity profile of cosmetic ingredients should be evaluated. Toxicological testing is used to determine the adverse effect of chemicals, cosmetic products, and personal healthcare products. Systemic exposure to cosmetic and personal healthcare products can stimulate cells or organs. Systemic toxicity includes two components, i.e., toxicokinetic (TK) and toxicodynamic (TD). The toxicokinetic study of cosmetic ingredients is necessary when the ingredients have the potency to penetrate the body and exceed a critical level. These studies focus on the toxicological mechanism and toxic effects of the chemicals on the cells, tissue, and organs. The toxicodynamic studies may help understand the molecular/cellular effects of the cosmetic ingredients and determine the concentration of chemicals/toxicants at the target site. The toxicodynamic and toxicokinetic studies provide complementary information, and combining their methodologies will help us build new approaches to safety evaluation.

A. Singh · V. Kumar · A. K. Gautam · M. R. Vijayakumar (✉)
Department of Pharmaceutical Sciences, Babasaheb Bhimrao Ambedkar University (A Central University), Lucknow, India

A. K. Yadav
Department of Pharmaceutical Chemistry, Hygia Institute of Pharmaceutical Education and Research Lucknow, Lucknow, India

A. B. Pant et al. (eds.), *Skin 3-D Models and Cosmetics Toxicity*,
https://doi.org/10.1007/978-981-99-2804-0_14

Keywords

Toxicokinetic studies · Toxicodynamic studies · cosmetics · 3D skin models · Safety assessment

14.1 Introduction

Typically, cosmetic products have very rare health hazards, but it cannot be stated that they are safer in use. Cosmetic products are used by all age group populations and the sensitive group of the human population like pregnant women, old people, and small kids. Therefore, the toxicity profile of cosmetic ingredients should be evaluated (Vinardell and Mitjans 2017). Scientific Committee on Consumer Safety (SCCS) recommends guidelines for the safety evaluation of the ingredients or raw materials used by cosmetic industries. The toxicological evaluations of the product, which are related to human health, should be performed before entering the market (Cancian et al. 2014). Toxicological testing is often used to identify the adverse effects of all chemicals or ingredients, such as preservatives, hair dyes, UV filters, and colorants used to formulate cosmetic products (Bernauer et al. 2018; Vinardell 2015). Therefore, to evaluate the toxicity profile of the chemical ingredients, a large number of human volunteers are required, but it would be unrealistic and unethical. Animals are used for toxicological studies because they are similar to humans in anatomical and physiological aspects. The animal models provide the preclinical data of the chemicals/cosmetic ingredients, which causes the involvement of a higher number of animals that creates unnecessary stress on animals (Ng and Yeong 2019). In this regard, the government of India prohibited the involvement of animals in testing the toxicity profile of cosmetic ingredients. It issued the guidelines in the drug and cosmetic act 1945 under rule 148-C (Akbarsha and Mascarenhas 2019).

14.2 Three-Dimensional (3D) Skin Models

The 3D skin model resembles human skin in structural and physiological aspects, making them a better replacement for animals used for in vivo studies (Niehues et al. 2018). These models are constructed by separating cells from human tissue and undergoing in vitro culture. These cells are mixed with biomaterials in a specific ratio and left to grow to form a 3D skin model. The 3D models can be used for in vitro studies and may act as effective tools for predicting the percutaneous penetration of substances. Full-thickness reconstructed skin models and epidermal skin models can be used to determine the penetration of cosmetic products. Therefore, reconstructed skin models predict the permeability of substances and serve as a better substitute for human skin. The 3D skin models also play a crucial role in estimating the release kinetics of the constituents of cosmetic products. These models also determine the percutaneous absorption and metabolism, such as N-acetylation, glucuronidation, and other reactions in skin tissue (Hu and He

2021; Schlotmann et al. 2001). The 3D skin models are two types, that is, scaffold-free 3D skin models and scaffold-based 3D skin models

Scaffold-free 3D skin model: In the scaffold-free 3D skin model, one or more types of cells assemble themselves and form nonadherent cell aggregates. Spheroids are scaffold-free 3D skin models that mimic the skin tissue on a small scale. Spheroids are nonadherent cell aggregates prepared by one or more types of cells (Randall et al. 2018). The cells are well assembled in a hard spherical shape and have definite morphology with different size ranges. This model is called microtissue produced by the gravitational force from monodispersed cells. The skin tissue is nourished with the growth media by immersing the tissue in media. This technique has few complications due to the absence of scaffolding that allows the self-assembling of cell colonies and forms clusters of microtissue called spheroids. Scaffold-free skin models are suitable for anticancer drug development and toxicology assessment. The uniform size and shape of the spheroid may be affected by the origin of the gel used for the spheroid formation. The gels used to prepare spheroids can be of plant origin or animal origin. The plant-origin gels are nontoxic, have significant stability at room temperature, and do not have the risk of animal contamination. Spheroids using agarose gel have various porous architecture and mechanical properties with the modulation in the agarose concentration because it is formed by heating up to boiling temperature and freezes after cooling. Alginate is also used for spheroid preparation, and a commercial product is available with the name AlgiMatrix™. It is a freeze-dried ready-to-use product, like a porous sponge with long-term viability. It is stable at room temperature, biodegradable, and nontoxic (Habanjar et al. 2021).

14.3 Methods of Spheroid Production

(a) **Hanging Drop Method:** The hanging drop method is used to prepare the spheroids using surface tension and gravitational force to produce spheroids of the desired size. This technique allows the aggregation of single cells to form droplets like spheroids.
(b) **Pellet Culture Method:** In this method, cells are concentrated by centrifugation. After the centrifugation, supernatant is discarded and the pellets are separated. The obtained pellets were resuspended in a culture medium.
(c) **Cultivation of Molded Lozenges and Liquid Overlay (Static Suspension):** In this technique, the spheroid formation involves the interruption in cell adhesion with biomaterials or gel. The cells spontaneously form the spheroids until the binding of the cell to the support is inhibited. This method is straightforward and makes it easy to monitor spheroid formation.
(d) **Spinner Culture Technique:** Cell suspension is mixed by stirring in the centrifugal flask. This technique has the advantage that the cell suspension is uniformly mixed by continuous stirring to form spheroids.

Scaffold-based 3D skin model: In this model, cells are cultured with the help of a support system. Hydrogel, hard polymer, and natural or synthetic support of animal or vegetable sources can be used as the support to culture the cells.

Some scaffold-based 3D skin models that are used for the assessment of cosmetic products are given as follows.

14.4 Human Skin Equivalents (HSEs)

HSEs are artificial skin models made up of keratinocyte cells and dermal fibroblast. The structure and function of these models resemble the native living tissue. HSE model can be beneficial in the screening of specific molecules. HSEs provide an alternative approach to evaluating substance efficacy and safety profile (Choudhury and Das 2021; Bellas et al. 2012).

14.5 Reconstructed Human Epidermis (RHEs)

RHE model contains human keratinocytes cells cultured on the acellular nonreactive substrate (Schlotmann et al. 2001). The commercially available RHE model contains a multistratified epidermis, differentiation markers of epidermal cells, and lipids such as phospholipids, cholesterol, and triglycerides. The main advantage of the RHEs is that it allows growing the epidermal cell in a serum-free medium. The epidermal cells are seeded on the nonreactive or inert substrate, which then grows at the air–liquid interface with definite moisture and temperature. The nutrient medium nourishes the cells of the basal layer through the filter substratum. The stratified epidermis is formed after 14 days of culture, which shows anatomical and physiological resemblance with the human epidermis (Tsai 2016). Some RHE models are described as follows.

(a) **EPiTRI** is the reconstructed human epidermis model that mimics the human epidermis. It is constructed by the Industrial Technology Research Institute (ITRI). Human keratinocytes cells are seeded into the nutrient medium for proliferation of up to 2 days and are allowed to grow at the air–liquid interface for up to 14 days. This reconstructed epidermis model is structurally similar to the native human epidermis and can be used to test the toxicity of cosmetic products (Liao et al. 2021).

(b) **EpiskinSM** is constructed by the EPISKIN SNC (Lyon, France). In this model, human adult keratinocytes reconstruct the differentiated multi-layered epidermis. To prepare the EpiskinSM model, human adult keratinocyte cells were inoculated on the surface of the collagen I matrix, which was coated with a collagen IV layer and fixed to the plastic chamber bottom. The plastic chamber is used to fix the epidermis in a certain position in the culture well. This design allowed the cells for the gaseous exchange and was ready to use for the assay. After 13 days of culture, the 3D epidermis was obtained. The main layers presented in the model were the basal layer, spinous, granular layer, and stratum corneum (Grandidier et al. 2007).

(c) **SkinEthic** helps to determine the skin irritation of chemical/cosmetic ingredients. It consists of keratinocyte cells cultured on an inert polycarbonate surface with a defined nutrient medium (Pellevoisin et al. 2018).

14.6 Full-Thickness HSEs

The full-thickness human skin equivalent model contains keratinocyte cells cultured on the dermal substrate with fibroblast cells. The total thickness HSE model has more resemblance with the native human skin in comparison with the RHEs. Recently, many advanced HSE models have been made that include varieties of cells such as neuron cells, Langerhans cells, and melanocyte cells. Melanoma cells and psoriatic epidermis have also been claimed (Tsai 2016).

14.6.1 Materials for Constructing HSE Scaffolds

14.6.1.1 Decellularized Extracellular Matrix (dECM)

Decellularized extracellular matrix is deputed as the scaffolding material for constructing a human skin equivalent model. It is prepared by the elimination of cellular components from the tissue. The epidermis removal can minimize the grafting rejection and leave some basement membrane and collagen that helps in the proliferation and reattachment of fibroblast and keratinocyte cells. The decellularization techniques can be more specific in addition to physical, chemical, and enzymatic removal of cellular material of tissue (Tsai 2016). The supports that are used to grow the cells are protein-based ECM and hydrogel-based support systems.

14.6.1.2 Protein-Based ECM

The biomaterials prepared using natural polymers have been used in manufacturing 3D surfaces for in vitro testing. Most natural polymers prepared for 3D skin models are obtained from the mammalian ECM. The biomaterials used for manufacturing 3D surfaces include Matrigel®, collagen, hyaluronic acid (HA), alginate, and gelatin (Choudhury and Das 2021). The Matrigel is formed by mixing gelatinous proteins derived from the mouse sarcoma basement membrane. It is the commercially available gold standard biomaterial with the Matrigel® brand name. It is the best available product in the market to develop the 3D surface for in vitro testing of the cosmetic product. It is the liquid extract at 4 °C, while at the physiological pH and temperature (37 °C), it is converted into a gel. It is commercially available and provides ease of application for in vitro testing. The biological activity is the reason for the success of Matrigel®, which allows the cell differentiation and development of organs under standard culture conditions. Non-mammalian ECM polymers are preferred to form biomaterials for 3D tissue culture. These polymers have the property to assemble themselves in 3D structures, but they show a lack of cell adhesion due to their origin. Therefore, the adhesive molecules or peptides should be added to get active material with biological properties (Habanjar et al. 2021).

14.6.2 Hydrogel-Based Support Systems

The hydrogel-based scaffold is the most useful for in vitro skin models. It is formed by hydrophilic polymers, which are attached through physical, ionic, or covalent interactions. The hydrophilicity of the polymer makes the hydrogel more specific for water absorption that penetrates between polymers and causes swelling. After swelling, hydrogels become rubbery that look like living tissue. Moreover, the polymers of hydrogels are attached with intramolecular and intermolecular physical attractions that maintain 3D structure (Habanjar et al. 2021).

Hydrogels have the property to mimic ECM and allow the substances like cytokines and growth factors to travel the gel, which makes it unique. It can be used to determine epidermal behavior. The hydrogel system has the crucial advantage that its physicochemical activity can be adjusted, and cells can be recovered, which provides convenience in molecular analysis, and their transparency allows for microscopy. It has biochemical and mechanical properties similar to the native ECM. Moreover, the cell-mediated degradability, stiffness of the gels, and materials used to prepare gel, such as collagen, fibrin, and gelatin, can be modified. The hydrogel is cheaper than the microfluidic device but more expensive than the spheroid system. The handling of the hydrogels is accessible, while the disadvantage associated with the hydrogel has batch-to-batch variability due to the involvement of different biological components, making it difficult to get the same result repeatedly. Hydrogels are two types, i.e., natural and synthetic hydrogels (Klicks et al. 2017).

(a) **Natural Hydrogels**: The natural polymers that are used to form hydrogels are categorized into two major classes, that is, polysaccharides (hyaluronic acid, alginate, and chitosan) and proteins (collagen and gelatin). Collagen polymer is used to construct hydrogels in which the dermal fibroblast cells are incorporated to mimic the dermal layer. Changing the concentration and preparation parameters of the hydrogel can modify the mechanical strength, structure, and biodegradable properties. The hydrogel architecture can also be controlled during polymerization by changing the ionic force, pH, and temperature. Additionally, higher density and reduced pore size were obtained with an increased collagen concentration, while the fiber diameter was unaffected. Furthermore, increased temperature and pH cause higher polymerization and reduced pore size and internal diameter, and thus modify the mechanical properties of the hydrogel matrix (Randall et al. 2018; Tsai 2016).

(b) **Synthetic Hydrogels:** The polymerization of synthetic polymers yielded synthetic hydrogels. These polymers can construct a 3D biomatrix to culture the cells. However, these polymers have an absence of basic biological properties. The production of synthetic hydrogels is easy and possesses flexibility in the testing of cosmetic ingredients. The synthetic polymers incorporated to form hydrogels are polycaprolactone (PCL), polyethylene glycol oxide (PEG), polyvinyl alcohol, and poly-lactide-co-glycolide (PLGA). Synthetic polymers have an active functional group, such as amine, acid, or alcohol. These functional groups actively participated in the chemical reaction to form ECM, which exhibits the desired feature to culture the cells. In physiological aspects, these polymers are associated with the drawback that they may form toxic byproducts. Additionally, they cannot produce biochemical signals essential for cell communication. Therefore, signaling molecules such as peptides, growth factors, and glycans need to be added to overcome the limitation of synthetic polymers (Choudhury and Das 2021).

Alvetex® is the commercially available hard-based scaffold that provides enough 3D space and volume to the cells to construct tissue. The cell can achieve the physiological shape because they are seeded in the presence of sponge shape structures. These scaffolds are made up of nonreactive substances that avoid byproduct formation. The cell–ECM interaction can also be determined because these polymers can reproduce the ECM structure and porosity. Synthetic hydrogels are prepared by gas foam technology or electrospinning (Habanjar et al. 2021).

14.6.2.1 Toxicological Testing of Cosmetic Products

Toxicological testing is used to determine the adverse effect of chemicals, cosmetic products, and personal healthcare products (Ng and Yeong 2019). Systemic exposure to cosmetic and personal healthcare products can stimulate cells or organs. Systemic toxicity includes two components, i.e., toxicokinetic (TK) and toxicodynamic (TD).

14.6.2.2 Toxicokinetic Study

1. Toxicokinetics can be described as the uptake and elimination of a chemical substance after entering the body. These processes are essential to understand the systemic toxicity of the substance. Therefore, in vitro or in vivo studies are required to determine the toxicity of cosmetic products. However, the use of animals is restricted for the testing of cosmetic products. The toxicokinetic study of cosmetic ingredients is necessary when the ingredients have the potency to penetrate the body and exceed a critical level. The toxicokinetic study includes absorption, distribution, metabolism, and excretion (Adler et al. 2011).
2. The absorption study is necessary for the cosmetic product as dermal penetration of chemicals can harm the tissue. The dermal or percutaneous absorption was performed by the in vivo animal model before ban animals for the testing of cosmetic products. Therefore, cosmetic industries choose an alternative approach to test cosmetic ingredients as the absorption study is compulsory and cannot be avoided. To study in vitro dermal absorption of cosmetic ingredients, some key points should be in consideration when no dermal absorption study is available for the particular compound, like the concentration or quantity of the product to be applied. If the cosmetic ingredients have the least dermal absorption and low

permeation, it is taken as the product cannot penetrate or enter through the epidermal route (Abou-Elwafa Abdallah et al. 2015).

3. When a chemical substance reaches the blood circulation, the metabolic profile of the product needs to be assessed. The metabolic profile of the compound gives complete information about the parent compound, whether it is being detoxified or converted into a reactive intermediate. In the case of aromatic amine hair dyes, assessment of the metabolic profile gives information about the hepatic metabolite or intermediates formed after the metabolism. The liver is the main organ for metabolism. The liver samples from human sources are restricted, but the 3D cultures are available to assess cosmetic products' in vitro metabolic profile. Some in vitro cell culture models are also available to test the chemical product; the keratinocytes (HaCaT) cells model is one of them. HaCaT cells were cultured into 96-well plates with the DMEM containing 10% fetal calf serum, penicillin, and streptomycin. After 24 h, the culture medium was replaced with the fresh medium, and the test compound was added. In this study, p-aminobenzoic acid was used as the positive control. The substrate was then subjected to incubation at 37 °C. After incubation, high-performance liquid chromatography (HPLC) analysis was performed to detect any change in the parent compound and/or the presence of metabolites (Manwaring et al. 2015).
4. Another study is reported to determine the metabolites by 3D tissue culture. For this, hepatocytes cells were seeded into 24 collagen-coated plates with the layer of GelTrex. Cultured hepatocyte cells were treated with the test compound and incubated at 37 °C for 24 h. After incubation, hepatocyte cells were centrifuged and analyzed to detect the loss of the parent compound and the presence of metabolite by HPLC analysis (Manwaring et al. 2015).

Toxicodynamic Study: These studies focus on the toxicological mechanism and toxic effects of the chemicals on the cells, tissue, and organs. The toxicodynamic studies may help understand the molecular/cellular effects of the cosmetic ingredients and determine the concentration of chemicals/toxicants at the target site. The toxicodynamic and toxicokinetic studies provide complementary information, and combining their methodologies will help us build new approaches to safety evaluation (Bois et al. 2017; Dellafiora et al. 2018).

14.7 Safety and Toxicity Assessment of Cosmetic Products

14.7.1 Assessment of Genotoxicity

Different tests can assess the genetic damage by detecting direct DNA reactive substances, such as the Ames test and in vitro micronucleus test for the genotoxicity evaluation. The reactive substances in cosmetic products that alter the genetic code can cause serious health issues in humans, so this needs to be assessed before it comes into human use. The Ames utilizes the prokaryotic cells that differ from the mammalian tissue in aspects like metabolism, chromosomal structure, and the DNA repair process. The Ames test uses two strains of bacteria, i.e., *Salmonella typhimurium* and *Escherichia coli*. This test work on the principle that the test sample reverts the mutation present in the bacteria. The second test for the genotoxicity assessment is in vitro mammalian micronucleus test. This test provides information regarding chromosomal damage potential after the exposure of the test sample (Vinardell and Mitjans 2017; Barthe et al. 2021; OECD 2016, 2020).

14.7.2 Assessment of Skin Sensitization

Most cosmetic products have topical applications, so the ingredients used for the cosmetic preparation should be biocompatible. Cosmetic products that have the potency to sensitize the skin may cause severe skin-related issues after topical application in a repeated manner. The adverse outcome pathways can describe the mechanism of skin sensitization. The skin sensitization reactions can be initiated by binding an electrophilic substance to the nucleophilic centers of skin protein by covalent bonding to a specific cell signaling pathway, i.e., antioxidant response element (ARE)-dependent pathway for the activation of immune cells such as T-cell proliferation. The ARE-Nrf2 luciferase test is the only validated in vitro method in KeratinoSens™, which is prepared using keratinocyte cells derived from the HaCaT human keratinocytes transfected with plasmid. This test assesses luciferase gene induction when electrophilic test compounds are applied (Adler et al. 2011).

14.7.3 Assessment of Dermal Absorption

Dermal absorption is the absorption of chemicals into the skin by passive diffusion. The in vitro dermal absorption study is suitable for evaluating the absorption of cosmetic products by human skin. Applying the chemical ingredients on the skin should be similar to human exposure. The substance should be applied 1–5 mg/cm^2 for solid ingredients and 10 μL/cm^2 for liquid materials. The test substance should remain on the skin for up to 24 h at 32 + 1 °C constant temperature to determine the passive diffusion of chemicals till the removal of chemicals with a definite cleaning protocol. The skin receptor fluid is taken at different time intervals throughout the study and evaluated to test the presence of chemical substances and/or metabolites. The purpose of the study is to evaluate the penetration of cosmetic ingredients into the skin through the dermal layer and assess the potency to be reached in blood circulation. The determination of dermal absorption is necessary for safety. The appearance of chemical substances in systemic circulation may cause adverse effects. The quantity of absorbed test samples is taken for the toxicological risk assessment and to determine the safety margin of products. The quantity of drug absorbed by the skin can be taken to assess the concentration of the test sample at the site of action (Barthe et al. 2021).

14.7.4 Assessment of Skin and Eye Irritation

Irritation reaction can be explained as the reversible degradation or inflammation of the skin after chemical/cosmetic product application. The inflammation automatically disappeared after a few days. Skin and eye irritation evaluation study of cosmetic ingredients is significant as it can be affected tissue with more prolonged use.

The cosmetic ingredients can cause tissue damage to the eye after application to the anterior surface, which reverts after 21 days. The commercially available reconstructed epidermis model (RhE) and human cornea-like epithelium (RhCE) can be used to determine the skin and eye irritation testing, respectively. The available reconstructed human tissue models have the same biochemical and physiological properties as skin and eye. RhE models are made up of keratinocytes cells of humans that are cultured to develop a multi-layered, highly differentiated epidermis. The test of cosmetic ingredients against skin irritation has been described in Organization for Economic Cooperation and Development (OECD) guidelines (OECD TG431). According to this guideline, the test compound is topically applied to the 3D RhE model. This model structurally and physiologically mimics the human epidermis. The penetration of chemical ingredients into the skin through the stratum corneum may damage the cells by releasing inflammatory cytokines. The inflammatory reaction that irritates can be occurred in two proposed ways, i.e., by the damaging stratum corneum barrier function and by the direct irritant effect on the skin tissue. The in vitro RhE test detects the cell damage by MTT (3-(4,5-dimethylthiazol-2-yl)-2,5-diphenyltetrazolium bromide) assay. The MTT assay detects the viable cell by converting tetrazolium dye into formazan crystals measured in cell/tissue extract. The irritant/chemical decreases the viability of the cells below the threshold (OECD 2021).

RhCE corneal model consists of alive human cells, which have been cultured for the construction of multi-layered, differentiated corneal epithelium. Both models are similar metabolic activity to human tissue and release many cytokines that have a more significant role in irritation and inflammation. The test sample is applied directly to the tissue surface. If the viability of the cells is greater than 50% and RhCE is greater than 60%, the test compound considers a non-irritant product. If the viability of RhE is $\leq$50% and the viability of RhCE is $\leq$60%, the test compound is considered an irritant (Barthe et al. 2021).

14.7.5 Assessment of Skin Corrosion

The irreversible degradation of the skin after applying a chemical/test compound is considered skin corrosion. This test involves the 3D RhE model. The test substance was applied to the 3D RhE model of human-origin keratinocyte cells. The principle of this model is based on the penetration of corrosive compounds through the stratum corneum, and the cytotoxic effects on the cells present below the stratum corneum. The presence of viable cells is evaluated by MTT assay. The MTT assay detects the viable cells based on the principle that the MTT dye converts the tetrazolium salt into the blue formazan salt, which can be measured in tissue after extraction. Suppose the chemical compound decreases cell viability below a threshold level (<50% viability after 3 min of exposure). In that case, it is considered a corrosive chemical, while the non-corrosive chemicals produce viability of cells above the threshold level (>50% viability after 3 min of chemical exposure) (Ng and Yeong 2019; Pistollato et al. 2021; Kandárová et al. 2006).

14.7.6 Assessment of Cytotoxicity

Cytotoxicity testing of cosmetic ingredients is used to determine the toxicity, contributing to preclinical studies. Therefore, many assays for the cytotoxicity study involve the incubation of cells with the reagents. The viable cells produce fluorescence by converting the substrate into a colored product detected by colorimetric equipment. The most commonly used assay for the viability test is the MTT assay and sulforhodamine B (SRB) assay (Nabarretti et al. 2022).

14.7.7 Assessment of Phototoxicity

Phototoxicity reactions may occur by skin exposure to ultraviolet or visible radiation after topical or systemic application of chemical products. For the assessment of phototoxic reactions, in vitro 3 T3 NRU model is used. This in vitro model is made up of fibroblast cells that are collected from the embryonic layer of Swiss mice. The phototoxic reaction is determined by using a neutral red dye. This dye measures the 50% mean inhibitory concentration (IC50) in the presence and absence of radiation (Nabarretti et al. 2022; Kapoor and Saraf 2007).

References

Abou-Elwafa Abdallah M et al (2015) Evaluation of 3D-human skin equivalents for assessment of human dermal absorption of some brominated flame retardants. Environ Int 84:64–70

Adler S et al (2011) Alternative (non-animal) methods for cosmetics testing: current status and future prospects-2010. Arch Toxicol 85(5):367–485

Akbarsha MA, Mascarenhas B (2019) Cosmetic regulation and alternatives to animal experimentation in India. In: Alternatives to animal testing. Springer, Singapore

Barthe M et al (2021) Safety testing of cosmetic products: overview of established methods and new approach methodologies (NAMs). Cosmetics 8(2):50

Bellas E et al (2012) *In vitro* 3D full-thickness skin-equivalent tissue model using silk and collagen biomaterials. Macromol Biosci 12(12):1627–1636

Bernauer, U., et al. (2018) The SCCS notes of guidance for the testing of cosmetic ingredients and their safety evaluation 10 th revision "the SCCS note of guidance for the testing of cosmetic ingredients and their safety evaluation–10th revision" SCCS/1602/18–final version

Bois FY et al (2017) Multiscale modelling approaches for assessing cosmetic ingredients safety. Toxicology 392:130–139

Cancian M et al (2014) Safety assessment of cosmetic products, with emphasis on the ocular area: regulatory aspects and validation processes. Braz J Pharm Sci 50:667–676

Choudhury S, Das A (2021) Advances in generation of three-dimensional skin equivalents: pre-clinical studies to clinical therapies. Cytotherapy 23(1):1–9

Dellafiora L, Dall'Asta C, Galaverna G (2018) Toxicodynamics of mycotoxins in the framework of food risk assessment-an in silico perspective. Toxins (Basel) 10(2):52

Grandidier MH, Roguet R, Leclaire J (2007) *In vitro* acute skin irritancy of chemicals using the validated EPISKIN model in a tiered strategy results and performances with 184 cosmetic ingredients. AATEX 14(Special Issue):351–358

Habanjar O et al (2021) 3D cell culture systems: tumor application, advantages, and disadvantages. Int J Mol Sci 22(22):12200

Hu X, He H (2021) A review of cosmetic skin delivery. J Cosmet Dermatol 20(7):2020–2030

Kandárová H et al (2006) Assessment of the human epidermis model SkinEthic RHE for *in vitro* skin corrosion testing of chemicals according to new OECD TG 431. Toxicol in Vitro 20(5): 547–559

Kapoor S, Saraf S (2007) Risk analysis tools for toxicological profile of cosmetics. The internet. J Toxicol 5:5

Klicks J et al (2017) *In vitro* skin three-dimensional models and their applications. J Cell Biotechnol 3:21–39

Liao CC et al (2021) Validation study of a new reconstructed human epidermis model EPiTRI for *in vitro* skin irritation test according to OECD guidelines. Toxicol in Vitro 75:105197

Manwaring J et al (2015) Extrapolation of systemic bioavailability assessing skin absorption and epidermal and hepatic metabolism of aromatic amine hair dyes *in vitro*. Toxicol Appl Pharmacol 287(2):139–148

Nabarretti BH et al (2022) A review of alternative methods to the use of animals in safety evaluation of cosmetics. Einstein (Sao Paulo) 20:eRB5578

Ng WL, Yeong WY (2019) The future of skin toxicology testing–three-dimensional bioprinting meets microfluidics. Int J Bioprint 5(2.1):237

Niehues H et al (2018) 3D skin models for 3R research: the potential of 3D reconstructed skin models to study skin barrier function. Exp Dermatol 27(5):501–511

OECD (2016) Test no. 487: *in vitro* mammalian cell micronucleus test

OECD (2020) Test no. 471: bacterial reverse mutation test

OECD (2021) Test no. 439: in vitro skin irritation: reconstructed human epidermis test method

Pellevoisin C et al (2018) SkinEthic™ RHE for *in vitro* evaluation of skin irritation of medical device extracts. Toxicol in Vitro 50:418–425

Pistollato F et al (2021) Current EU regulatory requirements for the assessment of chemicals and cosmetic products: challenges and opportunities for introducing new approach methodologies. Arch Toxicol 95(6):1867–1897

Randall MJ et al (2018) Advances in the biofabrication of 3D skin *in vitro*: healthy and pathological models. Front Bioeng Biotechnol 6:154

Schlotmann K et al (2001) Cosmetic efficacy claims *in vitro* using a three-dimensional human skin model. Int J Cosmet Sci 23(5):309–318

Tsai PC (2016) Development of three dimensional human skin equivalents based on decellularized extracellular matrices and tyrosine-derived polycarbonate polymers for in-vitro drug screening applications

Vinardell MP (2015) The use of non-animal alternatives in the safety evaluations of cosmetics ingredients by the scientific committee on consumer safety (SCCS). Regul Toxicol Pharmacol 71(2):198–204

Vinardell MP, Mitjans M (2017) Alternative methods to animal testing for the safety evaluation of cosmetic ingredients: an overview. Cosmetics 4(3):30

Resource Link

Sr. No.	Title	Resource link
1.	Skin models for cosmetic safety/toxicity studies: Bridging established methods and novel technologies	https://www.liebertpub.com/doi/10.1089/aivt.2014.0010 https://aip.scitation.org/doi/10.1063/5.0046376
2.	Artificial skin models for animal-free testing: a journey in past two decades	https://www.aerzte-gegen-tierversuche.de/en/information/animal-free-research-in-the-21st-century https://www.labbulletin.com/articles/artificial-skin-instead-of-animal-testing https://www.cnbc.com/2017/05/25/loreal-is-making-lab-produced-human-skin-to-curb-animal-testing.html
3.	Ready-to-use ex vivo human skin models for cosmetic testing: current status and challenges	https://genoskin.com/blog/2018/02/22/skin-models-cosmetic-testing/ https://www.liebertpub.com/doi/10.1089/aivt.2014.0010 https://www.altex.org/index.php/altex/article/view/2385 https://www.health.gov.au/sites/default/files/documents/2022/10/towards-alternatives-to-animal-testing-of-industrial-chemicals-in-australia-a-scoping-report_0.docx https://www.karger.com/Article/FullText/504063
4.	Advantages and limitations of human reconstructed skin for cosmetic R&D, efficacy and safety studies	https://qima-lifesciences.com/en/in-vitro-testing-pros-cons/
5.	3D bioprinted skin-on-chip: Novel skin equivalent models and their characterization	https://aip.scitation.org/doi/10.1063/5.0046376 https://pure.manchester.ac.uk/ws/portalfiles/portal/213187512/FULL_TEXT.PDF https://www.computerworld.com/article/3135055/harvards-3d-printed-heart-

(continued)

A. B. Pant et al. (eds.), *Skin 3-D Models and Cosmetics Toxicity*,
https://doi.org/10.1007/978-981-99-2804-0

(continued)

Sr. No.	Title	Resource link
		on-a-chip-could-lead-to-personalized-medicine-breakthroughs.html
6.	3D Bioprinting of skin tissue model	https://www.cellink.com/blog/the-skinny-on-bioprinting-human-skin/ https://pdfs.semanticscholar.org/1800/2a29e756b42cf1ba114e4ca6f32ccc687e62.pdf https://www.news-medical.net/life-sciences/Bioprinting-Applications.aspx https://www.aha.org/aha-center-health-innovation-market-scan/2022-06-07-3-ways-3d-printing-revolutionizing-health-care
7.	Solar radiations and phototoxicity of cosmetics: avenues of in vitro skin models	https://www.skincancer.org/risk-factors/photosensitivity/ https://www.msdmanuals.com/en-in/home/skin-disorders/sunlight-and-skin-damage/photosensitivity-reactions https://core.ac.uk/download/pdf/155248219.pdf
8.	Various efficacy based testing methods in 3D skin based model system	https://promocell.com/cells-in-action/using-3d-skin-models-to-test-cosmetics/ https://www.screenlifescience.com/3d-in-vitro-skin-model
9.	Improved animal component free medium for long term maintenance of human skin explants and cosmetic toxicity studies	https://journals.sagepub.com/doi/abs/10.1177/02611929211038652?journalCode=atla
10.	Regulatory requirements for the efficacy/safety assessment of cosmetics/nano-cosmetics products: Opportunities and challenges for new approach and methodologies	https://www.mdpi.com/2079-4991/10/5/979 https://health.ec.europa.eu/system/files/2020-10/sccs_o_233_0.pdf https://www.fda.gov/regulatory-information/search-fda-guidance-documents/guidance-industry-safety-nanomaterials-cosmetic-products
11.	Integrated approaches to testing and assessment (IATA) for cosmetic and personal health care products	https://www.oecd.org/chemicalsafety/risk-assessment/iata/ https://www.tandfonline.com/doi/abs/10.1080/17435390.2022.2085207?journalCode=inan20
12.	Approach for in silico validation of safety/toxicity data for cosmetics	https://cordis.europa.eu/project/id/266835/reporting
13.	Role of omics approaches in the toxicity/safety studies of cosmetics	https://www.fda.gov/science-research/bioinformatics-tools
14.	Application of 3D skin model systems in assessing the safety/toxicity of fragrances and flavors, hair dyes, preservatives, colorants, etc.	https://promocell.com/cells-in-action/using-3d-skin-models-to-test-cosmetics/ https://bioscience.lonza.com/lonza_bs/CH/en/download/content/asset/37016

MIX
Papier aus verantwortungsvollen Quellen
Paper from responsible sources
FSC® C105338

If you have any concerns about our products,
you can contact us on
ProductSafety@springernature.com

In case Publisher is established outside the EU,
the EU authorized representative is:
Springer Nature Customer Service Center GmbH
Europaplatz 3, 69115 Heidelberg, Germany

Printed by Libri Plureos GmbH
in Hamburg, Germany